SKETCH MAP OF EASTERN HIMALAYA

Numbered areas refer to chapters in the book.

Inset shows area of sketch map

A Quest of
Flowers

A Quest of Flowers

The Plant Explorations of
Frank Ludlow and
George Sherriff
told from their diaries
and other occasional writings
by Harold R. Fletcher

At the Edinburgh
University
Press

o

ISBN 0 85224 278 6

Printed in Great Britain
Photo-litho reprint by
W & J Mackay Limited, Chatham
from earlier impression
Reprinted 1976

The Bimbi La valley

Contents

o

Acknowledgments

o

The publisher gratefully acknowledges
financial assistance from the Carnegie
Trust for the Universities of Scotland,
the Stanley Smith Horticultural Trust,
and Mrs George Sherriff; and the loan of
original photographs by the Department
of Botany, British Museum (Natural
History), Mrs George Sherriff,
and Sir George Taylor

o

List of Plates

○

vii

Historical Introduction

BY SIR GEORGE TAYLOR FRS

○

It has been my good fortune to know most of the noted plant collectors of this century who have revealed the amazing richness of the flora of south-west China and the neighbouring Himalayan region. The first whose acquaintance I made was George Forrest, and his vivid verbal and written accounts of his field experience fired my youthful hopes of one day savouring the joys and risks of plant collecting in remote parts. As a post-graduate student at Edinburgh working on his material I was greatly influenced by his splendid collections; the specimens, beautifully prepared and most adequately labelled, were indeed an inspiration and left an enduring impression on me, and there is no doubt that their perfection was a standard to which I aimed in all my later collecting activities. Forrest's vast botanical collections and superb introductions to British gardens have been amply chronicled; suffice it here to say that this sturdy and indomitable Scot is commemorated in one of the most impressive chapters in the history of plant exploration.

During my years at the British Museum (Natural History) I frequently met with Frank Kingdon Ward, one of the most celebrated and tenacious of plant collectors, who became a very good friend and with whom in correspondence and conversation I shared vicariously his later expeditions. Ward, who travelled in austere fashion, had the happy knack of spotting élite plants suitable for gardens in this country, and his most noted introduction is probably the superlative Blue poppy. He was also gifted with a facile pen, and the descriptive narrative of his many expeditions is forcefully recorded in his many books. I acted as his sole executor on his death in 1958.

Latterly when I became Director at Kew I met Joseph Rock, and his annual visits were always packed with incident and far-

George Sherriff (*left*) and Frank Ludlow

ranging talk about exploration in remote areas of western China. Rock was a most remarkable man, for some years Professor of Botany and Chinese in Hawaii, who settled in Yunnan to devote his life to exploration of the country and intensive study of its history and culture. He had mastered the language and thus the literature of the region and published several impressive, beautifully-illustrated scholarly works. In addition he made substantial natural history collections and introduced a number of fine plants to cultivation, but he was forced to leave Yunnan when the wave of Communism engulfed the country.

It is to my greatest friends in the field of botanical endeavour, Frank Ludlow and George Sherriff, that I now come, and I find it extremely difficult to express my feelings adequately in regard to these men, staunch and wonderful companions, who contributed most significantly, with the help of one or two fellow-travellers, to our knowledge of the fauna and flora of south-eastern Tibet and who enriched our gardens with many lovely species. Their backgrounds were very different and it was a happy and fortuitous event which brought them together to forge an enduring friendship which led to the series of expeditions in the Eastern Himalaya.

Ludlow graduated from Cambridge in 1908, taking a BA in the natural science tripos, and during his course he read botany under Professor Marshal Ward, the father of Frank Kingdon Ward. On leaving the university he became Vice-Principal of Sind College in Karachi where he was simultaneously a Professor of Biology and a lecturer in English. He was indeed an outstanding naturalist but also one who was steeped in English literature and with a deep appreciation of poetry, which he could quote at length. In my notice of Ludlow published in the *Journal* of the Royal Horticultural Society in September 1972 I emphasized his penchant for reciting passages from Shakespeare and Tennyson, but he also had a fondness for the ditties of Gilbert and Sullivan and frequently for his own amusement and that of his friends he devised doggerel verse in the style of Gilbert generally with a theme of Sullivan in mind. Here are two examples in this idiom, the first prompted by publication of my book on *Meconopsis*:

> 'The offspring of *discigera*, or shall we say its progeny,
> Are placed by Dr Taylor in a section called Discogyne,
> Because the flattened disc upon the ovary's an impediment,
> To penetration by the rain, or any other element.
> *Torquata* too, from Lhasa, where the Dalai Lama rules the land
> Is placed within this section, which is very small you'll
>   understand

Though in loveliness its members are unequalled in the genus
One's as handsome as Adonis, t'other's beautiful as Venus.'

'*Concinna*'s mauve, *superba*'s white,
Paniculata's yellow;
Punicea red—a lovely sight,
Blue *grandis*—a fine fellow
Sherriff-i-i, I really think
'S the only poppy purely pink.'

The other was in a letter from Ha, Bhutan, in April 1949 where he
was enjoying wonderful trout fishing: 'I dare not fish for more than
an hour at a time for fear of catching too many; 4 and 5 pounders
are an everyday occurrence. The consequence is I am blasé. To
put things in a Gilbertian nutshell, here is a little poem which sums
up my contentions:

When anyone can roam this state,
And thrust his nose in any gate,
Then L & S sad to relate
Become no more than trippers,
If every time I cast a fly
A lusty trout is doomed to die
For lusty trout I cease to try,
Up goes the price of kippers.'

After the first World War, in which he was commissioned with
the 97th Indian Infantry, he went into the Indian Education Service
and became Inspector of European Schools, where his jurisdiction
extended from Bombay to the Central Provinces and from Karachi
to Derwa. In 1922 a request came from Delhi to submit the names
of any candidates for the post of Headmaster of a school to be
opened in Gyantse, in Tibet, where Tibetans of good family could
be taught the rudiments of western education. The work and
prospect of living for three years in a mediaeval surrounding
appealed to him, and he submitted his own name and was eventually
selected. Thus occurred a vital turning point in Ludlow's life. He
gave up the certain rewards of an assured career in the Indian
Education Service for the chancy prospects of a brief spell as a
headmaster of a school still to be founded. (In a Memorandum
bequeathed to me, along with his other papers, there is an account
of the difficulties he had in establishing this school and amusing
anecdotes of his experiences with his boys.) Clearly he sensed a
challenge to his adventuresome nature and, indeed, as a result of his
three years in Gyantse the spell of Tibet and its mountains held him
in thrall. From this decision to help the Tibetans stemmed the

realisation of the explorations in the eastern Himalaya. Although the Tibetan Government, fearful of Ludlow's influence on the young impressionable Tibetans, closed his school he was allowed access to Tibet, then very much a forbidden land, with chosen companions.

In 1927 Ludlow retired to Srinagar, in Kashmir, intending to travel extensively in the Himalayas collecting specimens, particularly of birds and only incidentally of plants for the British Museum (Natural History). In 1929 he was invited by Mr F. Williamson, then Consul General of Kashgar but previously British Trade Agent in Gyantse, to spend the winter with him and he jumped at this chance. His diary graphically describes his journey to Kashgar and, later, his trip to the Tian Shan. He left Srinagar in May and reached Kashgar at the end of September, collecting birds and plants but also, at the request of the Punjab Government and accompanied by a surveyor, investigating the state of the embankment impounding the 14-mile long Shyok Dam. When he first saw the dam in 1928 he thought it was in such an uncertain state that it would not retain the water for long unless it was strengthened. This premonition was confirmed before reinforcement could be made and the dam burst in 1929, fortunately without loss of life. Ludlow stayed at the Consulate with Williamson, meeting there Williamson's friend George Sherriff, and for five months greatly enjoying their companionship and especially the frequent forays with Sherriff on shooting trips. Together with Williamson (who was destined to occupy the key post of Political Officer in Sikkim) he and Sherriff discussed plans to explore Bhutan and Tibet in the years ahead, and these eventually came to splendid fruition as is recorded in the following pages.

Ludlow left Kashgar at the beginning of March 1930 for the Tian Shan Mountains, and his finely-prepared herbarium specimens from these mountains were amongst the first collections that I had to identify after my appointment to the British Museum. Ludlow returned to Srinagar in mid-November, having covered some 4000 miles and having obtained over 800 bird skins, 2000 butterflies, 6–700 birds eggs and about 200 different species of plants. I remember particularly two extremely attractive species — *Acantholimon venustum* and a magnificent *Eremostachys* — which I thought would be most desirable garden plants.

The meeting of Ludlow and Sherriff in Kashgar was the beginning of a long friendship which led these two comrades to explore together for over twenty years with outstanding success. They were two very independent individuals with differing backgrounds:

Ludlow, Sherriff's senior by fifteen years, very much in the scholarly academic mould and to whom anything mechanical was a closed book; Sherriff, equally talented in his chosen line, the precise efficient professional soldier who was an expert mechanic and electrician. The affecting and abiding magic of Tibet had touched Sherriff from his experiences on the Ladak-Tibet border, and the two men resolved to travel to Tibet whenever it became possible to do so. Their common love of out of the way places, of flowers, birds, fine music and field sports sealed the bond between them, and they shared a dislike for the conventional social scene. I cannot now recollect when I first met Ludlow and Sherriff, but it must have been in the early 1930s when they visited the British Museum to discuss plans for future expeditions. Afterwards, I became very closely associated with them and was privileged to join their expedition in 1938. Living together for weeks on end in isolated country, often in most austere conditions, is a real test of compatibility but on that journey, and over the later years, I was impressed by the great mutual respect they had for each other, their utter selflessness and harmony of views. I never knew of any friction or serious argument between them, and Ludlow actually dubbed himself Sherriff's *Fidus Achates*. Yet over the years of their long friendship they somehow could not bring themselves to address each other by their Christian names, and always used surnames. They concerted their series of expeditions with precise planning, so far as that was possible in the varying and often uncertain political climate and domestic conditions of Tibet at the time. I acted as their home agent and was responsible for the distribution of the seeds and living plants from all their travels.

Sherriff excelled at games at Sedbergh, where he delighted in the glorious country around the school, and on leaving he determined on a career as a professional soldier. So he went to the Royal Military Academy at Woolwich whence he was commissioned in the Royal Garrison Artillery early in 1918. Soon after going to France he was gassed and spent the rest of the war in hospital. In 1919 he was on the North-West Frontier as a gunner serving in the 117 Pack Battery in Mountain Artillery, and saw action in Wazeristan where he was mentioned in dispatches. He was an outstanding young officer, and his skill at games and generous nature made him very popular with his men. He entered the Consular Service in 1928 when he was appointed British Vice-Consul in Kashgar, eventually becoming Consul before leaving in 1932. He sensed that the Communist influence in Turkistan would lead eventually to the subjection of that country, but his opinion was not shared by his superior officers and he felt that these differences were so grave that he resigned. Thus he was able to animate the plans which he and Ludlow had for expeditions to the eastern Himalaya which, since they met in Kashgar, had become their lodestar. Sherriff's consular activities enabled him to travel widely in Turkestan and Ladak, and he kept exact diaries in which he recorded day-to-day events and his impressions of the countries and their people. On these and all his later journeys he showed supreme artistry with the camera, as the illustrations in this book reveal.

The outbreak of War put a temporary end to the Ludlow and Sherriff expeditions. Sherriff rejoined the artillery and for a time commanded an anti-aircraft battery at Assam, but he was brought back to political duties and based at Gangtok in Sikkim until in 1943 he succeeded Ludlow as British Resident in Lhasa, where he remained for two years. He made a unique series of colour ciné films of life in Lhasa, with marvellous portrayals of the colourful distinctive ceremonies, and these are a wonderful and fascinating record of a culture which has been extinguished.

In 1950 Sherriff retired to Scotland where at Ascreavie, Kirriemuir, at an altitude of 900 feet he and his wife created a garden strongly reminiscent of the Himalayan scene, where many of the lovely plants which had been the fruit of his expeditions flourish as luxuriantly as they do in their native habitat.

Sherriff's training and temperament made him an ideal organiser and his proficient planning, typical of his strict military practice, ensured that the primary object of the expeditions were achieved. He was responsible for arranging the commissariat, and even contrived to have vegetable seeds sown at intervals on the long journeys and to have porters collect the produce as it matured and bring it to the advance camps. Thus we frequently had tomatoes, lettuce, turnips and radishes in superb Russian salads prepared by Doud, our gifted Turki cook, who also excelled in baking succulent fruity cakes full of tasty ingredients but whose *tour de force* was an invigorating dish called *Tukpa* — a sort of rich vermicelli which took hours to prepare in the mixing, kneading and rolling of eggs and flour and the chopping of meat, the resulting dish being served piping hot as a delicious nourishing ragout.

Much foresight was exercised to ensure the comfort of the entire party. Sherriff knew success depended on all members being comfortable, contented, fitly shod, fairly paid, well-fed and when deemed necessary—and for the principals that was every night— suitably fortified with the highly concentrated Jamaica rum and 'Treasury Whisky' from the family distillery at Bowmore,

Islay. He spared no effort in his preliminary preparations and he bore by far the greater part of the expenses of mounting these costly expeditions. Thus the polyglot band of collectors, servants and porters were kept in good spirits largely by Sherriff's intuitive command. He also showed timely resource as a nurse, and I was immensely grateful for his sympathetic ministrations when, in 1938, I was floored by a puzzling and painful illness which immobilised me for about a fortnight. Following his care I was able to resume a full part in the activities of the expedition.

To keep in touch with the outside world Sherriff provided a massive short-wave radio and two recollections of significant events are particularly memorable. Naturally, at a time of international tension when we were isolated by weeks of travel from India we were concerned to know the trends of events at home, as there was a distinct likelihood that we might have to abandon the expedition and return to more serious duties. Surging over the radio when we were in camp at Melin on 30th September 1938 came the voice of Neville Chamberlain announcing his agreement with Hitler over Czechoslovakia, and this gave us borrowed time to complete the expedition. The other highlight which we relished was the commentary on the Test Match at the Oval in August when Len Hutton made his record score of 364. This episode did much to hearten me as I lay stricken with illness at Kyabden.

Nor was the need for agreeable reading matter overlooked. A pleasurable element amongst the store boxes was a small library, mostly of Nelson's Classics, including novels by Scott, Trollope and others. A thin paper copy of Kipling's poems was much used and so also was Palgrave's *Golden Treasury*. Opportunities for reading only came at night when we retired to our respective tents with candles, and I remember being particularly engrossed and burning a good many candles into the night over Wilkie Collins' *The Lady in White*. As space in the cases became necessary for our collections the books were discarded.

Both men were splendid marksmen, Ludlow especially with a catapult and Sherriff with a gun. The skill of being able to hit small birds with catapult slugs was a tremendous advantage in certain parts of Tibet where there was a complete ban on the taking of life and stringent veto on the use of firearms, so that the silent despatch of rare birds by means of a catapult added many valuable skins to the collection. Ludlow's prowess with the catapult was balanced by Sherriff's proficiency with the gun. His crack marksmanship in bringing down such birds as diving swifts was quite extraordinary.

The combined results of the Ludlow-Sherriff expeditions — botanical, horticultural, ornithological and etomological — are a magnificent contribution to our knowledge of the natural history of a region of breath-taking grandeur still greatly unexplored. Inevitably an end had to come to these rewarding expeditions, but it was sad that the Ludlow and Sherriff partnership was unable to put a seal on exploratory work in the Tsangpo area. This vast and topographically difficult country is surely the greatest existing sanctuary of horticultural and botanical riches still untapped, and without doubt there remain to be discovered many first-class plants of garden value. Let me quote from a letter from Ludlow dated 17th June 1949: 'It would be grand if I could conclude my research with an investigation of the Tsangpo Gorge, wouldn't it? Just think of the plants there, there were 20 different Rhodos at Pemakochung in four days and we didn't go more than a couple of miles from the monastery or ascend higher than 10,000 ft. Nature has run riot there. New species by the dozen flaunting their blooms asking for discovery and demanding a name.'

In this prefatory chapter it may be of interest to summarise the history of plant collection in the area chosen by Ludlow and Sherriff for their main collecting activities. Their expeditions are described in detail in the main part of this work and thus there will be a degree of factual repetition here to maintain the chronological sequence, but this will not detract in any way from Dr Fletcher's detailed account. Ludlow, Sherriff and I had in mind eventually to prepare an account of the series of expeditions to the eastern Himalayan area and as a long term project with the collaboration of the Department of Botany at the British Museum to elaborate the results into a catalogue of the plants of the Tsangpo drainage area north of the Himalaya and of Bhutan and outliers of Tibet to the south. Preliminary notes for this object were prepared by Ludlow, Sherriff and myself and as it now seems unrealistic to expect this work to proceed it might be useful to list chronologically the significant collections made in the region. In compiling this list with its ancillary comments I have relied heavily upon and quoted freely from Ludlow and Sherriff's voluminous notes and long series of letters. This account embraces the following areas in the Himalaya: (a) Bhutan; (b) the drainage of the Tsangpo, as far north as the 31st parallel and lying between the 89th and 96th meridians; (c) Tibetan-administered districts south of the Himalayan axis lying between the above meridians. A glance at the map will show that this area comprises roughly that part of south-east Tibet lying between Chomolhari in the west and

Namcha Barwa in the east, bounded to the north by the Ninchentangla and Yigrong ranges, and to the south by the Tibetan-administered districts of Chumbi, Monyul, Mago, Packakshiri, and Pemako. The whole of the Bhutan comes within this compass. But Sikkim is excluded and also the montane and sub-montane tracts of Northern Assam, inhabited by the Akas, Abor, Daphlas, Mishmis and other barbaric races.

1838–39. The first important collection of plants from the area was made by Dr William Griffith in Bhutan in 1838–39. George Bogle in 1774 and Samuel Turner in 1783 both passed through Bhutan when despatched by Warren Hastings to the Tashi Lama at Shigatse, but neither left any account of the floras of either Bhutan or Tibet. Dr William Griffith however, is in a totally different category. Already a botanist of international repute, he was attached as medical officer to a political mission led by Capt. R. B. Pemberton to the court of Bhutan in 1838. The mission entered Bhutan by way of Diwangiri in December, travelled northwards through Trashigong to Trashiyangse, and thence turned westwards through Biaka and Trongsa to Punaka, which in those days was the capital. Here they remained for a month, and then turned south making their exit from Bhutan to Buxa in May 1839. During the four and a half months Griffith spent in Bhutan he collected 1,500 species of plants, and so thorough and methodical was his work that at the present day it is possible to trace from his diaries and charts almost the exact locality in which each species was collected. In addition to his botanical and medical work Griffith also helped to collect birds, and, for this purpose engaged as his assistant a taxidermist called Monteiro. Several hundred skins were prepared, but exactly how many we do not know. Griffith, however, was no ornithologist and the labels written in French by Monteiro, were not very informative, and omitted all dates and locations. However, the amazing thing is that Griffith, despite his multifarious duties, was able to make such a large and comprehensive collection of plants in so short a period and at a time of the year when most of the plants he encountered were either dormant or just waking from their winter rest.

1849–50. A decade passed. J. D. Hooker was in Sikkim, and had already completed the first of his great journeys and had returned to Darjeeling with a collection amounting to eighty coolie loads. The horticultural world was astounded at the richness of the Sikkim flora and clamoured for further research and more seed, and so Thomas Booth, a nephew of the American botanist Thomas Nuttall, was despatched to India by his uncle for the express purpose of

collecting rhododendrons, orchids and ferns. And hereby hangs a
tale which deserves to be told, although strictly speaking the
activities of Thomas Booth do not concern us. At the present time it
is widely believed that Booth collected his specimens in Bhutan, but
this is not so. Not a single locality mentioned by Booth is to be found
on any map of Bhutan. Such localities as he gives occur to the east
of Bhutan in what is nowadays called the Balipara Frontier Tract.
Rhododendron nutallii, for example, unsurpassed in the genus for the
beauty and fragrance of its blooms, was discovered by Booth on
the banks of the 'Papoo Bootan', a river that rises in the Dafla
Hills and flows south-eastwards into the Bhareli River with
which it unites at approximately lat. 27°15′ long. 93°3′. Again,
Rhododendron windsori was collected at 'Roophry' which is the same
place as Rupa, a village on the Tenga river in lat. 27°13′ long.
92°24′ and the Tenga river itself is the type locality of *Primula
boothii* which is still recorded as having been collected by Booth in
Bhutan (*Trans. Roy. Soc. Edin.* LXL part I p. 281).

In Hooker's *Journal of Botany* and *Kew Miscellany* (Vol. 5 p. 353,
1853) Nuttall described twenty-two species of rhododendron collected by Booth in 1849–50 of which sixteen were new. (In this same
Journal Booth described *Rhododendron nuttallii*, which was also new.)
Dr Anderson, however, of the Botanical Survey of India ('Report
of the Political Mission of the Honourable Ashley Eden to Bootan'
p. 134) speaks rather disparagingly of Booth's efforts, and remarks
'the contributions made by Booth to our knowledge of the flora of
Bootan are very meagre', and goes on to say that Booth appeared to
have devoted himself to the collection of seeds, ferns and orchids.
Dr Anderson, it will be seen, was also under the delusion that
Booth collected in Bhutan though the maps of his day show quite
clearly the eastern frontier of Bhutan in exactly the same position
modern maps show it today, i.e. along the 92nd meridian.

And now for more than half a century there is no evidence that
any collections of importance were made either in Bhutan or Tibet,
though in 1888 H. A. Cummings of the RAMC, who served as
Assistant for India for a time in 1906 at Kew and later became
Professor of Botany at Cork, was medical officer with the British
Forces that defeated the Tibetans and advanced into the Chumbi
Valley. He made a collection of plants which are preserved at Kew.
In 1863–64 another political mission under the leadership of the
Honourable Ashley Eden proceeded to Punaka via Kalimpong,
Sipchum Ha and Paro. Many obstacles were encountered by this
mission and there is no record of any plants having been collected.
In the '80s and '90s a few plants were collected by various botanists

along the southern frontier of Bhutan, and in so-called 'British Bhutan' which for the most part means the Kalimpong district. It is probable that some of these specimens were actually gathered within Bhutanese territory, but no one collected during this period very far across the border, and it was not until the Younghusband Mission of 1903 — 04 that any noteworthy botanical collections were made.

1903–04. When Colonel Younghusband led his famous mission to Lhasa he had on his staff Colonel L. A. Waddell and Captain H. J. Walton, two officers of the Indian Medical Service, both of whom collected plants and birds. A complete list of Waddell's plants, totalling 160 species, was prepared by Prain and appeared in an appendix to Waddell's fascinating book *Lhasa and its Mysteries.* No list ever appears to have been compiled of the plants collected by Walton. Many new species were discovered by these two medical officers, and these have from time to time been described in the pages of the *Kew Bulletin.* Perhaps the most striking of these new plants was the Lhasa Poppy (*Meconopsis torquata*), which was discovered by Walton, or one of his collectors, in the vicinity of Lhasa. Plants collected on this expedition are preserved in the herbarium of the Royal Botanic Gardens, Kew.

1905–07. Sir Claude White, Political Officer of Sikkim, toured extensively in Bhutan. In 1905 he journeyed to Punaka to present the KCIE to the Maharaja, and returned to Sikkim via Lingshi Dzong and Phari. The following year he explored eastern Bhutan and crossed the main Himalayan range into Tibet by the twin passes Kang La and Bod La (Pö La). In 1907 he paid yet another visit to Punaka. White took more than a passing interest in plants and made several small collections. His name is commemorated in *Primula whitei* which, along with *Primula jonardunii*, now regarded as the equivalent of *Primula dryadifolia*, he discovered on the Kang La in 1906.

Major H. M. Stewart made a small collection in Gyantse in 1907, and two Lepcha collectors Ribu and Rohmoo, in the employ of the Royal Botanical Garden, Calcutta, collected in the Chumbi Valley and elsewhere in Tibet in the neighbourhood of Phari.

1913. The first man to reconnoitre thoroughly the lower Tsangpo valley in Tibet and to reveal something of the natural history riches of the region was Frederick M. Bailey, who was one of the most remarkable and least acclaimed explorers of this century. He was proud of his Edinburgh stock and was educated at the Academy of that city. He accompanied the Younghusband mission to Lhasa in 1903, and in 1913 with Captain H. T. Morshead of the Survey of

India he penetrated into south-eastern Tibet through the outlying province of Pemako. They were the first to confirm that the mighty Tsangpo did in fact cut through the Himalaya to emerge in India as the Dihang, a major tributary of the Brahmaputra. The painstaking report of this journey, printed as a confidential document by the Government of India, is a most impressive document and a remarkable tribute to his powers of observation and recording. On the journey Bailey collected fragmentary botanical specimens and also seeds. From his imperfect material Prain described two new species of *Meconopsis*, of which one was the celebrated Blue Poppy which bore his name but which, alas, is only a local variant of the previously known *M. betonicifolia* from south-west China. The *Rhododendron* which commemorates him was described from plants raised from seed which he sent to the Royal Botanic Garden, Edinburgh. Bailey also collected butterflies and birds.

1914–15. Considerable collections were made by R. E. Cooper in Bhutan in 1914 and 1915. He became aware of the richness of the Himalayan flora when under the tutelage of W. Wright Smith who was in Indian Government Service at the Royal Botanic Gardens, Calcutta, and who collected in alpine areas within Sikkim. Cooper made two journeys for desirable garden plants for A. K. Bulley who had also been a patron of Forrest and Kingdon Ward. He discovered a number of new species and introduced several plants to gardens in this country, though in the midst of World War I his material suffered from lack of attention. Cooper has written about his journeys in *The New Flora and Silva* volumes I and II, and in volume 18 of the *Notes from the Royal Botanic Garden Edinburgh*.

1920–50. Unquestionably these were the golden years of exploration in the area and resulted in the formation of many plant collections. Though some were small in size they were all of botanical interest. Such were those made by Cave in his visits to the Chumbi Valley and on the one occasion when he reached Gyantse. And so too were those made by the various Everest expeditions, although for the most part they are outside the scope of this review. These later small but valuable collections are all housed in the Royal Botanic Gardens at Kew whereas Cave's material is in the Calcutta herbarium. It was whilst in the employ of the Tibetan Government at Gyantse from 1923–26 that Ludlow first began to collect plants. All his specimens were plateau plants and were gathered either at Gyantse or in the vicinity. In all there were 200 gatherings. Unfortunately, he left untouched the rich flora of the Chumbi Valley in the mistaken idea that it was well-known.

Ludlow's plants are in the British Museum (Natural History).

The one outstanding plant collection of the 1920s, in fact more important by far than the previous ones, was that made by F. Kingdon Ward and Lord Cawdor in 1924–25. They entered Tibet via the Chumbi Valley in March 1924 and proceeded to Gyantse, which they reached in early April. Then they struck eastwards along the shores of the Yamdrok Tso to Tsetang and followed the Tsangpo river as far as Tsela Dzong, the capital of Kongbo. Leaving Tsela Dzong in May, Ward crossed the Temo La to Tumbatse in the Rong Chu Valley and made this delectable spot his base for the botanical exploration of Kongbo. Ward's principal gatherings were made on the Doshong La and Nam La on the main Himalayan range, and on the Temo La, Nyima La and Tra La on a subsidiary range on the left bank of the Tsangpo. In addition to working these passes Ward and Cawdor also collected on the Nambu La and Pasum Kye La and reached the great Tibetan highway known as the *gyalam* at Atsa Gompa. In November 1926 Ward and Cawdor began their remarkable exploration of the Tsangpo Gorge below Gyala. On their return journey to India in December and January they again crossed the Nambu La to Gyantse, and thence journeyed in a south-westerly direction across the Kumba La to Tsetang. From Tsetang they continued south across the frozen plateau to the trade mart of Tsona, whence they crossed into east Bhutan by the Pö La and reached the railhead at Rangiya in Assam in February 1925. A full and fascinating account of this great journey is contained in Kingdon Ward's *The Riddle of the Tsangpo Gorge*.

Botanically and horticulturally this was one of the most fruitful of all Kingdon Ward's expeditions, and forcefully demonstrated for the first time the astounding richness of the flora of south-east Tibet — a flora equal in its wealth and variety to the 'Eldorado' Forrest had discovered in north-western Yunnan. It was from seed gathered on this journey that *Meconopsis betonicifolia*, *Primula florindae* and *Primula alpicola* were first introduced into British gardens.

Hardly any less notable was Kingdon Ward's lightning journey of botanical and geographical exploration in south-east Tibet in 1935. Leaving Charduar, an outpost on the Balipara Frontier Track, on 26th April, he penetrated northwards into Monyul and Mago and thence crossed the Tulung La and Pen La, two passes on the Himalayan range each over 17,000 feet into Tibet. Descending the Chayul River he reached Chayul Dzong on 19th June and the large monastery of Sanga Choling ten days later. Tsari, a botanist's paradise where no cultivation is permitted, was reached via the Cha La and a visit paid to Migyitun, the last Tibetan village on the upper Subansiri. Leaving the Tsari district by the Bimba La, Ward turned north to Kyimdong Dzong and then crossed the Lang La to Molo and Tsela Dzong, whence he was on ground already familiar to him as a result of his 1924–25 journey. Crossing the Temo La he reached his base at Tumbatse on 27th July and descended the Rong Chu to Tongyuk Dzong. The sight of the unknown snowy Yigrong range had excited Ward's explorer instincts on 1924, and he now determined to ascend the Po-Yigrong river to its source and place it on the map. From Tongyuk Dzong he crossed the Sobhe La to Tsona Chamna at the head of the Yigrong Tso, and from there commenced his ascent of the great river. The track was always difficult, often dangerous, and involved dizzy cliff climbs and frequent river crossings by means of rope suspension bridges, but in seventeen days Ward found the source of the river in two longitudinal glaciers, and made his exit from the Yigrong by a pass called the Lochen La. Ward now turned south to Gyantse Dzong on the *gyalam*, proceeded southwards to De, and reached Tromda on the Tsangpo via the Ashang Kan La on 5th September. Still continuing south he crossed the Kongmo La into Tsari, the Dip La into Charme, and the Mo La into Chayul whence he retraced his footsteps along the route he had followed in the spring of the year to Charduar. On this journey, which lasted for six months, Ward marched 1200 miles and crossed thirty major passes. Constantly on the move, and with the minimum of transport, Ward nevertheless brought back a large number of valuable plants. The important geographical results of this expedition were many, but of more moment in this survey was his discovery of the rich alpine and forested areas of Tsari on the Po Yigrong. Ward's specimens are in the British Museum.

Although F. M. Bailey, Political Officer in Sikkim, paid an official visit to Lhasa in the summer of 1924 and collected a few plants in the process, it was during the 1930s that several collections were made in and around Lhasa. In 1935 there was the C. S. Cutting and A. S. Vernay Lhasa expedition, which also visited the important towns of Gyantse and Shigatse. Cutting and Vernay made a collection of about 200 numbers, amongst which were one or two new species which were described by C. G. C. Fischer in the *Kew Bulletin* of 1937. There was the journey to Lhasa in 1936 by F. Spencer Chapman, when he accompanied Sir Basil Gould on the latter's mission to Lhasa and collected assiduously both along the treaty road and at Lhasa itself. About 500 species were obtained, a list of which, prepared by C. G. C. Fischer, can be found in an appendix to Chapman's book *Lhasa the Holy City*.

In 1939 H. E. Richardson, whilst in charge of the British Mission in Lhasa, made a collection of approximately 400 gatherings which he presented to the British Museum (Natural History).

Sir Basil Gould, Political Officer in Sikkim, made official visits to Bhutan in 1938 and 1939, during which he collected 2,500 numbers. The bulk of the material was obtained along the Ha-Paro-Bumthang highway but Sir Basil also dispatched collectors into other areas. The collection was a valuable one and contained several new species. All the specimens are in the herbarium at Kew.

Although somewhat outside the boundaries of this review, yet impinging upon it, it seems desirable to mention here a very important collection of 3,500 specimens made by my friend and former colleague at Kew, Dr N. L. Bor, in the Aka-Sherdukpa country in the Balipara Frontier Tract of northern Assam in 1931–33. An account of this collection is to be found in India Forest Records (Botany) vol. 4 entitled 'A Sketch of the Vegetation of the Aka Hill, Assam'.

For the rest, the thirty years under review are dominated by the seven journeys of plant exploration made by Ludlow and Sherriff and their colleagues, during the course of which they botanised in all the areas traversed by those who had collected before them and, in addition, ventured into many previously unexplored regions both in Bhutan and south-east Tibet. The first was a five-month journey in 1933, chiefly to the Trashiyangse Valley in eastern Bhutan, to the Me La in the extreme north-eastern corner of Bhutan and then over the main Himalayan range into Tibet. The result was a collection of close on 750 bird skins and over 500 gatherings of plants which included several species new to science and the rediscovery of *Meconopsis superba* in western Bhutan, as well as a renewed determination to undertake a series of journeys gradually working eastwards through Tibet along the main Himalayan range, each succeeding journey overlapping its predecessor until the great bend of the Tsangpo was reached.

Towards this end in 1934 Ludlow and Sherriff visited the basins of the Tawang Chu and Nyam Jang Chu in Tibet, entering and returning by the Trashiyangse Valley in eastern Bhutan. 600 gatherings of plants, with a predilection for the more aristocratic genera, as well as almost 1,000 bird skins resulted from this five-month trek. In 1936, accompanied by Dr K. Lumsden, their objective was the exploration of the upper reaches of the Subansiri, the district Kingdon Ward had visited in the previous summer.

Again the route out and back was through eastern Bhutan, and during the ten months they were in the field they made nearly 2,000 gatherings of plants and sent home to Britain two crates of living material (which travelled in the cold room next to the butcher's shop on a P & O liner) and numerous packets of seeds. Although on this particular journey Ludlow and Sherriff's gatherings were somewhat limited, these contained such a large percentage of new species that this expedition to the Upper Subansiri must certainly rank as one of their most successful ventures. In the genus *Primula*, out of sixty-five species and varieties gathered fourteen were new, and with the genus *Rhododendron* fifteen out of sixty-nine were novelties. In *Meconopsis* there was a new species of great beauty (*Meconopsis sherriffii*) and a new yellow variety of *Meconopsis horridula*. In July 1937 Ludlow was occupied with work in Kashmir and so was unable to accompany Sherriff to the high massif in Bhutan called Dungshinggang by the Bhutanese and the Black Mountain by the Survey of India. Sherriff made 600 gatherings of plants during his four-month journey, paying special attention to the genera *Primula* and *Rhododendron*.

In 1938 Ludlow was free to join Sherriff, and I was asked to accompany the two friends on a ten-month journey within the drainage of the Tsangpo from the vicinity of Molo on the Lilung Chu down to Gyala at the entrance to the Tsangpo Gorge. The collections made on this memorable journey embraced every form of plant life and totalled nearly 4,000 gatherings, a vast quantity of seeds and many living plants which, probably for the first time in the history of plant exploration and certainly in this region of the world, were dispatched to Britain by air. The results of this expedition were satisfactory in every way, and Ludlow has described it as producing the largest and most comprehensive collection of plants that has ever come out of Tibet in one season, the collections being all-embracing and of great taxonomic and horticultural value. Alas, the outbreak of World War II prevented proper care being given to the living plants and the seeds so that the horticultural results were largely nullified, though a number of species were introduced and still survive in gardens. Unfortunately too the bombing of the British Museum severely damaged some of the herbarium material.

Whilst Ludlow, Sherriff and I were in south-east Tibet Kingdon Ward was collecting in Monyul, a district whose boundaries are ill-defined and which at the time of Ward's visit was being administered by Lhasa through the big monastery of Tawang. Ward followed the usual Assam–Tibet trade route from Charduar to

Dirang Dzong after which, with Dirang Dzong as his base, he spent the greater part of the flowering season exploring the Poshing La, Ze La, and Orka La passes. He returned from Assam in October with a valuable collection of 800 gatherings.

During the years of the war plant exploration on a large scale in the eastern Himalaya was quite impossible, though Ludlow and Sherriff seized any limited opportunities they had to collect. From April 1942 to April 1943 Ludlow was in charge of the British Mission in Lhasa, and he was succeeded by Sherriff who went to Lhasa with his wife and remained there until April 1945. The story of these Lhasa years is charmingly told by Mrs Sherriff on pages 227–46 of this book. Plants were collected at intervals between their official duties, chiefly around Lhasa but collectors were also sent as far afield as Reting, sixty-nine miles north of the capital where the rainfall is heavy and the vegetation more luxuriant in consequence. The flora of Lhasa and its neighbourhood is of great interest and contains a number of endemics, but by far their most important find was the Lhasa Poppy (*Meconopsis torquata*), until then very imperfectly known from a single gathering made by Walton or one of his collectors in 1904. It was rediscovered by Ludlow and Sherriff on the mountains overlooking the city.

The end of the war gave Ludlow and Sherriff the opportunity of continuing their botanical exploration of south-east Tibet, and in 1946 their objectives were the Po Yigrong range and the low-lying gorge country in Kongbo and Pome. Mrs Sherriff accompanied the party on this expedition and Colonel Henry Elliot joined as Medical Officer. They travelled in the winter to enable them to be on the rich collection grounds by early spring. From January to March 1947 the members of the expedition collected in the lower reaches of the Po Yigrong and the Po Tsangpo and visits were paid to Showa Dzong and Gompo Ne. In March the expedition split up. The Sherriffs made preparations to return to the Yigrong and Po Tsangpo whilst Ludlow and Elliot left for the gorges of the Kongbo Tsangpo below Gyala. Unfortunately a catastrophe now occurred which upset the carefully-laid plans. Sherriff, who had strained his heart badly in his exertions in 1938 and was also affected by his lengthy residence in Lhasa, now began to suffer so acutely from his old strain that he was compelled to return to India. In spite of this ill-fortune the fourteen-month journey produced close on 4,000 pressed plants, four crates of living plants which were sent home by air freight from Calcutta and a vast amount of seeds. The journey was the most ambitious of all the Ludlow and Sherriff expeditions and had it not been for Sherriff's unfortunate illness the results would have been more eminently successful.

This proved to be the last of Ludlow and Sherriff's Tibetan expeditions. In 1949 Ludlow planned to return to Tibet to work the gorge of the Tsangpo whilst Sherriff, mindful of his heart, planned to work at a lower altitude in the Mishmi Hills from a base near Rima. But both these projects came to naught as neither the Tibetan nor Indian Governments would accord the necessary sanction. The shadow of Communist China was already descending on the great Himalayan frontiers. Thus foiled in their plans to work independently two of the richest areas left to the field botanist, Ludlow and Sherriff decided to undertake another joint expedition and carry out as complete a botanical survey as possible in temperate and alpine Bhutan from west to east. Once again Mrs Sherriff accompanied her husband and Dr J. H. Hicks also joined the expedition both in a botanical and medical capacity. Fortunately Dr Hicks was at hand when Mrs Sherriff, as a result of a loose saddle-girth, was thrown from her mule, fractured her arm and had to return for medical treatment to India. During the seven months Ludlow and Sherriff and Hicks were in the field they made approximately 5,000 gatherings of plants for the herbarium, numerous bulbs and tubers and almost one hundred living plants which Sherriff had flown to Britain from Calcutta, and seed in such bulk that it was possible to distribute 20,000 packets. One of the most noteworthy plants discovered was a new lily named *Lilium sherriffiae* in honour of its discoverer, and another important find was the rediscovery of the beautiful pink *Meconopsis sherriffii* growing in great quantity on the upper Po Chu and Mangde Chu. This last expedition of the Ludlow and Sherriff series was certainly crowned with success and a magnificent climax to their years of endeavour.

The story of these Ludlow and Sherriff expeditions is chronicled in detail in the following pages. Like Forrest, and quite unlike Kingdon Ward, Ludlow and Sherriff were loth to put pen to paper and apart from the occasional brief article neither made any attempt to make the story of their exploits generally available. Ludlow did however publish a good many papers recording the ornithological results and also, in later years, on the botanical collections, while Sherriff in his retirement was often willing to lecture and to show his remarkable films. However, methodically and conscientiously they both kept a daily diary of all their travels and it is these records as well as their copious field notes which have formed the basis for Dr Fletcher's narrative. As both men regarded their expeditions as essentially scientific their diaries in large measure reflect this attitude, Sherriff in particular often-writing at considerable length on the plants in which he was particularly interested. Thus while it is not very difficult to obtain a clear idea of the plants they were collecting there is no real indication in their personal writings of the character of the country and the terrain from which the plants were taken. On our 1938 travels we took with us a copy of Kingdon Ward's *The Riddle of the Tsangpo Gorge* and also F. M. Bailey's confidential report on that region. Together these gave a very fine description of the country but alas are not now readily available. In the *Journal of the Royal Horticultural Society* vol. 77, 1947, I published an account of the 1938 expedition and my good friend Harold Fletcher has used small extracts from that article in chapter five. I was mostly concerned to describe some of the more interesting plants collected but I also tried to give some impression of the country and of the hazards of plant collecting in that region.

Chapter One

To The Pass of The Flowers,
N.E.BHUTAN

o

On 26 April 1933, George Sherriff and Frank Ludlow arrived in
Gangtok, the capital town of Sikkim, in preparation for their first
plant collecting expedition together. Gangtok, little more than a
small town on the mountainside, is one of the world's smallest
capitals. Two days before, in Kalimpong, the largest entrepôt of the
Indo-Tibetan trade, they had met the remarkable Dr John Anderson
Graham who, through his organisation of the St Andrew's Colonial
Homes for children of European and mixed parentage, made
Kalimpong familiar to many English-speaking people. Dr Graham's
youngest daughter was later to play an important rôle in the
explorations of Ludlow and Sherriff. Also in Kalimpong they had
discussed their itinerary for the next few months with Raja Sonam
Tobgye Dorje, the Prime Minister of His Highness Sir Jigmed
Wangchuk, KCIE, Maharaja of Bhutan.

The Minister, known to Ludlow and Sherriff, and always re-
ferred to by them by the name of Tobgye, had the greatest interest
in, and sympathy for, their explorations, and it was in large measure
through his influence with the Maharaja that they were able to
travel throughout Bhutan on so many occasions without restraint.
Tobgye's father had owned, in Western Bhutan, the lovely Ha
valley which runs parallel to the better-known Chumbi valley, a
narrow strip of Tibet interposed between Sikkim and Bhutan. He
had always been regarded with much favour by the Government of
India and had rendered considerable assistance to the Younghusband Mission of 1904 — the mission which eventually became
an armed expedition and penetrated to Lhasa before concluding
a treaty with the Tibetan Government. Tobgye had inherited from
his father not only the estate in Western Bhutan but much of his
character as well. Wise and far-seeing, he had the future progress
of Bhutan very much at heart. He realised that the mediaeval
system of Government was doomed and that his country must open

her doors wide to contacts with the outside world. Improved education, medical services, the development of Bhutan's forestry resources, were all projects he had begun to develop. Also, on his initiative, fruit trees — apples and pears — were introduced from Kashmir, and some of Bhutan's rivers stocked with Kashmir trout. In later years Ludlow maintained that in the Ha river, as well as in the Paro and Thimbu rivers, there was trout fishing unsurpassed in any country in the world; trout of 5–10 pounds were common-place and there were monsters of 18–20 pounds.

In Kalimpong, with Tobgye's assistance, Ludlow and Sherriff had concluded their plans. Hunting for plants — as well as for birds — they would work the Chumbi valley near Lingmotang from mid-May till the beginning of June. They would then cross the Bhutan frontier to the Ha area, keeping north of Ha and west of the Paro watershed. This was Tobgye's land. At Ha they would be joined by Tobgye, as well as by F. Williamson, CIE, Political Officer in Sikkim, and his bride. The party would then journey leisurely — and, as befits any honeymoon party, rather light-heartedly — to Bumthang, the summer residence of the Maharaja. Once in Bumthang the naturalists would be guaranteed several weeks of intensive collecting in the north-east of Bhutan, whilst waiting for permission from the Lhasa Government to proceed, by way of the Yamdrok *Tso,* (=lake) to the famous town of Gyantse which stands 13,260 feet above sea-level about half-way between the Indian frontier and Lhasa.

The next few days in Gangtok were fully occupied in making final arrangements for the forthcoming journey. Supplies had to be collected and belongings packed into suitable loads. There were also certain social duties to be undertaken, such as a visit to the Maha-raja of Sikkim, His Highness Sir Tashe Namgyal, KCIE, and his charming Tibetan wife. Moreover, as Ludlow and Sherriff were also on the quest for birds as well as for flowers, and as bird-hunting is never very satisfactory during a long and tiring day's march, the stay in Gangtok offered an excellent chance for bird collecting in the vicinity of the Residency where they were staying. In this occupation they were severely handicapped by the weather, and by leeches. Although the monsoon proper does not break until mid-June, they had heavy rain daily, generally in the mornings. On 1 May there was a terrific hail-storm which covered the ground with inches of hail and destroyed most of the flowers and foliage in the Residency garden. Such hail-storms are of frequent occur-rence at Gangtok in the spring and must be extremely depressing and disheartening for gardeners. And Ludlow and Sherriff were

Map 1. From Gangtok to Trashiyangsi

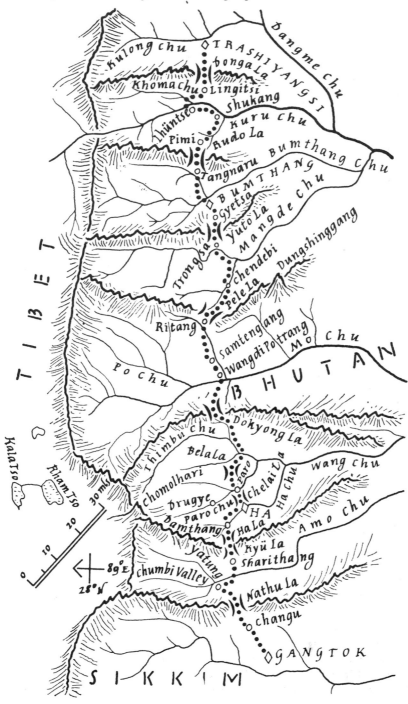

depressed and disheartened by the leeches — there were great swarms of them infesting the jungles near the Residency. 'We dared not move off the stony path. Immediately we did so we were attacked by myriads of these obnoxious creatures. Sometimes we shot a bird which dropped above or below the pathway. When this happened we made a dash for the bird and after retrieving it fled back to the path and picked off the leeches. If we could not find the bird immediately we had to leave it and beat a hasty retreat. We lost several specimens in this way.' [1]

To the Chumbi valley (Map 1)

On 5 May they were glad to leave Gangtok, by mule transport, for the Chumbi valley. Gangtok is some 6,000 feet above sea-level, and as they climbed steadily a further 3,000 feet the sub-tropical vegetation, dominated by *Ficus religiosa*, the pepul tree of the Hindus, was soon left behind and gave place first to a zone of bamboo and above this to a further zone of rhododendrons, great trees of *Rhododendron arboreum* 50–60 feet high, with many trees of the huge white-flowered *Magnolia campbellii*, twice that height, growing among them. Sadly, all flowers had been terribly battered by the recent hail.

The following day, on the way to Changu and its beautiful glacier lake, the first plant was collected and pressed for the herbarium. It was a small primula with a flat rosette of oval or oblong finely-toothed and conspicuously-veined leaves surrounding a cluster of pinkish-purple flowers each with an orange-yellow eye accentuated and surrounded by a zone of white. None of the party knew the plant — and neither did the primula authorities at the Royal Botanic Garden, Edinburgh. There it was at first attributed to the little known *Primula boothii*. Ludlow and Sherriff were to collect it on many future occasions, in Sikkim, Bhutan and SE Tibet, and were to send home living plants by air-mail. It was from the behaviour of these plants in cultivation that the first collection ultimately was identified as *Primula gracilipes* (L & S 1) [2] which first had been introduced into British gardens as long ago as 1886 and had been figured in the *Botanical Magazine* three years later, albeit under the name of *P. petiolaris* var. *nana*.

In view of the great interest both collectors were later to take in the genus Primula it is perhaps significant that of the first six plants collected, four were primulas. No. 2 was the well-known Himalayan Drumstick primrose, *Primula denticulata*, which first flowered in Britain over a hundred and thirty years ago, whilst nos. 5 and 6 were the at that time almost unknown, and closely related,

P. griffithii and *P. tanneri* respectively. The collectors remarked on the beauty of the large, deep purple, yellow-eyed flowers of *P. griffithii* which they found growing at 13,000 feet in the Chumbi valley in the shade of the conifer jungle; *P. tanneri*, with its pinkish-purple flowers, was at the lower elevation of 9,500 feet. Moreover the collectors endeavoured to introduce both species into cultivation not only in 1933, but in later years. They had little or no success with *P. tanneri*, but *P. griffithii* still is in cultivation mainly through the efforts of Major and Mrs Knox Finlay, who have grown it marvellously well for many years in their Perthshire garden, and who have distributed plants and seeds on numerous occasions. Plants they exhibited before the Royal Horticultural Society received the Cultural Commendation and the Award of Merit in March 1953. Those interested in primulas, especially in their taxonomy, were heartened and excited by the collecting of these three little-known species (1, 5, 6) but hardly foresaw the large numbers of rare and of quite new species Ludlow and Sherriff were to discover, and attempt to introduce into cultivation, in later years, and the enormous contributions they were to make to the biology and taxonomy of this group of plants.

The business of plant collecting, and of the preparation of the collected material into dried specimens for the herbarium, is a time-consuming one. For this purpose Danon, a Lepcha from a quinine estate near Darjeeling, had joined the expedition. He had worked for the Botanical Survey of India and appeared to know his flowers, even the scientific names of many of them. Though he had the occasional lapse, he proved a superb collector of plants and preparer of herbarium specimens. 'His eyes were everywhere and nothing living escaped them. He never spoke harshly to anyone, never quarrelled, never complained; he just roamed the forest or hillside with a plant-press on his back throughout the day.' [3]

At Changu the expedition had to halt for a week, held up by the heavy snows on the Nathu *La* (= pass) which would take them across the Tibetan frontier at 14,000 feet. Though frustrated by the delay they were not unduly depressed, for the Changu bungalow was reasonably comfortable, there was any amount of rhododendron wood to burn and thus no problem in keeping warm, and there were many interesting birds to study and collect. A strangely silent blackbird (*Turdus merula buddhae*), very powerful in flight and usually a denizen of rocky grassy hillslopes covered with dwarf rhododendrons and willows; a small Rose finch (*Pyrrhospiza punicea punicea*) in brown garb — there appeared to be a marked deficiency of red males; and a snipe (*Capella nemoricola*), many

5

specimens of which they were to see later, especially in early August, flighting like woodcock and uttering a croaking 'chur chur' call; these, and others, were added to the collection which was destined for the Natural History Museum in London.

The one mainly responsible for the preparation of the bird skins was the faithful Ramzana, a Kashmiri whose real name was Ramzan Mir. He was Ludlow's personal servant and, in fact, remained in Ludlow's employ from 1928 until 1950 when Ludlow left India. Ludlow originally engaged him to skin birds in 1928 for, as in 1933, Ludlow was far more interested in birds than in plants.

'Originally a tailor, he proved a most apt pupil and soon became far more proficient in the art of skinning than his master. Over the years he made upwards of 7,000 skins which are now in the British Museum. He possessed nimble fingers, made first-class skins, and worked quickly. On one occasion I remember him skinning fifteen birds in two and a half hours. A bird the size of a starling, say, would take him ten minutes. In addition to his skills as a skinner of birds he took a great interest in plants which he collected copiously. Slimly built, he delighted to spend the whole day on the mountain side where he was quite tireless, entirely on his own, and always he would return with a bulging press and often with his shirt stripped from his back and crammed with plants. Ramzana was quite illiterate, and like so many who can neither read nor write he had a wonderful memory and knew immediately when he had discovered a rarity or something new to our collections. Ramzana had another virtue; he was a really good cook, better by far than Ahmed Shaikh, his compatriot. He served me loyally and much do I owe him.'[4]

Ahmed Shaikh was the expedition's official cook. Ludlow had engaged him in this capacity in 1929 to accompany him to Kashgar and the Thian Shan. He was a fairly good cook but, being domineering and very argumentative, he did not mix too well with the other servants. And he was utterly uninterested in the work and collections of the expedition.

The enlightened Tobgye had had some of his more promising subjects educated in India and two of these Bhutanese youths he had placed at Ludlow and Sherriff's disposal. One of them was Pintso, who spoke respectable English and fluent Tibetan and Nepali. 'He was our interpreter and manager of transport and supplies

1. The Nathu La
2. Yatung, in the Chumbi valley

and we could not have had a better man. He got on well with everybody — rich and poor alike — being always jovial and loving a joke. Always *he* insisted on paying coolies and the owners of transport personally, and, on one occasion I remember, he paid a coolie woman twice the normal rate because she was going to have a baby, remarking that as she had been carrying a double load all day long she was deserving of double pay! There were roars of laughter of course, as well as approbation from the other coolies. Pintso was not interested in plants, or birds, and did no collecting whatsoever, but as transport officer and interpreter he was first class. He smoothed the way before us and as long as he remained in our employ we had no difficulties.'[5]

The other one of Tobgye's most excellent 'educated' subjects who accompanied the expedition was Tendup who, like Pintso, spoke very fair English — in addition to good Tibetan, Urdu and Nepali. But unlike Pintso he had an eye for a good and interesting plant. Being quiet in manner and extremely efficient in his work he was popular with everyone. Tobgye later chose him to undergo a three months' course of instruction in trout culture in Kashmir, and he afterwards successfully transported 20,000 ova across India to the little hatchery on the Ha river in West Bhutan where they prospered abundantly.

Not until 13 May was it possible for the expedition to cross the Nathu La and then only with difficulty, for the mules sank into the deep soft snow whenever they stepped off the narrow hardened snow-track which those on foot had prepared. On the summit of the pass there were still five feet of snow but the morning was bright and clear and the view quite lovely; to the south a low sea of cloud over the plains of India and to the north-east the cap of Chomolhari, 'Goddess of the Mountain Pass'. Only on one other high pass, during the next few months, would there be a similar superb view.

Once across the Nathu La and into the Chumbi valley the luxuriant Himalayan vegetation was soon left behind. Even so, there was much in the river valleys and on the Tibetan plateau to interest the naturalists. From the scrub jungle, at 10,000 feet, the first rhododendrons were collected; two forms of the immensely variable native of Sikkim, Bhutan and SE Tibet — *Rhododendron cinnabarinum*, one (7) with hanging bells salmon at the base paling to yellowish-pink at the tip, and the other (8) with bells of clear yellow; and the often-collected pale lemon *R. campylocarpum* (9) which ranges through Nepal, Sikkim, Assam and Tibet. Rhododendron was another genus to which Ludlow and Sherriff were to make important botanical and horticultural contributions, though not so much on this expedition.

From Chumbi to Bumthang

From the Chumbi valley they proceeded in leisurely fashion to Ha in West Bhutan, crossing into the Ha valley by the twin passes of the Kyü La (pronounced Chu) and the Ha La (14,000 feet) and halting for ten days at both Sharithang and Damthang, two encampments set amidst beautiful conifer and rhododendron forest on either side of the Massong Chung Dong Range. The forests were mainly of the beautiful Himalayan fir, *Abies spectabilis*, which occurs on the inner Himalaya from Afghanistan to Bhutan at elevations of 8,000 – 13,000 feet. With the abies there was a luxuriant undergrowth of rhododendrons, hollies, willows, viburnums and many another shrub.

At Sharithang (11,400 feet) they were housed in a large wooden hut with open windows along the south side. Though it was dirty its very size was useful for spreading themselves and their belongings, as well as their bird skins and flowers. Busily erecting huts for the arrival of Williamson and his party were some Bhutanese from the Ha valley. They too were very dirty, all dressed alike, all with short hair, so that it was difficult to know which were the men and which the women. But they were all extremely cheery and willing to help in any way, even to shooting birds and finding their nests. The encampment was thus a busy one. The road, too, was busy, with a good deal of traffic passing up and down carrying rice from Bhutan to Yatung in the Chumbi valley and returning with salt.

Of course a ten days' halt at a camp affords a splendid opportunity for collecting birds and flowers. But here birds were by no means easy to find, for the jungle was desperately thick and the constant noise of so many streams and rivers racing along made bird calls difficult to hear. Moreover, although plants were there in abundance it was a most difficult business botanising in the forest which was infested with ticks. However, half a mile above Sharithang, at 11,500 feet, in swampy ground in the forest, Sherriff found a real jewel of a plant; a primrose a few inches tall carrying a head of up to ten semi-pendent bell-shaped flowers so dark a claret in colour as to seem almost black until the sun caught them when they glowed a glorious rich ruby-red; this was *Primula kingii* (40) and the first record of it for Bhutan. There wasn't much of it in evidence, but on future expeditions Ludlow and Sherriff were to find it in masses — acres of it in fact — not only in the Ha valley but also in SE Tibet, and were to introduce it into cultivation. Little did Sherriff realise when he collected these first specimens

on 30 May that only he — and later his wife — was to cultivate this beautiful species with any measure of success.

On 1 June, on the journey to Damthang, ten miles distant, by way of the Kyü La and Ha La, two plants new to science were collected; a barberry, *Berberis lasioclema* (41) and a white-flowered apple, *Malus bhutanensis* (44). Damthang means 'The Muddy Meadow' — something of a misnomer for a spot where there were splendid grassy clearings in the forest both above and below the rest-house, which was built on the same spacious plan as that at Sharithang and lay near the junction of the two considerable streams which together form the Ha *Chu* (= river). The forest clearings had been made with fire and axe to provide grazing grounds for numerous herds of cows, yaks and zhos, the latter said to be a cross between the male yak and the common horned cow. These clearings apart, the woods, as at Sharithang, were very thick, in some instances almost impenetrable, making collecting, especially bird-hunting, very difficult. Even so, Ramzana, in camp, was always fully occupied preparing his bird skins. Collecting, of course, took up much of the time as did the drying and the packing of the bird and flower specimens. And sometime during the day (Ludlow and Sherriff usually rose at 4.15 a.m. and were abed by 8.15 p.m.) photographs had to be developed, and diaries, field notes, and sometimes letters, had to be written. Occasionally letters and papers had to be read, as on 6 June when a post arrived bringing the good news from Lhasa that the expedition might proceed from Bumthang to Gyantse, via Pomo Tso, Yamdrok Tso and Nangkartse.

As the Williamsons were expected at Ha on 15 June, Ludlow and Sherriff proceeded there on the 11th in the hope of a little plant collecting (shooting having to be curtailed) before the honeymooners arrived. The eight-mile journey to Ha was a pretty one, the land being cultivated for most of the way; wheat, barley and buckwheat were the principal crops, and very good they were, with lots of straw. Though the villages were small the houses were built on fairly massive lines. The two friends were given the comfort of Tobgye's guest-house delightfully situated on the left bank of the Ha river and opposite the three dome-shaped, very steep, wooded hills immediately behind the *Dzong* (= a fort, the headquarters of a district). The house was strongly built — the floor boards were 18 inches across and 7 yards long — with four rooms opening

3. *Primula kingii* (above)
4. *Primula obliqua*

from a central hall, all gaudily painted. Its comfort, and its hospitality, were almost too much for those seriously bent on a collecting expedition. Although herbs and shrubs were now fast coming into flower and although several new kinds were collected Tobgye was too kind and too lavish in his entertainment for organised collecting to be contemplated. In any case he was insistent on his guests occupying their days in other ways — in archery practice for instance. Archery is the national sport, about which the Ha people worked up great enthusiasm. The bow is a six-foot bamboo, the bow-string appears to be made from the fibres of some kind of nettle, and the target, a wooden board $3\frac{1}{2}$ feet by 15 inches, is placed at a range of 150 yards. Leather finger stalls are used as well as arm pads, otherwise finger tips and forearms become very sore. And woe betide the man who holds his bow upside down for *he* will be turned upside down — and flung into the river.

Most of one day was devoted to a serow and musk deer drive — and anything else that happened to be in the jungle. They climbed to the summit of one of three rounded hills overlooking the Ha valley to the west and took up very precarious positions in machans in trees. The machan Sherriff occupied was a pretty terrifying one, being dangerously situated in a tree on the edge of a precipice.

With the Williamsons' arrival all these entertainments, and many more, were provided for their amusement. The highlight was a great game of football — Ha versus the Political Officer's team. The game had been in progress only for a few minutes when, in some astonishing fashion, Sherriff bashed his head against an opponent's shoulder with such force that some of his front teeth were dislodged, his upper lip badly cut and his jaw damaged. The game was immediately abandoned and Sherriff wasn't fit for anything, least of all for serious collecting, for several days.

Despite such distractions and Sherriff's accident, several important gatherings of plants were made in the neighbourhood of Ha. In some instances it was the collectors' first acquaintance with species of which they were to see much more on future expeditions and which they were to attempt to introduce into gardens in Britain, either by seeds or by living plants. One such was that noble primula, *Primula obliqua* (129), with stout fleshy roots, overlapping reddish scales surrounding the base of the growing plant and forming a fat resting bud in the dormant season, and handsome head of 4–6 fragrant creamy-white or pale yellow flowers, the posterior petal elegantly bent backwards. It was revelling on moist well-drained slopes, covered with deep snow in the winter, in the shelter of scattered rhododendrons near the upper limit of the abies forest.

Another gem, in almost identical situations, was the exquisite *Corydalis cashmeriana* (131) — bright green deeply-divided leaves and sky-blue or turquoise-blue spurred flowers — pounds and pounds of whose tubers Ludlow and Sherriff later distributed to gardens in Britain where, for many years, only Major and Mrs Knox Finlay in Perthshire really solved the problem of their successful culture. Happily the plant now appears to be well established in several gardens.

There were also two species of meconopsis; *Meconopsis simplicifolia* (122), the dense tufted rosettes of which, beset with golden-brown hairs, were throwing up flowering stems of blue-purple flowers; and of much more significance, *M. superba* (93). Though overwhelmed by the beauty of the latter — flowering stems 3–5 feet high carrying large 4-petalled white flowers and arising from beautiful silvery-haired leaf-rosettes 2–3 feet across — neither Sherriff nor Ludlow was aware of the botanical importance of their find. Until this 14th day of June 1933, when specimens were gathered for the herbarium, *Meconopsis superba*, though in cultivation probably from seeds gathered by Col. F. M. Bailey on the Ha La in 1923, was known in its natural state only from the rather unsatisfactory original collections on which the description of the species is based. These had been obtained by a native collector at Ho-Ko-Chu in the Chumbi district of Tibet in 1884, and portions of them had been deposited in some herbaria in Britain, the type specimen being preserved in the herbarium in Calcutta. The splendid series of herbarium specimens which Ludlow and Sherriff now made enabled botanists to obtain a much more complete picture of this glorious plant, and the ample supply of seeds they later harvested gave gardeners a further chance to experiment with its cultivation.

On 21 June the entire party left Ha en route for Chanana via the Chelai La (12,500 feet), the transport being by ponies, mules and men and women, the women seeming to manage the 80 pound loads almost as easily as the men. On the following day they continued to Paro (7,750 feet). Soon after leaving the Chanana camp they were met and entertained by dancers sent by the local *Penlop* (= governor) who played and danced for half an hour and distributed saffron, rice and tea. Later, a dozen or more steel-helmeted soldiers of the Penlop arrived and preceded the party during its descent into the broad Paro valley. The Paro Penlop recently had built a most imposing three-storied guest house decorated in the usual ornate fashion of the country and occupying the centre of a flagstoned quadrangle at the four corners of which

were suites of rooms for guests. Here the travellers were once again comfortably housed in an atmosphere conducive to relaxation, and Sherriff, whose mouth had become a little septic, was hopeful that in such conditions he would recover quickly and completely from his accident.

One of these days of relaxation was spent sightseeing in the abundantly cultivated Paro valley up as far as Drugye. The wheat crop was being harvested and rice was being planted, the rice fields occurring as far as Drugye Dzong some 8,500 feet above sea-level. A mile short of Drugye, perched on a cliff 2,000 feet above the river, is the Taktsang (Tiger's Nest) monastery. The Paro Penlop had provided the party with mules and on these the tourists serpentined their way up a spur to inspect the *gömpa* (= monastery). There were several buildings scattered about the steep and precipitous hillslopes but attentions were confined to a monastery eerily placed on a limestone rock-face. A narrow tortuous track along a cliff-face linked the monastery with the bridle-path by which they had ascended. On reaching the gömpa they found themselves overlooking a chasm with a vertical fall of 600–700 feet into the ravine below. 'During the past ten years I have gradually grown accustomed to the amazing sites often chosen for Tibetan monasteries but the extraordinary situation of this Bhutanese monastery at Taktsang caused me greater astonishment than any I have hitherto seen.'[6]

Other days were spent in further sightseeing, in archery practice, in lunching with the Penlop and in more entertainments from his dancers. However, by 28 June Sherriff's mouth was much better, and in any case the time had come to journey eastwards. The Penlop and his guests rode up to the Bela La (11,650 feet) where he gave them all tea and rice and where the goodbyes were said. Everyone exchanged scarves with the Penlop. He held out his hands with a closed scarf in them whilst his guests placed an open one over his forearms before taking the one he offered. Then they parted and, following the custom, waved to each other and imitated wolf calls until the Penlop was out of sight.

They were now heading for Wangdi Potrang, four marches distant, and enjoyed a profitable time on the Dokyong La (10,500 feet) which proved a happy plant-hunting ground. Had the travellers been familiar with the British flora they would have been reminded of home by the presence of the round-leaved wintergreen, *Pyrola rotundifolia* (193), its lily-of-the-valley-like flowers spotted

5. Tobgye's residence, Ha Dzong
6. Tiger's Nest Monastery, Paro

15

16

within with purple, and by the bluish-purple-flowered Meadow crane's-bill, *Geranium pratense* (198). But no doubt they were much more impressed by the fragile beauty of the nodding white bell-flowers of that ally of the lily, *Streptopus simplex* (194). And they were so thrilled by the beauty of the magnificent white, egg-shaped or almost globular flowers of *Magnolia globosa* (192) that they took cuttings and optimistically planted them in a shady spot on the summit of the pass! Two rhododendrons at 10,400 feet fascinated them chiefly because the flowers were so strongly reminiscent of those of other genera. One, growing to a height of 20 feet, with correa-like, narrow, tubular flowers, was *Rhododendron keysii* (184) and greatly puzzled Danon; it was the form with red, yellow-tipped flowers. The other, a straggly shrub no more than 5 feet tall, with small camellia-like flowers, was, of course, *R. camellii-florum* (190), and a splendid form of it with deep wine-red fleshy flowers, as dark in colour in fact, and just as fleshy, as the port wine corollas of *Buddleia colvillei* (191) which grew about 1,000 feet lower down the pass.

Wangdi Potrang, some ten miles beyond the Dokyong La, was reached by a narrow path through the densest of jungles dominated both by ferns with chest-high fronds and by millions of leeches which, if the travellers halted for a second, covered their feet. The nine-hour march was thus tiring, strenuous and rather miserable. But Wangdi Potrang proved to be a picturesque village perched on a cliff above the left bank of the Mo Chu which was spanned by a splendidly-made cantilever bridge. The name has an interesting origin. Many years ago an old Shabdung arrived and found a child building mud castles. The child was called Wangdi and he told the Shabdung that he was making a palace (*potrang*). When the Shabdung built a Dzong here he called the place Wangdi Potrang or Wangdi's Palace.

The hot valley of the Mo Chu is comparatively dry and arid and the surrounding mountains are devoid of trees, except for *Pinus longifolia* near their summits, and are clothed only with grasses and shrubs. Naturally under such conditions and in such surroundings there were marked changes in the avifauna, and the characteristic birds of this hot dry zone were the Brown hill-warbler (*Suya criniger criniger*) and the Crested bunting (*Melophus lathami subcristatus*) with its typical call-note 'tweet-twe-twe-too'.

After a day in the heat of Wangdi Potrang the expedition moved on to Trongsa, five marches distant, halting at Samtengang in a pleasant camp on a hilltop among *Pinus longifolia*; at Ritang after a long hard march of fourteen miles in rain and mist; at Chendebi after another foul wet and misty thirteen-mile trek over awful slippery mud paths; at Tsangka after a third filthy wet day. Throughout these marches the travellers were again plunged into leech-infested rain forest, and at supper time at Chendebi, when the soup was brought in, there was a leech on the tray gaily swinging to and fro in search of a victim. In spite of the leeches, the rain, the mist, the mud and the long tiring marches, the flora was not neglected and the travellers added to their plant presses the great sweet-scented pale lemon blooms, deeply stained with dull crimson at the base within, of *Lilium nepalense* (223); the purplish-pink flowers of another lily, the low-growing *L. nanum* (225); and the apple-blossom-pink flowers, again scented, of *Rhododendron maddenii* (218).

The fifth day's march of eight miles to Trongsa (7,350 feet), on 7 July, was much easier, for the day was a fine one. For once butterflies, which the travellers were also collecting, were on the wing in large numbers. Even so, it was a relief to reach the beautiful camp site on the left bank of the Trongsa river, for Sherriff's mouth had again become septic; Ludlow's feet were blistered through treating leech bites, which wouldn't stop bleeding, with soda-nitrate crystals; and Williamson was in trouble with a septic arm, also the result of a leech bite. A day's rest, developing photographs, preparing bird skins, drying the plants, and admiring the river racing through a great deep gorge, probably would benefit everyone.

And so it proved, for the following day, 9 July, the entire party was fit enough to tackle the very steep seven-mile ascent to the Yuto La (11,200 feet) in rain and mist, gathering on the way further specimens of the epiphytic *Rhododendron camelliiflorum* (253), this time with paler, pinkish-red flowers, as well as material of the lovely large white-flowered *Clematis montana* var. *longluensis* (252). The descent on the eastern side of the pass, to Gyetsa (9,525 feet) produced a startling change in scenery and in vegetation, the woods of oak, maple, rhododendron and bamboo on the western side giving place to beautiful fir forests with open grassy glades.

Bumthang (9,725 feet), the summer residence of His Highness the Maharaja, was but a short march of six miles from Gyetsa and the arrival there of the distinguished visitors was greeted with

'a blaze of trumpets, bagpipes, drums and other instruments. . . . H.H. has the nucleus of an army in 25 young Bhutanese trained by a man who did some years with the 8th Gurkhas. They possess three bagpipes, two bugles, a cornet and two fifes. Add them to the skull drums, big drum and surnais of the bodyguard and you get a wonderful amount of varied music.' [7]

Collecting of any kind during the week at Bumthang was quite impossible, for the days were filled to overflowing with dinners, lunches, archery contests, football, racing, wrestling — even bull-fights, so anxious was the Maharaja that the best of Bhutanese hospitality should be lavished on his guests. Almost always His Highness and the Maharani, the two brothers Naku and Dorji, the sister Ashe Wangmo, and the half-sister Ashe Paldan were present, no matter the function, and at all times they were charming, thoughtful and kind.

On to Trashiyangsi

On 18 July the party divided, the Williamsons leaving for Lhasa and Ludlow and Sherriff proceeding eastwards. All were sorry to leave behind such good friends who had heaped gifts upon them and had gone to so much trouble in arranging for the naturalists' further journey eastwards and northwards. At the same time Ludlow and Sherriff were relieved at the thought that lavish hospitality was now at an end, that from now on their time was entirely their own and that they were free to collect to their hearts' content.

They headed eastwards for Trashiyangsi which, though the map did not say so, proved to be exactly a hundred miles from Bumthang and involved nine marches and a couple of days of rest. They found the map by no means accurate, the map names in many cases incorrect and often quite wrongly spelt. In fact for the next two months, until they reached Nangkartse on 13 September, they were not only making important botanical and ornithological discoveries but correcting the maps of the Survey of India at the same time.

The first march, one of fifteen miles, to Tangnaru (9,400 feet) was uneventful. The second, to Pimi (9,690 feet), thirteen miles distant, was a hard one, especially the latter part of it, and involved travel first through thick conifer forest, then on to open grassy hillsides and rhododendrons before reaching the Rudo La (12,600 feet) and several desirable plants, including primulas. There was *Primula dickieana* var. *aureostellata* (284), a denizen of boggy ground, specimens of which R.E. Cooper had found in this same locality in 1915; in fact this lovely plant, whose pale yellow flowers have a bright saffron-yellow star-like centre, is known only from this particular area of Bhutan. There was *P. prenantha* (291), a small, not very elegant, candelabroid species with one tier, or sometimes

7. Wangdi Potrang

two, of from 4–8 small yellow nodding flowers; it is native not only to Bhutan but to Sikkim, the Burma-Tibet frontier, Tibet and Assam as well. And there was a beautiful little plant Ludlow and Sherriff were to collect on several future occasions, for it is common throughout Nepal, Sikkim and Bhutan to the Mago district of SE Tibet, the dainty *P. pusilla* (294); small compact heads of purple or violet white-eyed flowers are held no more than a couple of inches above the damp rocky banks on which the species dwells. Most spectacular of all was *Meconopsis paniculata* (286), four-foot branched stems, carrying many lemon-yellow flowers, rising from great elegant flat leaf rosettes on the floor of the abies forest.

Part of the descent from the Rudo La to Pimi, in torrential rain and down none too safe rock steps along a knife edge, was extremely hazardous, especially for the ponies, and by the time the party reached Pimi they were all pretty tired and looking forward to a good night's rest. But the unprepared camp was in a swamp and the leeches were frightful; and though they took the precaution against the leeches of sprinkling brine around the legs of their camp beds the travellers spent a miserable night and next morning, 21 July, were happy to start the ten-mile walk to Shukang. It was literally a walk for the transport mules had disappeared during the night. Compared to the Pimi camp the one at Shukang (6,500 feet) was paradise — utterly free from leeches — and here the expedition halted for a day, developing films, drying specimens, shooting birds and collecting plants, including the first delphinium so far seen, *Delphinium carela* (298), low-growing and spreading its handsome large light purplish-blue flowers over the open hillside at 7,000 feet altitude.

From Shukang eastwards to Tangmachu (Takila) and the finding of a magnificent form of *Lilium nepalense* (310) with huge creamy-white blooms faintly tinged with pink at the base on the outside, and thus a very different form indeed from that which had been gathered at Ritang on 4 July. Thence to Lhüntse (4,520 feet), a short and very hot six-mile journey, though the 'Hot Baths' marked on the map just to the east of the bridge over the Kuru Chu were unknown, and certainly couldn't be found.

The sixth march from Lhüntse to Lingitsi (6,500 feet) was mainly noteworthy for the crossing of two rivers. The Kuru Chu had to be crossed by a frail-looking bamboo bridge fifty yards long. Having crossed the Kuru Chu, the Khoma Chu, also unfordable in summer, then had to be crossed by another extremely shaky bamboo bridge and again the mules managed only with the greatest difficulty for they only had a one-foot-wide plank on which to walk. The Kuru

Chu valley was desperately hot and it was a relief to reach camp at Lingitsi where there was a decent breeze.

The march from Lingitsi to Donga Pemi (10,000 feet) may have been only a short one of five miles, but for most of the way it was very difficult, the path through the thick forest being very steep and rough, often over tree-roots and rocks. But Donga Pemi was surrounded with delightful open grassy slopes, the camp was a good one, and as the country looked ornithologically interesting, they decided to halt here, for a day. And a splendid day it was, fine all the time, with beautiful cloud effects away down the valley, and a haul of twenty bird skins including those of a couple of owls (*Strix aluco nivicola*) and of the Spotted (*Bradypterus (Tribura) thoracicus thoracicus*) and Brown (*Bradypterus (Tribura) luteoventris luteoventris*) bush-warblers. In the evening letters and presents of ham, sausages, bread, cakes, vegetables and fruits arrived from the Maharaja and Maharani, and from Tobgye. Such gifts were to be of frequent occurrence on all Ludlow and Sherriff's future expeditions in Bhutan. His Highness apologised for the bad camping arrangements at Pimi, the only place where a camp site had not been prepared and shelters hadn't been built for the coolies and the other men. Here, at Donga Pemi, as at Lhüntse, Lingitsi, Shukang and as at some later camps, there were good shelters very easily made of plaited bamboo with roofs either of the same material or of long pieces of *Pinus longifolia*. They were a great boon as they kept out most of the rain — especially with tarpaulins above them.

Fortunately during these last two days leeches had not been a problem; but flies and lice and fleas had been a frightful menace. Every morning Ludlow had his clothes searched for lice, and Ramzana, when asked where they all came from, could only suggest that perhaps sweat was the source!

There was a steep climb the following morning, 27 July, to the summit of the Donga La (12,500 feet), first by a grassy ridge and then by a zig-zagging track over very rocky ground amongst trees. From the top of the pass the view should have been magnificent, but, as was so often the case, as the travellers arrived there the mist came down and only the cloud-filled valleys on all sides could be seen. However the mist did not hide the large numbers of Blood pheasants, of which four were shot, two of each sex. They proved to be *Ithaginis cruentus tibetanus*. Though this race is extremely common in Eastern Bhutan in summer, from 11,000–14,500 feet, especially in rhododendron scrub well above the conifer belt, the female, darker in colour than the male, was unknown until these two specimens were taken on the Donga La. From this point of view the day was of great interest. But not for this alone. The Donga La is the only known habitat of the very rare *Primula xanthopa* which Cooper had found in 1915 growing on mossy rocks on the floor of the abies forest. It is a fragile-looking species with thin deeply-toothed leaves thickly plastered below with golden-yellow farina and with a slender 4–6 inch stem carrying one to three, occasionally more, purplish-pink, yellow-eyed flowers. Ludlow and Sherriff were lucky enough to collect further specimens (329), the only ones since Cooper's original gatherings, which provided a more comprehensive picture of this primula than had been available hitherto.

The descent from the pass, over grassy hillsides, at first was easy, but as the path entered the forest and descended steeply over an execrable mud-track, the going became very tough. However, twelve miles beyond Donga Pemi they eventually reached their camp at Sana (8,400 feet), a delightful spot set in pine and fir forest.

The ninth march, on 28 July, brought the expedition to Trashiyangsi (5,800 feet). From the map the journey, at most, was one of eight miles. It turned out to be one of over thirteen miles the first five of which were very difficult, along a muddy track through the dense forest. Afterwards the track was much better, passing through more open and partly cultivated land before descending to the main river, the Dangme Chu, which had to be crossed, as had the Trashiyangsi Chu a little later. The Trashiyangsi river is a fine one and Ludlow and Sherriff were anxious to explore it as far as the Tibetan frontier. The people of Trashiyangsi are not at all like the Western Bhutanese; they are smaller, not so finely made, and more Tibetan in appearance. They gave the good news that Shingbe was very close to the Me La, which is on the Tibetan frontier and which Ludlow and Sherriff were bent on visiting; that it could be reached easily in four days; and that, from a day's march below Shingbe one can get across country, even with mules, to Singhi Dzong, which is of some importance as a place of worship and well on the way to the Pomo Tso which Ludlow and Sherriff had to pass on their way to Gyantse.

Two days were spent in the Trashiyangsi camp making arrangements for two weeks of rations for the party, drying and packing the bird skins and sending them off to Bumthang and thence to Yatung in the Chumbi Valley, washing and drying clothes, and developing films. Plants were not neglected and among other things

8. *Meconopsis simplicifolia*

two species of didymocarpus, one with pale purple flowers, *Didymocarpus pulchra* (346), the other with lilac, *D. oblongata* (356), as well as the pure white *Lilium wallichianum* (341) and the cream rather waxy-flowered and very sweet-smelling *Schima khasiana* (343), were gathered.

To the Pass of the Flowers (Map 2)

On 31 July the party turned north up the Kulong Chu river into botanically unknown country on the first of the four marches to Shingbe and the Me La. The day was the best either of the naturalists had experienced in Bhutan, hot sun being tempered with a pleasant breeze, and they voted the Trashiyangsi river valley the finest either had seen. The valley journey to the camp at Shapang (6,500 feet), between densely tree-covered and very steep hills and through wet muddy rice fields, was an easy one of only eight miles. Moreover the camp was one of the best they had had; it was pitched close to the river, which at this point was over fifty yards across, and awaiting the travellers were a few tethered cows to provide fresh milk — sent on the instructions of the Maharaja. The locals were wearing a novel and most excellent form of hat made of thick felt fitting close to the skull and with four or five points, each about four inches long, down which the rain dripped.

The second march, even shorter than the first, to the camp at Tobrang (7,500 feet) — also beautifully situated and also with cows in attendance — was equally easy. In contrast, the next two days were strenuous and tiring. 'I have seldom felt so exhausted' wrote Ludlow, on 2 August, in the camp at Lao (9,200 feet). 'We were continually climbing up 50 feet and going down 40 feet [a slight exaggeration. Ed.]. The road was as bad as it could be and very wet and muddy. Moreover it was hot and steamy.' By way of some compensation the gorge of the Kulong Chu was magnificent, the river running in dramatic falls for over two miles and leading up to a series of stupendous ragged rock peaks over 20,000 feet high. 'This is just the country we are looking for — no trees and plenty of rock. But we can't get there' was the frustrated entry in Sherriff's diary that evening.

At Shingbe (12,750 feet) the following evening he wrote at some length. '. . . This was one of the finest marches we have had, up a
> narrow steep valley. Whereas yesterday's was very steep and
> very bad, and after climbing 100 feet dropped 50 feet, today's
> kept steadily up, a much more satisfactory way of getting
> uphill. It is a pity the jungle is so dense. The waterfalls on the
> way up were magnificent, but we never get a full view of one.
> From camp we can see quite close to the end of a glacier in a

valley to the north, while the one leading to the Me La is rather E of N and the path goes up over a glacier-moraine covered with firs. Ludlow and I have very mild arguments, today's being the height of this place [Shingbe]. Although the aneroid makes it 14,200 feet I know it is about 700 feet too high and would place Shingbe at 13,500 feet, higher than we've been since Ha. Ludlow insists that it is not more than 12,500 feet at the most. We are both certain and I can't see how to settle it, as we have no boiling point thermometer. I must always have one in future.' Sherriff entered 14,200 feet in his diary and later changed to 13,500 feet; Ludlow entered 13,500 feet and later changed to 12,750 feet which was confirmed the following year.

At Shingbe there were neither inhabitants nor houses, only a couple of log shelters with a few Tibetans with their yaks there for the grazing. None agreed on the nature of the boundary between Tibet and Bhutan. Some affirmed it was the Me La, some the river one-and-a-half miles beyond the pass, and still others the Cho La which they reported was fifteen to twenty miles beyond the Me La and considerably higher. No matter; the expedition was to spend the next week at Shingbe exploring the Me La — which in Tibetan means 'The Pass of the Flowers' — and was hoping for a rich harvest.

The Me La (14,950 feet) was three miles distant from Shingbe and involved a fairly steep ascent. Just above camp the main valley divided into two, one branch leading to the north, the other a little south of east. This latter led to the pass. Soon the firs were replaced by junipers which struggled on for some distance before giving way to open grassy hillslopes. Sherriff described it in his diary for August 4 and 5. 'At last we have come to a part where the flowers are both interesting and plentiful. Some of the primulas, androsaces and saxifrages are beautiful and I hope one or two will prove to be new. The views were lovely beyond the pass and in the valley just below the Me La. Collected what we could and then rushed home to try to photograph them, arriving just in time for heavy rain and a thunderstorm.' (August 4). 'I went up myself to a spot close to the Me La and took the flower photographing apparatus with me and there collected and photographed twelve species. Before leaving camp took a further eight. Then home and spent till 6 p m developing — a good twelve hours' work. The Tibetans' remarks about my

9. The Me La (*overleaf*)

photography are amusing. They told our men that it had rained hard till we came here, but that I was like a lama, and with my box of tricks was putting off the rain. They also thought the reason we were collecting butterflies was to take them home to make new dyes for our clothes. The present dirtiness of our clothes may have warranted this remark.' (August 5)

Sherriff's next diary entry, for 6 August, struck a different note. He had spent six hours on the Me La wandering about the hillside collecting plants, and disturbing Snow-cock in the process, before returning to camp with his collections. 'The day was completely spoilt for me by discovering that Danon [the Lepcha collector] has been very lazy and has not been changing the drying paper. He has made an awful mess of all the fine flowers collected here. I haven't had time to keep a check on him and he has taken advantage. I thought he was a better man than that. I photographed a little Tibetan girl who is up here with some yak herds from Tibet. She is a pretty thing and very capable, being sent, as I saw yesterday, by herself to the Me La to bring in the horses. I gave her a Woolworth necklace which pleased her immensely. I find on looking over my photos that with the damp a few have stuck together and have had to be thrown away. Truly this is no easy country to work in in the summer. I think we will both be very glad indeed to get out into the drier plateau of Tibet in three or four weeks' time.'

Conditions for taking photographs no doubt were difficult and conditions for developing them no less so, but no one has taken more splendid photographs of the Himalayas, and of the plant life of the Himalayas, than has Sherriff.

The Me La lived up to its name of 'The Pass of the Flowers' and among the 'fine flowers' collected were *Ranunculus, Trollius, Doronicum, Cremanthodium, Draba, Potentilla* — all with flowers in every conceivable shade of yellow; *Lactuca* and *Aster* in pale and dark lavender-blue; rich purple-flowered *Astragalus* and *Delphinium*, including the hitherto unknown species *Delphinium ludlowii* (368); pale yellow *Aconitum* and one specimen only of a large violet-flowered kind which proved to be a species new to science, the dwarf *Aconitum fletcherianum* (378) which on future expeditions Ludlow and Sherriff were to collect on many occasions regarding it as one of the supreme jewels of the genus; purplish-pink, crimson-rose and rich wine-purple *Pedicularis*; deep blue and purplish-blue *Gentiana*; creamy-white and slate-blue *Swertia*; *Meconopsis* in several shades of blue, especially the sky-blue *M. grandis* (387) which was a first record for Bhutan; *Arenaria* in rose-pink and white; creamy-

Map 2. To the Me La, and Return to Gangtok

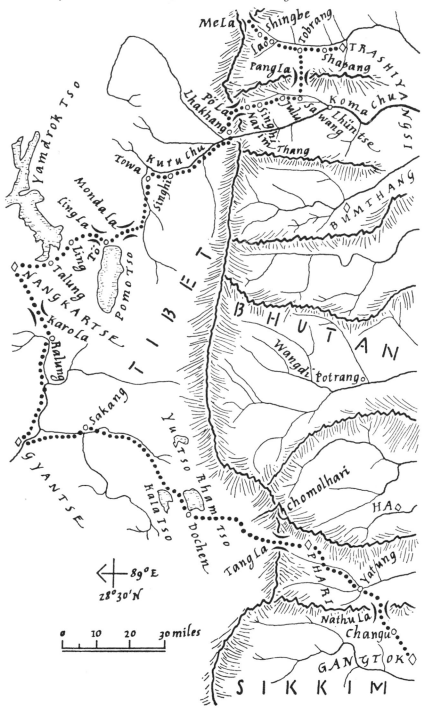

white *Cassiope* and pink *Diapensia*; *Notholirion bulbuliferum*, (411) and *Lilium nanum* (436) with flowers of purplish-pink; *Saxifraga* in many hues of yellow and orange-yellow, as well as in white and deep red, including three species never previously recorded, *Saxifraga erinacea* (376), *S. montanella* (423) and *S. haematochroa* (445).

But it was the primulas which were in greatest abundance and which most captured the collectors' imagination to the extent that from this time forth they developed a scientific interest in the genus and helped to solve many a taxonomic and distributional problem. On the Me La they gathered nine different kinds and one, later named *Primula jigmediana* (397) in honour of Sir Jigme Wangchuk, His Highness the Maharaja of Bhutan, proved to be a new species which appears to be confined to the Me La, where only Mrs Sherriff and her companion Dr John Hicks have since refound it and that in 1949. It is a strange little plant whose taxonomic position in the genus is uncertain. The slender flower stem is no more than a couple of inches high, arises from a compact rosette of spatulate leaves less than half an inch long, and carries a single nodding, or erect, funnel-shaped flower nearly half an inch across and thus very large for the size of the plant, and blue-mauve in colour. It was found in rocky ground not far removed from the species it probably most resembles, *P. sapphirina* (391, 419), another very dwarf plant, with a head of up to four pendent violet-blue flowers, which was discovered as long ago as 1849 by Sir Joseph Hooker during his explorations of the Sikkim Himalaya where it is not uncommon and whence it extends into Bhutan and the adjoining parts of Tibet. Ludlow and Sherriff made every effort to introduce *P. jigmediana* into gardens in Britain and although plants were raised by several gardeners and flowered at the Royal Botanic Garden, Edinburgh, in 1936, they did not survive anywhere. Neither did the living plants which Sherriff sent to Britain by air in 1949. Ludlow and Sherriff were no more successful in their efforts to introduce *P. sapphirina*, at any rate not until 1949 when, from the seeds they harvested, plants grew and survived for three or four years in at least one garden, that of Mr and Mrs J. Renton in Perth.

A third interesting primula from the Me La was *Primula jonardunii* (386) which was discovered in June 1905 by J. C. White, at that time Political Officer in Sikkim, whilst on a tour of Bhutan. He collected it on a pass known then as the Bod La but which should now be called the Pö La. Although Cooper also collected it in Bhutan, Ludlow and Sherriff's more comprehensive specimens, as

10. Tibetan girl, Me La

well as others they were later to gather both in Bhutan and in SE Tibet, showed that this cushion-plant with its stout rootstock, its evergreen leaves like those of the White Mountain dryas, *Dryas octopetala*, and with its rose-crimson, dark purple or white-eyed, flowers, is really but a form of *Primula dryadifolia*, wide-ranging throughout Yunnan, Szechwan, SE Tibet and Burma; a form with the flowers immersed among the leaves and with the flower-tube more hairy within than is usual.

All the other Me La primulas Ludlow and Sherriff were to collect again, both in Bhutan and in SE Tibet, on their future expeditions; the minute, solitary-flowered *P. tenuiloba* (395); the robust, tight-headed, deep purple, yellow-eyed *P. capitata* subsp. *crispata* (366, 451); the nasty-smelling *P. calderiana* (385, 412) with glorious deep purple, occasionally white (412) flowers; the magnificent *P. macrophylla* (400) again with flowers of rich purple, or sometimes of lilac; the very distinct creamy-yellow-flowered form of *P. dickieana* (401); and *P. involucrata* (393), the white flowers being occasionally tinged with purple. In 1933 all these species were well-known to botanists, but the degree of variation exhibited by almost all of them in nature was not realised until Ludlow and Sherriff's magnificent, and remarkably annotated, future collections were made available.

It was obvious to the two naturalists that a week's collecting in the region of the Me La at any one season of the year could only tap the richness of the flora and that they would have to return sometime in the future. And return they did — on two occasions.

On the other hand the birds were rather disappointing. Some, of course, occurred in considerable abundance, and were expected to. The Nepal wren (*Troglodytes troglodytes nipalensis*), the Monal pheasant (*Lophophorus impejanus*) and the Snow-cock (*Tetraogallus tibetanus tibetanus*) were all fairly common among rocks and boulders in the pass. The most frequent Rose finch in Bhutan and SE Tibet, at least in summer, *Propasser thura thura*, and the almost equally common Dark Rose finch, *Procarduelis* or *Carpodacus nipalensis*, were, naturally, much in evidence. But, strangely enough, Ruby Throats, which Ludlow also expected to find, apparently were quite absent. The Snow partridge (*Lerwa lerwa*) was there in abundance, all over the La, and a number were shot and many more could have been for they are among the tamest of game birds.

After a fine dry spell the weather had now broken, and in any case it was time to return to Tobrang whence it was planned to march west across the Donga La Range by a hitherto unexplored pass called the Pang La. The journey to Tobrang was uneventful save for the fact that even at 10,000 feet leeches were a pest — on taking off his stockings Sherriff found four of them 'hard at it' — and that several specimens of a gorgeous swallow-tailed butterfly were netted — and later described as a new species, *Bhutanites ludlowii.*

Into Tibet

On leaving Tobrang on 13 August they marched due west and immediately began to climb steeply and at first the going was good, the track leading up a grassy ridge. It then entered thick evergreen and deciduous forest which gradually led into the rhododendron zone and the ascent became very difficult owing to the quagmires into which everybody was continually floundering. Eventually, after climbing 4,000 feet in six miles, camp was pitched on a spur surrounded by rhododendron jungle, and a very wet, very depressing and uncomfortable camp it was. To keep out the rain from their roughly built shelter huts the Bhutanese peeled off the bark of conifers in six-foot strips and with this covered the roofs. By so doing they made a complete ring round the bole and, of course, eventually killed the trees.

It was still raining when they broke camp next morning and ploughed their way through mud in rhododendron jungle for 1,500 feet before turning north and reaching the summit of the Pang La (14,000 feet) before noon. Throughout the morning thick clouds surrounded them and they saw nothing of the surrounding country. But they did see some interesting birds. Near the summit of the pass, in rhododendron scrub, they met with the rare shortwing, *Heteroxenicus stellatus*, in considerable numbers. The birds were on the topmost branches of rhododendron bushes and without any effort Ludlow and Sherriff quickly obtained five specimens. This was the only occasion this shortwing was seen on this 1933 expedition. The other birds of the pass were more common, including Blood pheasants, monals, and the bush-warbler, *Homochlamys major*. For so high a place flowers were disappointing — no doubt because there was no permanent snow — and the only plants of interest were the yellow-flowered *Cremanthodium thomsonii* (461), the minute *Primula pusilla* (464) which had been gathered on the Rudo La almost exactly a month before, and a new and very beautiful and distinct aconite, some 12 inches tall, with large deep purple flowers, *Aconitum sherriffii* (462).

From the Pang La the expedition plunged and slithered steeply down a ridge for 4,000 feet with, as is usual in these dense Bhutanese forests, no view of the country. Eventually, after a twelve-mile march they reached a clearing in the forest where evidently there had been cultivation in the past and where they now found a good camping ground. The Khoma Chu was below them, and over it, five miles away, Sawang (7,700 feet) to which they marched the next day, 15 August, making a steep, almost perpendicular, descent through the thick jungle to the Khoma Chu, crossing the river by the type of cantilever bridge found all over Bhutan, and then trekking through hot steamy forest on the river's right bank.

From Sawang on to Julu (9,800 feet), 'a tiring march along an execrably muddy track through thick forest. Leeches were common and the weather foul.'[8] Thence, continuing up the right bank of the Khoma Chu, through birch, fir and rhododendron jungle along a track too difficult for riding the mules, to Tosumani (10,500 feet), shooting, on the seven-mile walk, the Spotted-winged crossbeak, the Scaly-breasted wren and Blandford's Rose finch. Just beyond Tosumani they left the main river and branched northwesterly up the stream leading to Singhi Dzong. In places the ascent was fearfully steep, with the stream in cascades, but at intervals there were small flats where the stream meandered peacefully through willow and rhododendron glades, and on one such flat lay Singhi Dzong (12,500 feet) at the junction of two glacier streams from the north and west. The Dzong was little more than a small ramshackle building more worthy of the name 'hut' than anything else and close by, and built into the cliff, was the monastery, habitation of two or three lamas. The whole place was rather bleak and desolate, surrounded by towering cliffs and peaks and hanging glaciers and, down below, forests of stunted *Abies spectabilis*.

The day's march had belied the Tibetan proverb which says something to the effect that a horse is no horse unless it can carry a man up hills and a man is no man unless he leaves his horse and walks downhill. One could hardly blame any animal for not carrying a man on this day's march; one could hardly blame any man for preferring to walk. And the leeches had been a constant plague.

> 'Today I had what I hope will be my last leech bite. He got through my long boots, somehow. When Ludlow saw it he rather laughed and said that I should wear breeches, not shorts and then they wouldn't get at me. And went on to say how lucky he had been [recently] with leeches. However I had the laugh on him, for when he took off his boots and socks there was a fine fat one, in his sock, which had had its fill. A leech injects something first to make the blood run or keep it from coagulating and this seems to affect Ludlow more than me. It was some hours before the bite stopped bleeding.'[9]

The next day's journey to Narim Thang (13,900 feet), where the expedition intended to spend the next two weeks, was an easy one of only three miles, up the smaller of the two glacier streams which ran northwards and through rapidly diminishing fir forests to the Narim Thang plain where there was only rhododendron scrub and where the stream left the plain in two beautiful cascades. The plain is about three quarters of a mile long and some 600 yards broad and is surrounded by fine snow peaks. Two wooden huts had been prepared for the use of the party and they all settled themselves in for the rest of the month. But Sherriff was not optimistic. 'I fear we have come to rather a poor place for our last fortnight. There seem to be very few birds about and so far I have only found rather uninteresting flowers. But Cooper was the same until suddenly he came across *Primula eburnea* and *P. oreina*, both growing here. I think we may possibly be too late for them but will have a good look anyway.' [10]

And they had their 'good look' — and a profitable one — early next morning when they explored, northwards, towards the Kang La (16,200 feet) four miles away. A steep ascent of 600 feet led to a glacier lake and here, after much wandering around, they came across masses of *Primula eburnea* (476), 'the gem of the whole of the eastern Himalaya'.[11] It grows only under rocks, sheltered from wind and rain but open to the sun, and is certainly a plant of the most ethereal beauty with a head of 6–12 ivory-white funnel-shaped sweetly scented flowers. Cooper discovered it in 1915, and plants raised from the seeds he collected flowered in various gardens in Britain in 1918–19, those flowering for Mr A. K. Bulley receiving the Award of Merit from the Royal Horticultural Society in 1919 albeit under the name of *P. barroviana*. By 1933 the species had been lost to cultivation and Sherriff and Ludlow were anxious to reintroduce it. 'Found more *eburnea* today and am trying to make arrangements to get seeds home. It means a man coming up in October but we can make his work easier by tying on little bags to collect seeds as they ripen.' [12] And again: 'I tied on 25 small and 13 large bags, over a total of about 80 heads of fading flowers. Also brought away some seedlings and hope between them to get something home.' [13]

It was a brave effort on Sherriff's part but unfortunately he did not succeed — not on this expedition anyway. He had to wait until 1949 when he was able to fly plants to Britain. These flourished, for a time, in several gardens, some grown by Mrs Renton of Perth receiving the Award of Merit in 1950.

They were also successful in finding Cooper's other discovery at Narim Thang in 1915, *Primula oreina*, and their specimens (485) were sufficient to show that *P. oreina* is nothing more than a minor

variation of *P. jonardunii* which, in turn, is but a form of *P. dryadifolia*.

Otherwise the flora was rather poor and the avifauna not much better. They saw lots of Snow pigeons and Blood pheasants and on the Kang La a flock of grandalas feeding on insects and on the glaucous indigo-blue berries of a dwarf vaccinium. In addition they shot both the Brown and White-breasted dippers, three kinds of Rose finch and a wagtail. 'But bird life is decidedly scanty', Ludlow commented. 'I cannot help feeling that most of the high altitude forms have forsaken the wet southern slopes of the main range and will be found on the northern slopes of the Bod La. Bod La, by the way, is a misnomer. It should be the Pö La or Tibetan Pass.'

Throughout the sojourn at Narim Thang the weather, to use Ludlow's favourite adjective, was 'execrable'. 'Half an hour's sunshine in the morning, with masses of clouds playing hide and seek amongst the peaks, then the usual dense mist and rain in the afternoon. Rain, Rain, Rain.' On the nights of 23 and 24 August snow fell to a depth of 4–5 inches.

On 29 August they were not sorry to leave their wet and depressing camp, not sorry even to leave Bhutan and to cross the main range into Tibet, first by the Kang La and then by the Pö La. The former is a rocky knife-edge and the four-mile crossing took eight hours, a journey which prompted Ludlow to comment: 'I found the ascent to the Kang La very trying. I am afraid I feel high altitudes far more than I used to' [he was to be very much at home in them for the next sixteen years, for all that!]. On the other hand, the eight-mile journey over the Pö La (16,300 feet) to Hamo (13,500 feet), a small grassy plain at the head of the conifer forest, was comparatively easy.

They were now in the Hamo Chu valley heading for Lhakhang Dzong (10,000 feet) ten miles away and their descent of the Hamo Chu proved an extraordinary one which Sherriff well described.

'We left from a place just above the conifer (fir) forest: there the climate and vegetation were not far removed from the wet Bhutan climate and vegetation. We had had lots of rain there too. The cliffs on the right bank [of the Hamo Chu] are very nearly perpendicular and very high indeed. Many side streams come in but their valleys are all so narrow that most come

11. *Primula eburnea*
12. Glacier lake above Narimthang. *P. eburnea* grows under the black cliff on left

down as waterfalls. One fall we saw was, we thought, about 800 feet sheer. The water took 13½ seconds to reach the bottom but then it was all spray by that time. We crossed to the north face and there the conifer jungle was thick. There was some lichen on the trees but already a great difference could be seen. Suddenly, when we rounded a shoulder, we were straight from the wet to the dry zone, where there were no conifers, little grass, and only shrubs as vegetation. The suddenness was extraordinary. . . . Crops are ripe and being threshed now. There are many wild peach trees in the cultivated areas but I doubt if they will ever ripen. Gooseberries are not bad, though barely ripe. White [J.C. White of the Indian Political Service] remarks that there is a large trade in dried apricots; there are no apricots at all — only peaches. We managed on the quiet to get three birds and a number of butterflies, but we must be very careful now. The dzongpen is a nice old thing, keen on flowers. I made use of that by asking him to collect some delphinium seeds and send them to Gyantse. He promised to do this, and in return we will send his photo to him here, via a friend from Lhasa.' [14]

The delphinium in question had been gathered on the journey over the Pö La; it was a glorious affair with numerous large Mediterranean-blue flowers on stems up to 3 feet tall, and was later identified as *Delphinium grandiflorum* (499). It was by no means the only desirable and interesting plant discovered in the region of Lhakhang. There were, for instance, two which proved to be new to science; an aconite of great beauty with large purple flowers and floral bracts, *Aconitum bracteolatum* (509), as well as a greenish-white-flowered gentian, *Gentiana lhakhangensis* (515).

The end of the Journey

From Lhakhang Dzong the journey in Tibet was of no very great importance either from the botanical or ornithological points of view; in fact during the next month only twenty or so plant specimens were collected for the herbarium. But the actual journey was not without interest. Leaving Lhakhang Dzong on 3 September, they ascended the Kuru Chu for some forty miles to Towa Dzong (12,500 feet) which they reached two days later, a journey which, because they were having to change transport continually, almost at every village they passed, both irritated and amused the tired travellers. 'Changing transport is proving a perfect curse. Yesterday we took 13 hours to reach here [Towa Dzong]; started at 6.0 a m and arrived 7.0 p m [from Singhi Dzong, a distance of eight miles]. Luckily it was a day of fine weather. The changing is really rather amusing; it would be very funny were it not for the awful delay. Everyone in the village turns out. They look at and examine all our things: then there is endless talk which seems to lead nowhere at all. People shout to others miles away, for horses. Somehow or other these eventually turn up, and with some coolies, donkeys, cattle, yaks and horses off we eventually go again. Yesterday, at Lala, it took 1¾ hours for this to be done; otherwise we should have been here in daylight. The coolies are mostly women and they go on with the work pretty well. Everyone seems happy and they all joke and laugh about the weight of the loads and the size and shape. There is a most intricate business every morning casting lots as to who, or which village, will take which loads. Everything is laid out in a line, lots are cast, and then with a rush the men and women dash to the loads they think will be best, and off they go with a laugh. The women always amuse me with their blackened faces. They have, most of them, little round black plasters on their temples, put on by lamas to relieve pains in the head. Then sometimes they will cover all the face, but the eyes and mouth, with a black paste — to assure a good complexion! — and then they look exactly like nigger minstrels, especially as they are always smiling and laughing.' 15

Fortunately at Towa they were able to arrange with the dzongpen to have the transport changed only at Lalung, the Monda La, Ling and Nangkartse — one of three routes to the latter place and the one they proposed to take.

After a very easy march to Lalung (14,700 feet) on a marvellously fine day when the sky was a most lovely colour of bright clear blue they camped in a pretty little willow garden which, they were told, had last been used as a camping site by a European, J. C. White, shortly after the Tibetan war and thus 27 years previously. A messenger from the Maharaja of Bhutan was awaiting them with letters and presents of apples, vegetables and scarves. His Highness asked Ludlow and Sherriff to return to Bhutan another year, any time they desired in fact, and hoped that when they did return they would collect seeds, as well as plants. This was a most generous invitation of which they were to take good advantage.

Of course they hoped that this fine weather would hold for the crossing of the unexplored Monda La (17,200 feet) whence the views of the beautiful peak Kula Kangri (24,784 feet) and its satellites would be magnificent. But alas! The day of the crossing was 'perfectly foul', the wettest since leaving Bhutan, and they saw nothing. On reaching the Tö monastery, on the eastern shore of the Pomo Tso (16,200 feet), which probably had not been visited by Europeans before, and after a gradual descent over rolling open downs, they were greeted with a severe hailstorm which later gave place to snow. Happy were they to shelter in a couple of yak hair tents and to try to warm themselves at a fire of yak dung — a poor substitute for the great fires of rhododendron wood they had made at previous camps.

On the following day, 11 September, the march to Ling (14,700 feet) by the Ling La (17,000 feet) could hardly have been more similar. In the early morning low clouds began to clear off the Himalayan peaks and there were glimpses of Kula Kangri, Chenrazi and other peaks. But then, on the easy ascent of 1,000 feet to the Ling La, mist and rain gathered again and the views from the pass, especially looking over the Pomo Tso, were bitterly disappointing. The descent into the grass-covered Ling valley pasturing tens of thousands of sheep as well as thousands of yak, was, as yesterday, easy, and the tents had hardly been pitched at Ling when a terrific hail and rain storm flooded the camp.

Still another beastly wet day for the journey to Talung (14,700 feet), a picturesque village of some fifty houses, with the usual monastery perched on an isolated rock, in the middle of the great Talung plain. To the east was the Yamdrok Tso which, even in this awful weather, looked very fine, with many duck around the edges and, close by, hundreds of Greylag geese.

At last, on 13 September, a nearly perfect day, with marvellous cloud effects, for the ten-mile trek to Nangkartse Dzong (14,700 feet), with a patch of the Yamdrok Tso blue and sparkling in the brilliant morning sunshine. On such a warm sunny day Nangkartse was pleasant enough but it was easy to imagine how disagreeable and bleak a spot it would be in winter, surrounded by extensive flats and with tearing winds sweeping across the plain. In the distance could be seen the monastery where abides the famous Dorji Phamo (Thunder-bolt Mother of Pigs). She is the only woman who may have a meeting with the Dalai Lama, her fame resting on the fact that she can turn herself into a pig! Ludlow and Sherriff were now on the route Williamson and his bride had taken to Lhasa, and for the first time in their travels they were asked to show their passports.

Sadly, the fine break in the weather did not last for long and as usual when they had to cross a high pass, this time the Karo La (16,600 feet), on the way to Ralung (14,500 feet), wind, rain and sleet hit their faces. It was all rather remarkable; they had crossed some twenty major passes and the only two on which the sun had shone on them had been the Nathu La and the Me La. And now, wind, rain, sleet and snow, and thunder and lightning accompanied them for most of the two-day march of just over thirty miles over the wide cultivated plain on which stands the town of Gyantse (13,260 feet). Here Ludlow and Sherriff stayed for a week, and here they enjoyed uniformly good weather, before returning to India by the Phari road to Kalimpong where they arrived on 7 October.

This first Ludlow and Sherriff joint expedition proved a highly significant one; significant not only because the British Museum was enriched by close on 750 valuable bird skins and the major herbaria of Britain by over 500 dried plant specimens, several of them of species previously unknown to botanical science; but significant also because it spurred the two friends on to greater efforts. With the death of George Forrest in Western China in 1932, and with Joseph Rock's retiral from plant collecting in that area in 1934, a great era of plant exploration and plant introduction had been brought to a close. Frank Kingdon Ward, it is true, was at the height of his activities. He had collected in Tibet and Bhutan in 1924–25; in Burma and Assam in 1926–28, and again in 1933; in NE Upper Burma and on the Tibetan frontier in 1931. He was to visit Tibet, Assam and Burma in 1935–39; Assam again and the frontiers of Tibet in 1946; and Upper Burma, once more, in 1953. Even so, the 1933 expedition of Ludlow and Sherriff was a landmark in the history of plant collecting in, and plant introduction from, the Himalaya.

Their first expedition at an end Ludlow and Sherriff now decided on a plan of campaign for the future. They planned a series of journeys during which they would work gradually eastwards through Tibet along the main Himalayan range, each succeeding journey overlapping to some extent its predecessor, until they reached the great bend of the Tsangpo. Thus progressing gradually eastwards they hoped to obtain valuable information regarding the distribution of plants, to make collections of plants for herbaria, and to introduce to British gardens, by means of seeds and living plants, as many desirable plant species as possible. And as the avifauna of the country they proposed to visit was almost totally unknown they knew that there would be no problems in amassing a valuable scientific collection of bird skins.

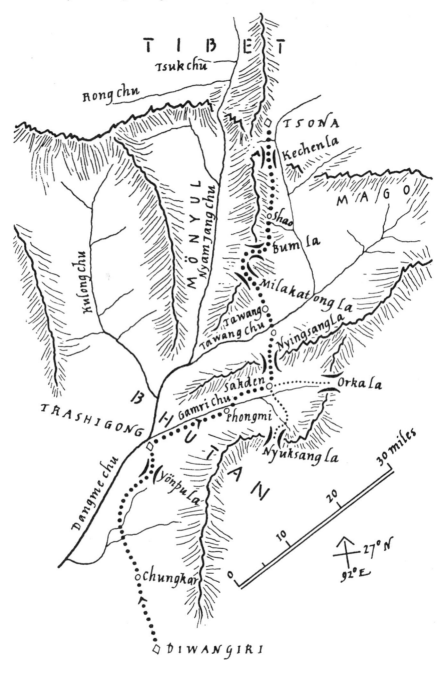

Map 3. The Journey to Tsona

T I B E T

Tsukchu

Rong chu

TSONA

Kechen la

M A G O

o Shao

M
Ö
N
Y
U
L

Nyam Jang chu

Bum la

Milakat ong la

Kulong chu

Tawang
Tawang chu

Nyingsang la

Sakden

Orka la

TRASHIGONG

Gamri chu

Phongmi

B
H
U
T
A
N

Nyuksang la

30 miles

Dangme chu

Yönpu la

20

10

27° N

92° E

o Chungkar

0

◊ DIWANGIRI

Chapter Two

Frustrations and Triumphs,

TIBET & E. BHUTAN

o

In accordance with a plan made at the end of their 1933 expedition, Ludlow and Sherriff decided that in 1934 they would work, as intensively as possible and in the time available, the basins of the Tawang Chu and Nyam Jang Chu in Tibet and then revisit the Trashiyangsi valley in East Bhutan, which seemed to offer further scope for botanical and ornithological exploration. His Highness the Maharaja of Bhutan readily gave permission to return to Bhutan, but it wasn't until late in May that sanction was received from the Tibetan Government to visit the province of Mönyul.

The start of the 1933 expedition had been rather disastrous, having been held up for a week at Gangtok because of heavy snows on the Nathu La. The start of this one was no less so. The season was well advanced, making it necessary to reach the collecting grounds without delay. The quickest route — the Diwangiri-Tawang Trade Road — penetrated the fever-belt of the eastern Duars. Ludlow and Sherriff decided that the risk was worth taking, and the expedition left Srinagar on 13 June, reaching Rangiya, on the Eastern Bengal Railway, on the 17th. To pass through the fever zone in daylight, they planned to proceed to Darrang, in the foothills, by means of lorry and car, and thence by pack-mules to Diwangiri, in one day. The plan miscarried. The day before they reached Rangiya the monsoon broke and rendered the 2,000 foot ascent to Diwangiri, on the Assam-Bhutan frontier, impossible, and the Diwangiri stream an unfordable torrent. They took refuge for four days in the neighbouring Menoka tea estate until the torrent subsided. Eventually Diwangiri was reached on 22 June after a nightmare journey. The torrent had to be forded thirty-five times. 'Some of the fords were so bad that we had to unload, swim our mules across, and manhandle the loads through the dense jungle for 100 yards or more when we would be brought up suddenly by a precipitous cliff face which was absolutely

impassable. Luckily the river continued to drop the whole day. But as we ascended higher and higher the river got narrower and narrower so we always had a rapid torrent to cross.' [1]
The nine-mile march took almost nine hours.

They were now on the road to Trashigong and Tsona. They were also in the land of the leeches, great red things which squeezed their way through the eyelet holes of Ludlow's boots and covered his socks with blood. Biting flies were also a pest, raising large blood blisters which were frightfully itchy. However, some fifteen miles beyond Diwangiri, on the way to Chungkar (6,400 feet) they had a great stroke of luck that made these inconveniences temporarily forgotten. On a great cliff, close to the right-hand side of the road, Sherriff spotted a number of handsome lilies, which he managed to reach with great difficulty. They were clearly allied to *Lilium nepalense* but very different indeed to the two forms of this lovely species which he and Ludlow had collected in 1933, at Ritang and near Lingitsi, having wide trumpet-flowers of pure pale yellow; they proved to be *L. nepalense* var. *concolor* (553). But this was not all. In the perilous business of collecting the lily Sherriff found two primulas which he hadn't seen before; 'I hope they may be new ones.' No one else had seen them before either, although other travellers had passed by that same cliff, and possibly collected on it, including William Griffith as long ago as 1838. And they were indeed new species.

They were growing entirely in moss; there was no soil whatsoever. One was a most elegant plant with large pale violet flowers delicately dusted with farina, the slenderly cylindric and slightly curved flower-tube being close on 2 inches long and the limb nearly 1½ inches across. As a consequence of the elongate tube, probably the most elongate in the entire primula genus, the head of flowers had a peculiar spreading character. Sherriff immediately recognised that he had found a very fine and unusual plant and in addition to taking material for the herbarium, as well as seeds, he also collected a living plant, potted it, and with splendid horticultural skill carefully nurtured it for several months before his journey home. The plant survived the sea-voyage to Britain, arrived at the Royal Botanic Garden, Edinburgh, in June 1935, and flowered the following month. In the meantime the seeds, received in Britain much earlier, had germinated well and a plant raised and flowered by Mr T. Hay received a Preliminary Commendation from the Royal Horticultural Society in September 1935 and the

13. *Primula sherriffae*

Award of Merit the following April. It was named *Primula sherriffae* (552) in honour of Major Sherriff's mother. Since that time *P. sherriffae* has proved an excellent plant for the cool greenhouse. Moreover its geographical distribution has been extended in a most interesting fashion by Kingdon Ward's discovery of it in the Mishmi hills, right across the plain of Assam to the south. He found it at Sirhoi, where, apparently, it grows in greater abundance than in Bhutan for he describes 'a cliff face which was so plastered with *P. sherriffae* that it loomed up in the mist like the white cliffs of Dover.' [2]

The other new species, *Primula ludlowii* (554), was to all intents and purposes a miniature and depauperate form of the other with a solitary hairy flower, and was of no horticultural merit.

Growing beside these two new species there was another primula which Sherriff failed to see on that memorable day of 24 June, and it is not inappropriate, here, to anticipate events. On 23 February 1936, Sherriff wrote to Sir William Wright Smith at the Royal Botanic Garden, Edinburgh: 'Today I visited the cliff face where we got Nos. 552 and 554. Neither was in flower; in fact neither showed signs of much life. But on the same face and growing amongst 552 and 554 were a number of what must be *Primula filipes*. It is in full flower now and some flowers have faded. Colour varies from white to pale heliotrope. At the same time I looked again for seeds of 552 and 554. I found some of 552 — last year's of course. You remember that I could find no seeds of 554 last time. This year I have found a few. . . . The pods of 554 are so low, right at the very base of the small leaves, that they can't be seen unless the plant is pulled out and examined closely.'

This is not the end of the story of the findings on this particular cliff in 1934. In order to reach the two primulas Sherriff had to support himself with a branch of a twenty-foot tree which was carrying magnificent, white, scented flowers and great elliptic or oval leaves, some as much as 10 inches long and 7 inches broad. Specimens were taken and they proved to be those of another new species, a species of luculia which was later named *Luculia grandifolia* (555). Until the discovery of this new species the genus luculia was represented by but three species, all well-known in cultivation as fine greenhouse plants with sweetly-scented flowers; the rosy-pink *L. gratissima* with a wide distribution in Nepal, Sikkim, Bhutan, Assam, Burma and South China; *L. pinceana* confined to the Khasia hills of Assam and with creamy-white flowers; and *L. intermedia*, the flowers darker in colour than those of *L. gratissima* and a native of Burma and Yunnan. Thus, to add to these three species a fourth just as beautiful was a tremendous piece of luck, especially as the collectors were able to introduce it into cultivation, a story which it is again convenient to anticipate.

In August 1949 Ludlow saw this new species again, in Central Bhutan at Tsangka between Trashiling and Trongsa. But here it was not tree-like in habit; it was a shrub 6–8 feet tall, inhabiting south-facing hillslopes at 7,500 feet (17,040). Again in 1949, in November, he met with it once more, this time in Sikkim, in the Palace Garden at Gangtok (19,868). It was to this shrub, said to have been introduced from Bhutan, that in March 1951 Ludlow sent his servant, Pak Tsering, to collect fruiting specimens (21518). When these specimens arrived in Britain seeds were obtained from them and in this fashion was the species introduced into gardens in Britain, receiving the Award of Merit from the Royal Horticultural Society when exhibited in June 1955 by the Society's Gardens at Wisley. Let us now return to June 1934.

The expedition was en route for Trashigong, a little over thirty miles from Chungkar and a matter of four marches, the third one involving the crossing of the Yönpu La (8,200 feet). Life was still being made pretty miserable by the eternal leeches and blister-flies and the exhausting heat. Except for the Yönpu La the travellers were always at altitudes well under 7,000 feet and in Bhutan, in summer, it is almost impossible to experience good temperatures under 8,500 feet. Moreover, under such circumstances it was rather depressing to climb 3,000 feet, first through *Pinus roxburghii* forest and then through oak and rhododendron jungle, only to lose all the height the following day.

Understandably little collecting was attempted although a few plants appealed so much that they simply had to be put into the plant presses; the epiphytic orchid with solitary four-inch white and purple flowers, *Dendrobium falconeri* (567) — it was through a plant imported from Bhutan that this beautiful orchid first flowered in Britain in 1856; and two marvellous lilies. *Lilium wallichianum* (564), just coming into flower, was growing in great profusion, and in another two weeks the hillside would have been a memorable sight with countless large creamy-white funnel-shaped blossoms. The other lily, *Cardiocrinum giganteum* (573), was at its most magnificent, its great white flowers, in this instance with blood-red streaks along the centre of the tepals, carried on stems up to 10 feet tall. It was revelling in dense shade in the thick forest, competing quite successfully with a host of trees and shrubs forming the undergrowth, near the summit of the Yönpu La.

At Trashigong (3,250 feet) they were in a dry part of the Manas valley, very reminiscent of the Mo Chu at Wangdi Potrang which they had visited in 1933, with the hillsides here and there cultivated but for the most part fairly barren and clothed only with a thin forest of *Pinus roxburghii* and a few other stunted trees and shrubs.

In and around Sakden

From Trashigong the expedition turned eastwards, heading for Sakden, up the unexplored Gamri Chu valley, passing through areas of rice, buckwheat, maize and lac cultivation and making excellent camps in splendid weather. 'The huts made for us yesterday and
today are perfectly wonderful buildings. Nothing but bamboo is used and they are strong, airy and light. We had a heavy shower of rain last evening but they proved pretty watertight after the first few minutes. The inhabitants [of Phongmi (5,450 feet)] are very mixed. Some are from Diwangiri, some from Trongsa, and probably some from the Tawang area. There are many "Takpas", the people who wear a hat shaped to the head, made of felt and having tails hanging down in three or four places round the edge. They wear a queer little felt patch hanging over their bottoms, presumably to sit on.' [3]
The fourteen-mile walk to Sakden (9,700 feet) on 2 July gave cause for some concern and the entry in Sherriff's diary that evening tells the reason. 'A difficult day for everyone as the march was so
long: 14 miles does not seem long, but add to it 4,000 odd feet of climb and many descents of 500 feet, and climbs of 500 feet in between and it is a good day's march. Ludlow was feeling pretty poor in the morning when we left and had no breakfast. He came straight in here [Sakden] and went to bed with fever, though not much temperature. He has had neuralgia, and pains in his legs. Gulla [Ramzana's assistant with the bird skins] gave out half way and now has a temperature of 102°. I'm afraid he must have got malaria in Menoka or Diwangiri.'
Sherriff's fears were justified. Gulla indeed did have malaria, and so had Ludlow. In fact apart from Sherriff and the two Lepchas the entire party collapsed with malaria at Sakden and there were fears that the expedition might have to be abandoned. However, each man was treated drastically with 30 grains of quinine bihydrochloride per day, and sometimes with a nip of Sherriff's 40 per cent-over-proof rum, and after a halt of eight days all were fit enough to continue the journey although from now until October one or other of the party complained of short bouts of fever.

14. *Cardiocrinum giganteum*

With Sherriff and Danon (the Lepcha collector) fit, the enforced halt at Sakden gave them the chance to explore the area. On the descent of the Munde La (9,600 feet) to Sakden, they had gathered another epiphytic dendrobium, the form of *Dendrobium fimbriatum* with the normally deep rich orange flowers blotched with dark reddish-purple and known as *oculatum* (581). More important, they had added several rhododendrons to their collections; the deep rose form of the tree-like and variable *Rhododendron arboreum* (586); a form of *R. keysii* (588) with coral-coloured tubular flowers tipped with golden-yellow; a pale rose-pink form of the usually straggling *R. camelliiflorum* (589); the straggly epiphytic pale yellow *R. micromeres* (590); the leggy but handsome creamy-white trumpeted *R. dalhousiae* (582); and most interesting of all 'one rhododendron [583], white with red streaks up the corolla segments which is almost certainly Cooper's 3937, found at Punakha at 5,000 feet in flower on 29 May 1915. It is a fine flower but I doubt if we could get the seeds. I saw it growing on rocky ground about ½ mile above the bridge over the main Gamri Chu. . . .' Thus wrote Sherriff on 2 July and he was correct; the plant *was* Cooper's 3937 — *Rhododendron rhabdotum*, a plant so beautiful that it has received both the Award of Merit and the First Class Certificate from the Royal Horticultural Society. Thus, having found such interesting and desirable plants on the approach to Sakden Sherriff was hopeful of finding many more within a few miles of the malaria-stricken camp.

Sakden stands in the midst of an extensive undulating plain surrounded by thickly wooded hills, and Sherriff and Danon explored two small areas of this country. One of these, to the south-west, was the Nyuksang La (c. 14,000 feet) the summit of which was nothing but rhododendron jungle with numerous forms of the immensely variable *Rhododendron lepidotum* (634) in hues of pink, crimson, purple, yellow, greenish-yellow and white, as well as with forms of *R. campanulatum* (605, 616, 619, 622, 624), some with white flowers tinged with pink, with or without a purple blotch, others with rose-pink flowers with or without dark crimson spots, and all showing variations in the foliage. That these two rhododendrons should be so variable is not surprising in view of their wide geographical range, the former from the NW Himalaya to SW China, and the latter from Kashmir to Bhutan.

This rhododendron zone was very good for other flowers, especially for primulas in which Sherriff was now deeply interested, and three of the species he collected proved to be new primula records for Bhutan. *Primula waltonii* (598) was growing in great profusion,

Meconopsis grandis ('Sherriff 600')

its sweetly-scented funnel-shaped flowers the colour of port wine and carried on stems 1–2 feet high making a brilliant show. First collected by Captain H. J. Walton on the hills above Lhasa in 1904, it wasn't seen again until twenty years later when it was found in Tibet, at Gyantse and on the Gokar La, by F. M. Bailey, and on the Mambu La by F. Kingdon Ward. *Primula gambeliana* (608) previously had been known from one gathering in Nepal and from several in Sikkim. Sherriff now found it growing in moss under rhododendrons and although the deep grape-purple flowers were practically over, the almost round or heart-shaped leaf-blades carried on leaf stalks up to four times their length clearly revealed the identity of the species. The third species new to Bhutan was *Primula stirtoniana* (632), a minute affair with solitary pale bluish-violet or lavender-blue white-eyed flowers immersed in small rosettes of leaves. A single plant is nothing much to look at. However the species never grows solitarily but in good-sized clumps on sheer cliff faces and as such is quite lovely. Until this finding of it, *P. stirtoniana* was known only by a handful of gatherings in Sikkim. Ludlow and Sherriff showed that it is exceedingly common in Central Bhutan, growing in massive clumps, usually, over a wide area.

Two other primulas which Sherriff and Danon gathered had previously been discovered by Cooper in 1914–15; the elegant *P. strumosa* (597, 612) with sturdy flower-stems a foot or more tall carrying several yellow, orange-eyed flowers, sometimes nearly an inch across; and the rather fragile looking *P. bellidifolia* (635) growing, like *P. eburnea*, below overhanging rocks and sheltered from direct rain, and with rather small, deep purplish-violet, slightly deflexed flowers. In 1915 when Bees Ltd first brought plants of the latter to the flowering stage from Cooper's seeds they were called *P. menziesiana* and under this name received the Award of Merit from the Royal Horticultural Society. However, Cooper's plant is nothing more than a rather robust form of *P. bellidifolia* which was described from plants found in the Chumbi valley as long ago as 1877 and which Ludlow and Sherriff were to find on several future occasions both in Tibet and Bhutan.

On the Nyuksang La Sherriff found a plant which surpassed in beauty all the primulas and every other plant on the pass — a most magnificent form of *Meconopsis grandis* (600), which he and Ludlow had recorded from Bhutan for the first time in 1933. It was occupying open stony ground beside *Primula waltonii*. Well might Sherriff have echoed the remarks of Kingdon Ward when *he* collected Bailey's Blue poppy, *Meconopsis betonicifolia*, near Tumbatse in SE

Tibet in 1924; 'Among a paradise of primulas the flowers flutter out from amongst the sea-green leaves like blue and golden butterflies.' Perhaps the collection of *M. betonicifolia* is the achievement best associated with Ward's name, for it is now firmly established in cultivation both in Britain and overseas. And it could well be that the discovery, and the introduction to cultivation, of 'Sherriff 600', as this marvellous plant is now known in horticulture, will be ranked as Sherriff's greatest achievement. It is a finer plant by far than *M. betonicifolia*; and it is a finer plant by far than the form *M. grandis* from Sikkim which grew for many years in the Rock Garden at the Botanic Garden, Edinburgh, and which carried only a solitary nodding flower on a 12 or 18-inch flower stem. 'Sherriff 600' grows to twice that height, sometimes higher, and bears several glorious deep blue flowers often as much as 6 inches across.

Before leaving Sakden Sherriff took one of the men to the pass and placed seed bags on the fading flowers of the meconopsis and of *Primula waltonii* (598), *P. gambeliana* (608), *P. strumosa* (612) and the dwarf *P. glabra* (602). The intention was for the man to return in the Bhutanese eighth month to collect the bags and, it was hoped, their enclosed seeds. Sherriff also potted seedlings of *P. waltonii*, *P. gambeliana*, *P. stirtoniana* and *P. bellidifolia* — all of which he hoped to nurture during the rest of the expedition.

In addition to the Nyuksang La Sherriff and Danon paid a fleeting visit east to the Orka La (13,900 feet). 'I found some good flowers . . . . and one could easily spend a month or so in that area. A new meconopsis and some primulas and a beautiful little corydalis and androsace were found. The best spot seems to be on the big round cliff to the south of the pass; on the north side of this there were many flowers.'[4] The meconopsis was the low-growing dark purplish-violet-flowered *M. lancifolia* var. *concinna* (642); the corydalis the exquisite delft-blue *C. cashmeriana* (644) which Sherriff had found for the first time in 1933; the androsace the purplish-pink *A. adenocephala* (643); and the primulas were similar to those of the Nyuksang La.

Ludlow's fever, of course, handicapped his quest for birds. Nevertheless he had a most exciting find, a new species of fulvetta, or tit-babbler, later named *Fulvetta ludlowi*, which was quite common, and very often remarkably tame, in bamboo and rhododendron forest, between 7,500 and 11,000 feet on the extreme eastern frontier of Bhutan. It is chocolate-brown in colour, the head rather

15. Trashigong
16. *Primula stirtoniana* with *Saxifraga clivorum*

53

darker than the mantle, and the white throat is heavily streaked with brown.

From Sakden to Tsona

The enforced stay at Sakden had been a profitable one and the expedition was to return in the autumn to collect seeds and to study the avifauna more closely. But now, 11 July, everyone was fit again and anxious to move on. They turned north, crossed the Se La Range, which is part of the boundary between Tibet and Bhutan, by the Nyingsang La (12,200 feet), and descended through silver-fir and deciduous forest to Muktur (8,250 feet), a Tibetan village of 30–40 houses. Thence to the monastic town of Tawang (10,200 feet) — a disastrous day when the coolie transport and riding ponies were totally inadequate and when, according to custom in this part of Tibet, the transport had to be changed at every village; seven changes in six miles on a day of incessant rain tried the patience of everyone. Tawang consisted of a walled 'gömpa' with between 550 and 600 lamas, not one of whom was willing to co-operate with the expedition — a new experience for the two natura-lists. Disgusted with the behaviour of the lamas, and with the weather — it rained incessantly and the surrounding country was constantly in mist — they were thoroughly glad to head for Tsona which is north of the main range and thus considerably drier.

The mountain passes between Tawang and Tsona held a rich flora and at the end of the first day's march, at Shao (13,300 feet), in a clean and comfortable room belonging to the headman of the village, Sherriff wrote: 'I have never seen so many primulas in one day.'[5] On the Milakatong La (14,200 feet), where they marked a very beautiful white form of *Meconopsis horridula* (659) for future seed collecting, they gathered six different primula species, in-cluding the beautiful and deliciously fragrant creamy-white *Primula obliqua* (655); and, approaching Shao, they had been blessed with an unforgettable sight, between 20 and 30 acres of disused fields absolutely chock-full with the pale yellow flowers of the so-called Himalayan cowslip, *Primula sikkimensis* (669). This march to Shao was quite a nostalgic one for the rocky grassy hillslopes covered with dwarf willows and with great pink drifts of the dwarf *Rhododendron hypenanthum* (661) reminded Sherriff of the heather moors of his native Scotland.

During the sixteen-mile journey from Shao to the Tibetan village of Tsona, involving the ascent of the Kechen La (15,600 feet), they collected very little for they were in a hurry and it was beastly cold and wet. But they stopped to admire the great masses of the deep sky-blue *Meconopsis simplicifolia* (672), the young shoots and seeds of which were eaten by the local people; drifts of a fine form of *Gentiana nubigena* (670) with deep purplish-blue trumpets; six-inch clumps of the minute *Primula tenuiloba* (674), the small blue-violet flowers when in such masses being quite beautiful — incidentally this was the first time this primula had been found in Tibet; and *Primula dryadifolia* (671) with its massive spreading low-lying cushions brilliantly studded with large grape-purple blooms.

Tsona (14,300 feet) in 1934 was a filthy village of close on a hundred hovels hardly fit for human habitation. There was excellent grazing for yaks and sheep, and some cultivation, a good deal of barley being grown and cut green for cattle fodder in winter. Here the expedition camped for three days, drying out its belongings and making arrangements, especially for transport, for its journey to the Mago district of Tibet on 19 July. In this the dzongpens, who lived a couple of miles below Tsona near some hot springs, were very helpful especially after they had been given their presents; a pair of gloves, a silver-lined papier-mâché bowl, a length of Kashmir tweed, a piece of Bhutanese cloth, and some saffron. The saffron, brought from Kashmir, caused great excitement for it was quite unobtainable at Tsona.

Something else also caused great excitement — not to the dzongpens but to Ludlow and Sherriff. Hanging from the cliffs north of the camp were dozens and dozens of aged clumps, some of them two feet across, of *Paraquilegia grandiflora* (678), delicate bluish-green deeply cut foliage blending beautifully with the large deep-violet, sometimes lilac, fragile, ever-trembling, flowers. The travellers were very familiar with this species, one of the world's most lovely plants, but they had never seen it in finer condition than here.

A month in the Mago district (Map 4)

On 19 July the expedition, with all members rested and with all baggage dry, set out for a month's stay in the Mago district. In the course of four days four passes of over 16,000 feet had to be crossed; the Rala La; the Gu La; the Dza La and the Tulung La. The hills were bare of almost everything but grass and the odd juniper and only occasionally, so persistent was the mist, were the snowy peaks of the main Himalayan range to the south, visible. Only on the descent from the Tulung La, after a final 300-foot very steep ascent along a bare knife-edge, were the flowers of any particular interest and even then not so abundant as had been anticipated, chiefly, it seemed, because the shaly hillside was almost continually on the move and the shale, gradually falling down to the river-bed, covered everything. However, Ludlow and Sherriff discovered one plant which was certainly new to them and which, at some distance, they at first mistook for a paraquilegia; and indeed, hanging from the cliff faces and with numerous azure-blue flowers on short stems, it was reminiscent of the beautiful Tsona plant. But a moment's examination convinced them that they had found a meconopsis new to their collections; *Meconopsis bella* (708) which Cooper had recorded from Bhutan in 1914.

Having crossed the Tulung La they were now in the valley of the Goshu Chu, marching towards the twin villages of Nyuri and Dyuri (11,600 feet) and, moreover, marching through miles and miles of the pale yellow *Primula sikkimensis* 'thicker than butter-cups in an English meadow'.[6] Apart from a fine deep wine-red lousewort, also in great profusion, *Pedicularis megalantha* (712), the plant of greatest interest was 'a most magnificent thalictrum. This is one of the finest flowers I have seen and would make a most beauti-ful table decoration. I shall take a pot or two of it, if the seeds are not ready.'[7] They had found, in rocky and shady places in the forest, the elegant, cut-leaved, mauve-violet-flowered *Thalictrum chelidonii* (710), a not uncommon plant in the Central and East Himalaya. The seeds were not ready when the expedition returned to Dyuri but Sherriff was not dismayed. 'I am trying to make arrangements here for a married lama(!) — a Bhutanese — to collect seeds for me. He will be bribed with butter . . . and I will send a man, in October, from Sakden, for the seeds from him.'[8]

The map of this area left much to be desired and gave little idea of the nature of the country. Instead of the expected plain in the Goshu Chu valley there were sheer cliffs everywhere, rising up to 4,000 feet from the valley-bed, whilst in the neighbourhood of Nyuri and Dyuri the gorges were quite stupendous, the twin villages being perched on the sides of precipitous hill faces, facing each other, and so close that the natives hailed each other across the gorge. This was the country of the Mönbas, shy, timid people, more like Takpas than Tibetans, who wore the tight fitting felt caps, like the Takpas, but with as many as a dozen 'drip-tips' instead of the four of the Takpas. And suspended from the crown of the head, instead of from the neck, the women wore some kind of necklace. They had no fields to cultivate and apparently grew very little of their own food. But they did possess large numbers of yak herds, which at this time of year were high up in the grazing lands, and thus never lacked for milk much of which they made into butter which they exchanged mainly for maize from Tawang and for salt from the Lobas of the Dirang district south of the Tse La. They also had a few hens which scratched out a pretty miserable existence

Map 4. A Month in the Mago District

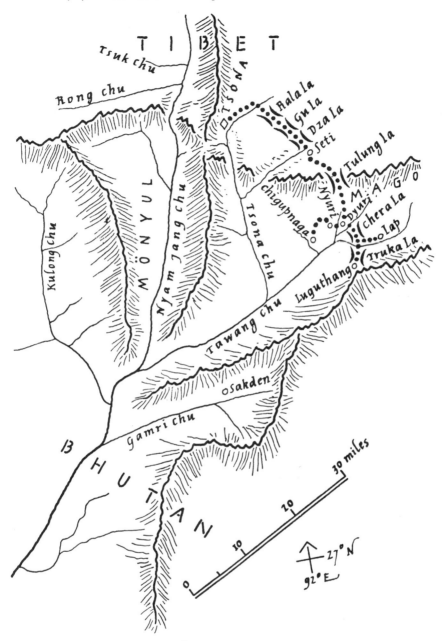

and provided a few eggs. There weren't more than thirty houses in the combined villages, and being without windows they were permanently dark within. The reason for this, apparently, was fear of the Lobas who visited them twice a year with their loads of salt. If there was light in our houses, the Mönbas said, the Lobas would see our belongings and would steal them. By the same token, if potatoes, for instance, a few of which the Mönbas *did* grow, were planted round the houses, the Lobas would steal these also; and for this reason the potatoes were grown in secret places in the jungle and thus out of sight of the Lobas.

In this district of Mago Ludlow and Sherriff planned to spend two weeks working the headwaters of the Tawang Chu. They made the ascent of the Chera La (13,500 feet) into the Gorjo Chu valley and the only plants which excited them were *Primula waltonii* (724), this time with pink, not port-wine coloured flowers; *P. bellidifolia* (720); *P. atrodentata* (721) like a small version of the Drumstick primrose, *P. denticulata*; a very dwarf mauve primula, no more than two inches high and closely akin to *P. pusilla*, which proved to be a new species which was named *P. occlusa* (725) and which has been found since only twice, by Kingdon Ward in 1938, on the Ze La on the Assam-Bhutan frontier and on the Orka La just within the Bhutan border; and best of all a magnificent slipper-orchid, *Cypripedium himalaicum* (722), the pale green ground-colour of the flower being boldly streaked with red and the labellum suffused with blood-red.

They marched for five miles up the Gorjo Chu valley to Lap (14,200 feet) and apart from the dwarf deep grape-purple *Lilium nanum* (726), a slaty-blue-flowered ally of the gentians *Swertia kingii* (727), *Meconopsis lancifolia* var. *concinna* (728), *Aconitum fletcherianum* (729), and one or two gentians and androsaces, they were again disappointed with the flora. On the evening of 28 July both Ludlow and Sherriff were very depressed. 'We have pinned our faith to Mago for the past year as a place where we should find a veritable rock-garden and it has turned out a failure' (Ludlow). 'It is a great disappointment, after coming all this way, to find that the upper Tawang Chu area is useless. At this height [over 14,000 feet] and in July and first half of August there should, one would think, be flowers all over the place. Flowers there are but more of the Tibetan type than we want. The Nyingsang La-Tse La Range seems to form more of a barrier than we had expected. It was this ridge which we had expected would make the area so particularly good. Anyway, we are wrong. I was out all yesterday, going over the ridge and a little down the Lopha side but there was nothing of any interest at all. Ludlow went up to the north with the same result' (Sherriff).

As the avifauna of the area was as disappointing as the flora, Ludlow and Sherriff returned very dejectedly to Dyuri on 30 July. For a change, the day was a fine one and from the summit of the Chera La there were views of the surrounding country. The snowy peaks to the west of Dyuri and Nyuri looked very inviting and they decided to spend the following week among these hills. But they hadn't reckoned with the dour uncommunicative and unco-operative attitude of the Mönbas who strongly asserted that it was not possible to visit these mountains by yak transport. If this were so, Ludlow and Sherriff reasoned, the journey would have to be made by coolie transport; and as there weren't enough coolies for the entire party the expedition would have to divide, one half of it exploring the hills to the west of the twin villages, the other half marching elsewhere with yaks.

Thus on 1 August Ludlow and Sherriff went their separate ways; the former, with Pintso, Ramzana and Danon, returning to the Gorjo Chu valley with the intention of ascending to Luguthang, a grazing ground on the Tse La Range, and to the south of it; the latter, with Amed Shaikh, Gulla and Kurtip, to explore the snowy peaks overlooking Dyuri. Sherriff soon realised that the Mönbas, who had professed to have no knowledge of the country beyond the fact that yaks couldn't go there, did, in fact, know the country well. Moreover he frequently came across well-trodden yak paths, and quite obviously the journey could have been done by yaks. The Mönbas had said that butter and milk were unprocurable in these hills yet Sherriff found lots of both. He also found a village of half a dozen houses, Bulung, which quite clearly the Mönbas knew though they professed not to. Why was it that these strange, shy, and really very kind and likeable people were so anxious that Ludlow and Sherriff and their men should not roam about the countryside ? After several days of trying Pintso found an answer. The Mönbas were afraid that the expedition would try to dig up pearls, or gold or other valuable stones and by so doing disturb the earth spirits!

Some twelve miles west of Dyuri, Sherriff spent four days in a camp at Chigupnaga (14,000 feet), a summer grazing ground with three or four huts occupied by yakherds; and in foul weather, with heavy rain and mist making it impossible to obtain a good view anywhere, he endeavoured to explore the countryside. *Meconopsis bella* (806) was much in evidence on all the cliffs; 'it is a beauty and

and should do well at home in rockeries'; and much in evidence also, above the conifer żone, either on open hillsides or amongst rhododendrons, was the tall-stemmed golden-yellow-flowered *M. paniculata* (747). There were some primulas, but nothing new; large clumps of the two miniatures, the grape-purple *Primula stirtoniana* (751) and the blue-violet *P. pusilla* (803); drifts of the magnificent *P. macrophylla* (761, 802) and fat cushions of the bright magenta *P. dryadifolia* (763); and growing in damp ground whenever there was a clearing in the forest, in fact in any open space above the tree-line, *P. sikkimensis*. But there was little else and on 3 August Sherriff sent a message to Ludlow telling of his disappointment.

In the meantime Ludlow was some eight miles from Dyuri, at Luguthang (13,500 feet), a small Mönba village of nine or ten houses at the junction of the stream from the Truka La with another stream coming from the east, both waters emptying into the Mago Chu. And from this village, on 3 August, *he* sent a message to Sherriff. '*Botany* N.B.G. Just the usual things we have seen and nothing of any interest whatsoever. *Birds* N.B.G. also . . . . *Butter-flies* nil. *Weather* bloody with a horizon of about 200 yards . . . . *Plans.* There is no point in stopping here and I propose to leave on the 5th. I may get in the same day or the morning of the 6th.'

That same evening Ludlow entered into his diary this note. 'I am afraid the Mago district must be given up as a failure. A pity, for we have wasted the best part of a season on it and it is too late now to hope for much in other areas. We may get a few late things between the Pö La and Me La but we cannot expect much. Birds also have been most disappointing, but, unlike flowers, we can expect better results later on. I sat in a Mönba hut on top of the pass yesterday and in consequence have got lice — disgusting.' Whilst he was writing so dispiritedly he did not realise that the two yellow saxifrages he had collected that day were new to science; they were later named *Saxifraga sphaeradena* (807) and *S. montanella* (809).

On 8 August the expedition began the return journey to Tsona by the route it had taken to Dyuri, and took a week over it. A lama accompanied it for the first two marches so that he could be shown the plants from which Sherriff was anxious to have seeds collected. 'I have shown him and told him to collect seeds of thalictrum, meconopsis, and primula and tomorrow will show him the other small mec (708) [*Meconopsis bella*] near the Tulung La. I

17. *Meconopsis bella* (above)
18. *Primula soldanelloides*

think he will manage to collect all. He knows something of meconopsis and when they should be ready, as they eat the seeds and also the shoots before the flowers appear.'[9] On the Tulung La, above Zangthang, they gathered three gentians new to their collections and the next day, 10 August, three others from the Dza La, one of them, *Gentiana urnula* (789), being especially fine; it was growing in tufts on the summit of the pass, in loose shale and with practically no soil, the large white flowers being streaked with slaty-purple. This was Ludlow's 49th birthday which he celebrated by shooting a couple of Brown accentors (*Prunella fulvescens tibetanus*), by buying a sheep, and by opening a tin of paté de foie gras.

At Tsona the margins of the fields were stained violet and deep blue with the tens of thousands of the tall *Delphinium grandiflorum* and of the invasive annual *Aconitum gymnandrum*. And in the rocks there was still the sublime paraquilegia. Sherriff spent the whole of the morning of 13 August trying to dig a living plant from out of the rocks. It was a difficult task for the tap-root is very long and wriggles its way downwards and in any direction through clefts and cracks. Such bits of plants as he was able to prize from the rock he packed into two boxes with stones, charcoal and earth. He also covered several lots of fading flowers with seed bags and requested the dzongpen to have the seeds collected and sent to Sakden. Whilst Sherriff was thus engaged Ludlow strolled around with his butterfly net and collected several specimens of the lovely *Parnassius imperator*.

Return to the Me La (*Map 5*)

On a morning of perfect weather, 15 August, the expedition left Tsona, with 24 yaks in charge of four men, en route for last year's pleasant and profitable haunts at Shingbe and the Me La. It was a journey which involved the crossing of six major passes, including the Me La itself. The route led up the valley west of Tsona through rolling undulating country with, at intervals, flat plains over which grazed countless yaks. The path gradually ascended to the Nyaba La (15,250 feet) and then turned northwards over open hillside before climbing steeply to the Gorpo La (17,750 feet), a pass on the Donkar Range which is flanked to the north by a snow massif with several glaciers falling away to the south-east and which, on such a day as this, offered magnificent extensive views of the Tibetan plateau stretching far to the east with, here and there, dark thunderclouds towering over the ranges. A few high alpine flowers new to the expedition's collections were gathered, including a pretty bright golden-flowered cremanthodium with a dark velvety green involucre, *Cremanthodium humile* (823); a pale lavender very sweetly-scented

Map 5. Return to the Me La

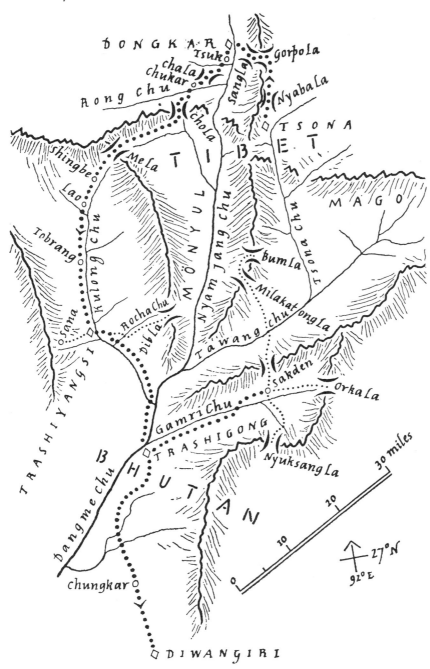

corydalis (799); a form of *Meconopsis horridula* (826), rather dwarf
at 17,500 feet, with the petals invariably tinged with pink; and
clumps of several saxifrages, including the golden-yellow, brown-
spotted *Saxifraga punctulata* (820) which was also above 17,000
feet in damp stony ground, and the nodding golden bell-flowered *S.
nutans* (822) which was usually never found higher than 15,500 feet,
the same elevation as that strange shrubby legume, from Siberia and
Mongolia, with its thick hairy and spiny branches and cream and
pink flowers, *Caragana jubata* (825).

From the camp west of the Gorpo La, where two inches of snow
fell overnight, they made for the Sang La (17,100 feet) on another
splendid morning of marvellous views, this time of the Nyam Jang
Chu valley stretching away to the north, and of Dongkar, where
they were to rest for a day, immediately below them. The sun on
the pass was warm and parnassius and fritillary butterflies took
wing, and gentians opened wide their flowers; *Gentiana algida* (828),
the white flowers streaked and spotted with greenish-purple, and
G. aglaia (829) of a wonderful bright delft-blue, were very conspicuous, and very attractive, growing together at 16,500 feet.
With the sun on their backs Ludlow and Sherriff lingered on the
summit of the Sang La searching for plants and butterflies and
allowing the yak transport to advance ahead of them. Not until
they made the 3,000 foot descent to the bridge over the Nyam Jang
Chu and found all the yaks off-loaded and the baggage piled up by
the roadside did they realise that whilst they had been happily
engaged on the pass there had been trouble at Dongkar.

Ludlow tells the story. 'The yaks from Kishong, unaccustomed
to bridges, had taken fright. One had cast its load and my tent
had fallen into the river and been swept away. Pintso and
several Bhutanese traders had started off down the river to
retrieve it, but so far had been unsuccessful. Ahmed Shaikh
and Ramzana were excited and incoherent and instead of
giving information started to pour into our ears a long story
as to how the tent bobbed up and down in the water, struck a
rock here, and was washed into the bank there. For a long
time we thought that both our tents had gone and began to
wonder how we should carry on without them. But about
5 p.m. Pintso turned up with my tent. He and his men had
found it two miles downstream of Dongkar somewhere in mid-
stream jammed up against a boulder. Pintso had swum out on a
line and retrieved it but not without great difficulty and considerable risk. Two or three times he was completely sub-
merged in the swift current. He lost the rope and his fellow

companions thought he was drowned. But eventually he was washed ashore. A stout effort. But P. ought not to have risked his life for the sake of a miserable tent.' [10]

The camp at Dongkar (13,350 feet) was a most charming one, pitched in a walled garden with firs, poplars, juniper and buckthorn, wild asters and *Delphinium grandiflorum*, and a sward of vivid green grass; both Sherriff and Ludlow fell in love with the place. Although situated in a deep gorge there were fine views up and down the river and to the snow peaks to the east and west. The cultivated fields, too, were an agreeable sight after the bleakness of the higher regions and the depressing lack of cultivation in the Mago district. Apart from the countless spires of the delphinium there wasn't much in the way of exciting flowers. But there *was* a fine gentian-like plant with blooms of pale lilac, growing in very dry stony ground, which proved to be something quite new — *Lomatogonium dialatum* (834). And there was an interesting aconite with grape-purple flowers (827) which Sherriff and Ludlow were not to see again until 1949, and then at Tsampa in Bhutan. This also proved to be a new species and eventually was named *Aconitum hicksii.*

On leaving Dongkar, the expedition descended the Nyam Jang Chu which, some two miles from Dongkar flows through a steep narrow gorge, so narrow indeed that at one point a number of huge fallen boulders completely obscured the river from view. Shortly beyond this the Tsuk Chu flows in from the west. Ascending this valley, and passing the customs village of Tsuk and its barley fields, they made the steep ascent to the Cha La (15,300 feet), the crossing of which brought them into the large Rong Chu valley and to the village of Chukar (13,600 feet). This route, not previously travelled by Europeans, is famous as the one by which the saintly Lopön or Guru Rimpoche entered Tibet from India, bringing Buddhism with him. He stopped between the Me La and Cho La — and this latter pass probably should be written Chö, which in Tibetan means 'holy', or 'religious', because it was the pass, the holy pass, over the main range chosen by the Guru Rimpoche. Over the Cho La he came on to Chukar and when he reached the Cha La he was met by all the people who, with hands folded and supplicating, begged him, unsuccessfully, to remain among them. Hence the name Cha La, the 'hand' pass, 'chha' being the honorific Tibetan for hand.

Though they were still north of the main Himalayan range the vegetation of the Rong Chu valley showed the rainfall to be considerable. 'Immediately we crossed the pass we knew we were in a good place for flowers', Sherriff wrote on 18 August. 'It is curious how many pink *Mec*[*onopsis*] *horridula* there are about here — from Tsona to Mago to a certain extent but more so from Tsona to Dongkar and on here. Besides *M. horridula* there are a number of the Tulung La little meconopsis [*M. bella*]. Two gentians were in seed and were collected on the Cha La [*Gentiana phyllocalyx* (844) and *G. tubiflora* (853)]. Down this side [descending to Chukar] we both thought of *Prim*[*ula*] *eburnea* and within a few minutes came across it. It was growing in a typical situation under overhanging rocks and in clefts of rocks. It is certainly a beauty [848]. Beside it, and in even more pronounced clefts and more overhanging rocks, was a, to us, new primula [847]. It is also a beauty, very delicate and very sweetly scented. . . . With these two, and in similar situations, was a third primula — the Seti primula collected on the way to Mago [*P. bellidifolia* (849)]. . . . I took photos of all.' The primula new to them (847) was *P. littledalei*, a delicate-looking species , with a long stout rhizome cleaving the rock, round-toothed leaves which are thickly plastered below with white meal and which, when they wither, persist at the base of the plant, and with lilac or pale purple flowers. Ludlow and Sherriff later found it on several occasions in SE Tibet, as well as several of its close allies, all beautiful plants.

On the day's halt at Chukar Sherriff tied seed bags over some of the flower heads of the primulas, of *Meconopsis bella*, and of *Thalictrum chelidonii* which was also here. He arranged for a local man to keep an eye on the plants until October when he proposed to send someone from Trashiyangsi to retrieve the lot. Even at this early stage in his explorations Sherriff would go to no end of trouble to attempt to introduce a worth-while plant into cultivation.

The march on 20 August took the expedition to a camp (14,300 feet) on the north side of the Cho La and a little under 2,000 feet from the summit, the final ascent of which — an absolute knife-edge — was made the following day. *Meconopsis bella* and *M. horridula* were once again much to the fore, as was *M. grandis* (875). There were several aconites new to the party; indeed there were two new to science, the blue-violet *Aconitum parabrachypodum* (861) which Ludlow and Sherriff saw on many future occasions both in Tibet and Bhutan, and the pale greenish-purple *A. rongchuense* (862) which they never found again. There was also a new species of delphinium with fine purple or blue-violet flowers, *Delphinium bhutanicum* (873). They thought they had discovered a new cyananthus with most handsome white blooms but it turned out to be a very beautiful albino form of *Cyananthus lobatus* (872). There was nothing new in the way of primulas though they gathered further specimens of *Primula gambeliana* (874) which they had seen for the first time at Sakden, and of the two tight-headed species, *P. atrodentata* (868) and *P. capitata* subsp. *crispata* (869). The views from the pass were quite superb; from the south, and round to the west, several snow peaks of considerable size from which descended numerous glaciers; and immediately below, the great lovely valley, its slopes well clothed with fir and rhododendron, into which they had gazed from the Me La, last year. The expedition was now across the main Himalayan range, and in two days' time, 22 August, had crossed the Me La into Bhutan and was at its old camping ground at Shingbe.

The valley between the Cho La and Me La had impressed Shcrriff greatly. 'I believe that had we spent all our time between here and Dongkar we would have got nearly every flower we have collected in all our wanderings. A few of the Sakden ones might not have been found but there would have been others of the early flowerers to make up for that. But it can't be helped now.' [11]

However, Ludlow and Sherriff were now back in Bhutan and near the Me La — 'The pass of the flowers' — and their first ascent revealed a treasure new to them. 'We found a new primula (878). It was growing on rock-ledges in wet moss. It is a pretty little thing, rather delicate looking. I call it a primula though it is unlike others in having a rather irregular flower, the lower petals being longer than the upper. But as far as we could make out it is a primula and, if so, will almost certainly be an interesting one, if not quite new. *Rheum nobile* was growing all over the hillside [the large straw-coloured bracts] being very obvious against the dark green vegetation. *Prim. roylei* [*calderiana*] is in seed and I will bag some tomorrow. P[*rimula*] *congestifolia* [*dryadifolia*] is also near the top and has just about [finished flowering]. . . . My difficulty now is to remember where the flowers are, the seeds of which I want to collect.' [12]

No. 878 was indeed a primula *and* an interesting one. It was *Primula soldanelloides*, a dwarf species the flowering stem of which, two inches high at most, carries a single glistening-white cup-shaped flower about half an inch across and thus rather huge for so small a plant. It was not new for it had been discovered, in Sikkim, as long ago as 1849. But this finding of it on the Me La did constitute a new record for Bhutan.

During last year's stay at Shingbe the weather had been kind. Now, it rained almost incessantly. Even so, Sherriff and Danon spent a lot of time on the pass, 'bagging' some plants and marking many others for seed collecting later — in October; *Lonicera, Berberis, Thalictrum, Geranium, Primula, Meconopsis, Gentiana, Delphinium, Aconitum, Swertia, Saussurea, Lactuca, Dracocephalum, Codonopsis* and *Lilium nanum*.

Observations on birds

Whilst Sherriff was thus occupied Ludlow 'strolled about with a gun'. For some time he had been interested in the Scaly-breasted wren-babbler, *Pnoepyga albiventer albiventer* which in 1933 he had thought was rather a rare bird but which, when once he had learned its habits and call, he now realised was very common in Bhutan, occupying a wide altitudinal range. In the rainy season it may occur at over 12,000 feet, whilst in the autumn it can be found as low as 2,000 feet. Its favourite haunt is the neighbourhood of streams in thick forest where there is plenty of undergrowth. When thoroughly alarmed and frightened it utters a shrill piercing whistle but its normal note Ludlow likened to 'an ill-mannered person loudly sucking his teeth!'. From now on Ludlow was to see a great deal of this wren-babbler and to secure a large series of carefully sexed specimens with the object of solving the long standing problem as to whether or not the coloration of the individual is any guide to sex. He concluded that the colour of the underparts is no criterion of sex and that *albiventer* and its cousin *pusilla*, of which he also secured a splendid series of specimens, are dimorphic.

Pintso, who had been sent to Trashiyangsi for coolies, returned with them on the evening of 27 August and the expedition was ready to leave Shingbe the next day. The height of the flowering season was now past, of course, and during the ensuing three months all energies were to be applied to collecting birds and seeds and to marking plants for seed collection. The expedition moved slowly, making prolonged halts in the Trashiyangsi valley, and again visited Sakden; and during this time close on 750 bird skins were prepared.

At Tobrang (7,500 feet) Sherriff and Ludlow spent a week bird-hunting in the dense jungle, an extraordinarily difficult and unpleasant occupation. 'Often one cannot shoot a bird because it is

> bound to drop into a spot where one cannot possibly retrieve it. Many birds are lost in the dense undergrowth. Then there are leeches in all the best places and the best places are the deepest, darkest, densest nulla beds where you cannot see more than 5 yards ahead. A 410 bore is valueless in such places. If you fire you blow your specimen to pieces; and if you don't fire you never see your bird again.'[13]

In an effort to procure as many skins as possible the services of Pintso and Danon were sometimes recruited. Sherriff recounted one such occasion at their camp in the Rocha Chu valley, at 10,000 feet on the Dib La, the knife-edge pass clothed with fir forest to the summit. 'As this was our last day here, and we still wanted some "binair dum" birds [*Pnoepyga*] we split up. Pintso took the 12 bore, Ludlow and I the 410's and Danon took the 22 rifle. The result was hardly as good as we expected. Pintso blew one to bits by being far too close; Ludlow and I saw none; and Danon who saw four missed them all as he did not understand the aim of the rifle. But I met him and gave him some instruction. After that he and I heard one, and stalked it. During the stalk he managed to place the bird between us. There is no stopping Danon if he sees a "binair dum", and he fired, getting the bird and bits of me beyond. No damage done though. Pintso is rather peeved that he always gets less than Danon if they are given guns. So he has taken out the 22 rifle and swears he will bring in some pnoepygas somehow. Danon's eyes are so sharp that he is great value when we are after birds. The sight of a tailless wren is almost too much for him and he gets terribly excited. If I ever come again I will most certainly take Danon with me.' [14]

It was to the babblers, warblers and wrens that Ludlow and Sherriff paid special attention during the next weeks and by so doing elucidated several problems. The problems of sex and colour in the Scaly-breasted wren-babbler was one such. The problem of the identity of *Tesia cyaniventer* and *Tesia olivea*, the Slaty-bellied ground-warbler, was another. This ground-warbler exhibits two colour phases. There is what may be termed a *pale* phase; birds with the crown concolorous with the back and with the pale slaty underparts, and with the base of the lower mandible yellow or orange-yellow. And there is what may be termed the *dark* phase; birds with glistening golden crowns contrasting with the olive-green of the back and with the dark slaty underparts, and with the base of the lower mandible deep orange or orange-red. What value is to be attached to these pale and dark phases ? Are they of sexual significance ? Are they seasonal ? Are they due to age, to individual variation or to dimorphism ? Are they of specific or subspecific importance ?

In an attempt to answer these questions Ludlow and Sherriff collected a large and carefully sexed series of twenty-six specimens. The first twenty were of the pale phase, the rest of the dark. All twenty specimens of the pale phase were obtained from temperate forest between 6,000 and 8,000 feet — not a single dark form was encountered within this zone. The six specimens of the dark phase were all taken in tropical forest at, or below, 3,000 feet, where none of the pale form was seen. Without breeding experiments these facts of course prove very little but they do very strongly suggest the existence of a high altitude pale bird and a low altitude dark one. The difference in colour is not sexual for in the series there were adult male and female dark forms and adult male and female pale forms. The colour is not seasonal for both phases occur both in summer and in winter. It is not a question of age for in the series there were both pale and dark juveniles. Individual variation cannot be the cause for there was never any sign of intermediate coloration — specimens were either definitely pale or definitely dark. Dimorphism seems unlikely for if it had prevailed it would hardly have been possible to have collected twenty pale forms in succession without once meeting the dark form. Thus, the conclusion the collectors reached was that there is a high altitude pale species or subspecies with a temperate summer distribution, *Tesia cyaniventer*, and a low altitude species or subspecies with a tropical or semi-tropical distribution, *T. olivea*.

Another carefully-sexed series of thirty specimens of the Chestnut-headed ground-warbler, *Tesia* (or *Oligura*) *castaneocoronata*, gave considerable hitherto unrecorded information about its morphology and habits. The males are larger than the females. They are also sometimes brighter below; otherwise there is no difference in colour in the sexes. In both sexes the young bird is dark olive-green, with a brownish tint above and chestnut below, and from this plumage the bird moults directly into that of the adult — bright chestnut crown and nape and bright yellow underparts, the upper parts, wings and tail of a dark olive-green. It inhabits higher ground than *Tesia cyaniventer* in summer, being frequently met with at 10,000 – 11,000 feet along with *Pnoepyga albiventer albiventer*, whilst at Diwangiri, in November, it was as low as 2,100 feet.

Several skins of both sexes of a Long-tailed wren-babbler were at first thought to be those of a new species. However they proved to be those of a rare race of *Spelaeornis souliei* from NW Yunnan which was to be named *sherriffi*; the head and back are warm brown each feather being tipped with white; the tail and wings are barred with wavy black lines; the throat and breast are white, some of the feathers being faintly tipped with black, whilst the belly and flanks are light chestnut. This race is more arboreal than most wren-babblers — *and* rather slower in its movements — never being found below 10,000 feet and usually hopping about bamboo stems and mossy tree-trunks, sometimes uttering a subdued 'cheep'. Apparently Sherriff's Long-tailed wren-babbler, *Spelaeornis souliei sherriffi* is confined to Bhutan. Further east in NW Yunnan, west of the Mekong river, the typical form *Spelaeornis souliei souliei* occurs whilst east of the river it is replaced by *Spelaeornis souliei rocki*.

One of the most interesting skins procured was that of the rare Ward's trogon, *Harpactes wardi* — they shot a female bird in the thick oak forest on the Dib La, at 8,000 feet. This trogon, a dark vinous bird appearing almost black when at rest and in flight showing yellow underparts and some crimson on either side of the tail, was discovered by Kingdon Ward in the Seinghku valley, north of Fort Hertz, in Upper Burma, in 1926, and in Burma it is now known to be fairly common in the mountains north-east of Myitkyena. In 1929 it was also found in considerable numbers in north Tonkin. Thus Bhutan represents a great extension westwards of its previously known range. And Diwangiri, where in November Ludlow and Sherriff found McClelland's Laughing thrush, *Garrulax gularis* — chestnut brown back, primrose-yellow underparts and bright orange-yellow legs, and known to occur from Assam across Northern Burma to Laos — must be pretty nearly the western limit of this bird's range.

During his hunts Ludlow was constantly impressed by the shyness of most jungle birds and finally concluded that this must be due to their many enemies. 'They probably have more enemies than

> we realise as we seldom see much ground vermin. But martens, weasels, snakes, squirrels, etc. must all take heavy toll of birds and eggs and this explains, I think, why the slightest noise, such as the cracking of a stick, suffices to send many birds helter skelter into the heaviest thicket. Some birds, such as *Garrulax albogularis*, which are quite tame appear to owe their immunity to their gregarious habits. They are always in large parties and with 20 or 30 pairs of eyes to look everywhere there is little likelihood of an enemy taking them unawares.' [15]

Return to Trashigong

Most of September, 8–29, had been spent at several camps at varying altitudes on the Dib La (13,000 feet), some fifteen miles east of Trashiyangsi, chiefly in the hunt for birds. On 1 October Danon left the main party for a week's seed collecting on the Me La, whilst the rest walked twelve miles west of Trashiyangsi to Sana (8,300 feet), there to stay for a week during which time their main concern, apart from seeds and birds, quite unexpectedly proved to be an ailing Tibetan, a Khamba. The diaries of the two friends tell the story. 'There is a poor Tibetan all alone here [Sana] in a shelter. He is ill and has been unable to leave the place for the last 10 days. He has no food and is in bad pain. It looks as if he had a very bad abscess behind his right shoulder-blade. We can at least feed him though what the treatment is I don't know.'[16] 'The Tibetan is no better today; in fact the poison has gone down to his right arm and hand and left leg too, it seems. We have put on four large hot compresses today but there is still no sign of the abscess coming to a head. A Mohammedan is a rotten man when it comes to helping a fellow creature. Our people get all food, and ample, practically free. They said, of course, they would feed him. When we looked to see what he was getting, it was dry rice alone.'[17] 'The Tibetan is rather better today and some pus came out of the abscess this morning. We intend to cut it this evening with a razor blade.'[18] 'In the evening opened the Khamba's abscess. Never have I seen such a large one. We must have taken a full bottle of pus out of it. He stood the pain very stoically and, of course, got much relief.'[19] 'The Khamba had a good night and seems better. There is still a lot of pus to come out though. The difficulty will be to prevent the incision healing up before all the pus has been got rid of.'[20] 'We dress it four times a day and still any amount of pus pours out. The hand is a little less swollen but still big and very painful. I wonder if he will be able to walk down with us in two days' time. We cannot leave him here or the whole thing will just start off again.'[21] 'The Tibetan is much better, able to walk a little and in much less pain. There is still a lot of pus and two cores came out today.'[22] 'The Khamba is a good deal better and we have been insisting on his taking as much exercise as possible. After lunch we sent him off to a hut ¾ mile down the road. We leave for Trashi-yangsi tomorrow and we want to get him along the road as far as possible. I expect he will have to be carried most of the way. He *must* accompany us for his wound needs constant dressing. There is still a flow of pus from it but this is getting less and less and his two arms and shoulders are now practically the same size. Also, he is in a much happier frame of mind. I hope he will be quite fit again in 3 or 4 days — at any rate by the time we reach Trashigong.'[23]

Such consideration was no isolated example. Wherever they travelled in Bhutan, Tibet and elsewhere, Ludlow and Sherriff

19. The Laughing thrush (*Garrulax gularis*),
    Fulvetta ludlowi and *Spelaeornis souliei sherriffi*

always treated those in need with similar kindness and as a result their somewhat limited medical services were eagerly sought. On their next expedition they brought a doctor with them to attend to the sick and needy.

On 9 October the expedition retraced its way to Trashiyangsi where it met Danon who had just returned from the Me La with a good collection of seeds, as well as with flowers of two fine autumn-flowering gentians. *Gentiana amoena* (1001), brilliant blue and tinged with purple, was in large masses above the tree-line in open grassy ground at 14,500 feet; in ten days' time it was also to be seen in great profusion on the Nyuksang La, Sakden, where the locals ate it, either dead or alive, and professed it to be 'very sweet'. *Gentiana gilvostriata* (1002) was also in abundance, its large sky-blue trumpets, paler within and banded with brownish-purple without, making a brilliant splash of colour at the edge of the conifer forest at 12,500 feet.

After a day at Trashiyangsi, labelling and packing up seeds, as well as bulbs of *Cardiocrinum giganteum, Lilium wallichianum, L. nanum* and *Notholirion bulbuliferum* — a very fine form of the latter over 4 foot tall, with as many as sixteen rich purple, green-tipped flowers on the one stem — they began their three-day march in the Trashiyangsi valley to Trashigong, and gathered several desirable plants on the way. There were two epiphytic orchids; one, *Coelogyne ovalis* (1048), with flowers predominantly of old gold with the throat of the labellum striped and ribbed with bright brown, was growing on some species of oak; the other, a form of *Cymbidium longifolium* (1051), was epiphytic on rhododendrons, its long spikes of basically lemon-green flowers heavily streaked with brownish-red making it very conspicuous. And there were a couple of marvellously scented shrubs; the pale pink *Luculia gratissima* (1059) and the old favourite, the white *Jasminum officinale* (1058) which, though it has been hardy in cultivation in Britain for over 400 years, was much at home in this hot valley at the surprisingly low elevation of 4,500 feet.

Seed collecting

From Trashigong the expedition made its second journey up the Gamri Chu valley to Sakden where, on 19 October, the party split up with the object of collecting seeds over as wide an area as possible. Danon departed northwards for the Milakotong La (14,200 feet) and the adjacent Bum La (15,000 feet) near Shao, hoping to obtain seeds of primulas as well as of the white *Meconopsis horridula*. One Takpa was sent even further northwards to Tsona and another north-east to Mago. No sooner had they departed when the

Mago lamas arrived. Though they had failed to collect seeds of the white poppy they brought with them seeds of *Meconopsis bella* and of *Thalictrum chelidonii*. For his part, Sherriff climbed the Nyuksang La south-west of Sakden, and was bitterly disappointed with the seed harvest. 'I found some of primula 602 [*glabra*]. The other one, 608 (or 641) [*Primula gambeliana*] I could not find. All have been eaten by sheep or yaks. No. 612 [*P. strumosa*] could not be found either. It had also been completely cleared by animals. . . . . The yaks have done a lot of damage to all flowers as I found lower down where the *Meconopsis grandis* (600) and the new primula 598 [*waltonii*] were in such profusion before. All had been trampled down and eaten and with great difficulty I collected a packet [of seeds] of each.'[24]

Another day Sherriff left camp at 5.15 a.m. to make the four-hour walk to the east to the Orka La and returned with a fair amount of seeds of *Primula gambeliana, Rhododendron anthopogon* (1091) and an androsace. The seeds of the rhododendron proved to be a lucky gathering. Many plants have been raised from them and one seedling, grown by E. H. M. and P. A. Cox of Perthshire and called 'Betty Graham', received the Award of Merit from the Royal Horticultural Society in 1969.

Danon returned from Tawang with a good haul of seeds including those of several primulas and rhododendrons, of *Meconopsis lancifolia* var. *concinna* (1080), of *Cyananthus inflatus* (1087) and of the new *Berberis buchananii* var. *tawangensis* (1089). Unfortunately the sheep had been at the lovely white *Primula obliqua* (655) which was so badly wanted and there wasn't a seed to be found. Danon paid another hasty visit to the Orka La and returned with seeds of *Meconopsis lancifolia* (1095), of *Primula atrodentata* (1094) and of the new species gathered in July, *P. occlusa* (1093). When the two Takpas rejoined the main party they brought nothing with them save seeds of paraquilegia mixed with capsules of *Meconopsis horridula*.

From a seed collecting point of view the expedition virtually was now at an end although Sherriff and Danon still had to visit the Chungkar cliff on the Diwangiri road where, in June, Sherriff had found the two new primulas and the new luculia. This they did on 7 November. 'We climbed up the cliff and found signs of many primulas, both 552 [*sherriffae*] and 554 [*ludlowii*]. However we could not find one seed of 554 and although there were lots of seed pods of 552 most of them were last year's. However we collected about 30 heads. It grows in the most impossible places, on sheer cliff, just hanging on to ½ inch of moss which is absolutely dried up. I took a few of the plants of both 552 and 554. The lily [*nepalense* var. *concolor* (553)] was even harder to get. We could see it nowhere at first but eventually saw a few. Danon with great difficulty got three bulbs and two seed pods. A few others were seen but we could not possibly reach them without a rope.' [25]

It was not quite the end of the journey as far as birds were concerned. With the cessation of the rains on 18 October the leeches miraculously disappeared and Ludlow and Sherriff were able to creep through the forest undergrowth in comparative comfort — at least when once the huge spiders' webs which were spun everywhere had been destroyed by brandishing a small tree-branch. Leisurely they worked their way down to the evergreen zone which, so far, they had not explored. They were surprised to see how greatly the numbers of resident birds were now reinforced by numerous autumn migrants and especially surprised to find at Diwangiri many Spotted-winged grosbeaks (*Mycerobas melanoxanthus*), the male black-throated, the female yellow-throated, feeding on the tops of trees in thick tropical forest. In the summer they had been seen between 10,000 and 12,000 feet in conifer and birch forest in East Bhutan. With the approach of autumn, towards the end of September, they begin to congregate into very noisy flocks. Then, apparently, as the autumn advances, they descend to lower altitudes, even as low as 2,100 feet as at Diwangiri.

With the departure from Diwangiri on 14 November the five-month trek was over. It had begun badly, with delays due to bad weather and to illness, and with disappointing collections, and had ended rather successfully, certainly from the ornithological point of view. Nine hundred and seventy bird skins had been prepared for the British Museum, 750 of these during the last three months, and valuable scientific information had been gained regarding the morphology, distribution and habits of a good many species. Of the 600 beautiful plant specimens which had been prepared for the herbarium many represented new records for Bhutan or Tibet and thus threw new light on the distribution of previously known species, whilst at least a dozen and a half were of plants quite new to botanical science. More efforts had been made to collect seeds than last year and more seeds had been collected. And from these seeds one of the great plants of the century was soon to be introduced into British gardens — *Meconopsis grandis* 'Sherriff 600'. This alone made the expedition eminently worth while.

Chapter Three

To TSARI,

A Tibetan Sanctuary

o

The first entry in Ludlow's diary of his 1936 expedition, dated 9 February, reads: 'Left Srinagar for another trip to Bhutan and SE Tibet. Our objective on this occasion is that portion of SE Tibet which includes the districts of Chayul, Charme and Tsari — in fact the upper reaches of the Subansiri. We have had this trip on our minds ever since 1934 and we were rather dismayed when we learnt a few days ago that Kingdon Ward had just returned from these very districts . . . . Apparently he crossed into Tibet via Dirang Dzong, the Tulung La and Pen La and then went east to the bend of the Tsangpo, returning to Tsari in the autumn. It is too late to change our plans and our only hope is that as he was travelling quickly he left a lot of ground untouched. At any rate, I don't suppose he has exhausted the botanical possibilities of the country.'

Ludlow immediately wrote to Kingdon Ward explaining the situation to him and a couple of weeks later received a helpful and encouraging reply by return airmail. Ward outlined his itinerary and recommended for exploration those areas which he was convinced would yield the most profitable results. He emphasised the hurried nature of his own visit and the richness of the flora and gave the assurance that he had collected not a tithe of the wonderful plants which he had seen.

Because of the inhospitable and hostile nature of the Dafla and Abor tribes inhabiting the various tributary valleys of the Subansiri south of the main Himalayan range, Ludlow and Sherriff did not consider it feasible to attempt to reach their collecting grounds from the south. Instead, they would approach from the west, through Bhutan. They entered Bhutan on 14 February, via the familiar little frontier post of Diwangiri where they stayed for a week, collecting birds in the thick tropical forests surrounding the village and reorganising their baggage and stores in preparation for their journey, in the first instance, to Trashigong on the Manas River. They were a party of ten, including Dr Kenneth Lumsden the expedition's medical officer, and they were to be away for over ten months. Not surprisingly therefore there were eighty loads of various kinds to be carried around.

On 22 February the expedition left for Trashigong, seven marches distant. Sherriff botanised on the Chungkar cliffs, which now were as dry as dust, whereas in 1934, towards the end of June, they had been covered with soaking wet moss. *Primula filipes* (1126) was in fine flower, heliotrope flowers where the plants grew in the shade and bleached almost to white when in the sun. On all sides the flaming red flower trusses of *Rhododendron arboreum* were a splendid sight. By 28 February the expedition had reached the grassy downs on the summit of the Yönpu La (8,300 feet) and was rewarded with entrancing views of the snowy range of peaks to the north. In 1934 when Ludlow and Sherriff had travelled over this pass, *Cardiocrinum giganteum* had been in magnificent flower. Now, the only plants of any interest to the travellers were *Primula filipes* which had been quite common all the way from Chungkar, and in one place only, on wet mossy rocks within reach of the splash from cascades, *P. gracilipes* (1147) with most beautiful heliotrope flowers over an inch across.

Though the plants of the Yönpu La were disappointing there was some consolation in the fact that here, for the first time in Bhutan, Ludlow saw Beavan's bullfinch (*Pyrrhula erythaca erythaca*) which, at the time, was generally considered to be a rare bird and indeed was represented in the National Collection in Britain only by seven male specimens. Ludlow shot two females and a male and was very elated at procuring so apparently rare a prize. However, some weeks later, when the Manas basin had been left behind and they had entered that of the Subansiri, Ludlow saw large numbers of this bullfinch north of the main range. Quite obviously its real home is east of Bhutan; Bhutan and Sikkim represent the western fringe of its distribution, hence its rarity there.

The expedition reached Trashigong on 2 March and was received with great hospitality by the dzongpen whom Ludlow and Sherriff had met in 1934. He had once been a mendicant monk and had travelled widely in Tibet and Bhutan. Then, abandoning his vagabond life, he had settled in Eastern Bhutan, had proved himself an able administrator and had become Governor of an important frontier district. Now, he was anxious to be of all possible help. He presented the party with a *maund* (= 80 pounds) of fresh

Map 6. Mera, and by the Pö La to Tsona

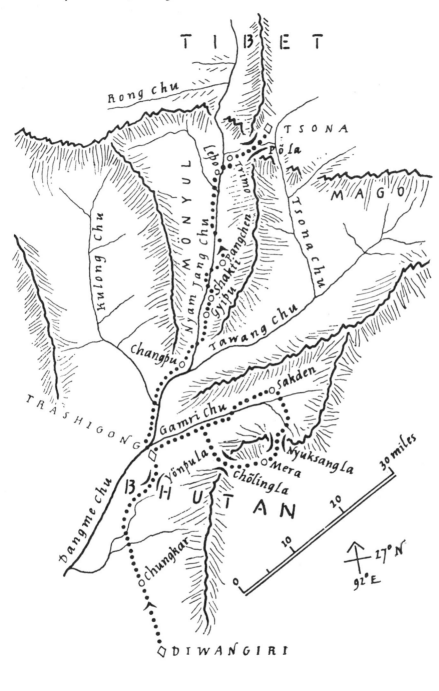

butter, and as His Highness the Maharaja presented them with two maunds they were all well supplied with fats for many months, during which time the butter never became rancid although no kind of preservative had been used in its making.

Already Dr Lumsden had been very busy. On the Yönpu La a middle-aged woman was brought to him, from several miles away, on a stretcher, with appalling suppurating burns from knees to ankles; two weeks previously she had had a fit and had fallen into the fire. Saline baths for several days greatly improved her condition. During the three days at Trashigong, Lumsden was fully occupied examining and treating patients of all kinds; many had goitre and leprosy and there were several cases of cataract.

As it was still too early in the year for flowers and as there was little to be gained in reaching the upper Subansiri much before the end of April, they made a short diversion eastwards up the valley of the Gamri Chu where, after marching for some ten miles, the party divided. Ludlow and Lumsden continued up the valley to Sakden on what, to Ludlow, was now a familiar route, whilst Sherriff moved south-east over the Chöling La (11,100 feet) to Mera with the intention of rejoining the others at Sakden by crossing the Nyuksang La (13,600 feet).

Sherriff was greatly impressed by Mera, 'the prettiest part of Bhutan I have ever seen'.[1] Like Sakden, it was a village of some sixty or so houses, in a beautifully wooded valley. The East Himalayan fir, *Abies spectabilis* (1174) covered whole hillsides, mostly on the north faces, with splendid stems over 100 feet high and often with a girth of 15 feet at breast height. Fine big trees of *Tsuga dumosa* (1180), the Himalayan hemlock, 80 to 100 feet tall, were also frequent on these same faces. On the south slopes, on the other hand, the Drooping juniper, *Juniperus recurva* (1175), was more common than the fir, beautiful 45-foot specimens frequently with a 10-foot girth four feet from the ground. In fact the trees, at this time of the year, were far more impressive than the flowers. Much the most beautiful and interesting — and common — of the latter was a blue primula (1166), akin to *Primula gracilipes*, which Ludlow and Sherriff, quite independently, and excusably, at first mistook for a colour form of this species. 'About 1,000 feet

> below the pass [Chöling La] Tendup came across the primula 1167 [*gracilipes*] . . . . Then we reached the snow, going up a north face, and at the first big patch which was half melted,

20. *Primula whitei*
21. Sanglung, Nyam Jang Chu

in rhododendron and abies forest, came across 1166, which I
at first mistook for the same primula, only blue . . . . on the
south side of the pass we found masses of 1167 with a good
deal of 1166 with it. Both were growing together.'[2] And at
Sakden Ludlow 'found the pink primula . . . . and also a blue
form. The two forms grew cheek by jowl, and it looks as
though here is a case of colour dimorphism.'[3]

However, the more they saw of the blue primula the more they
realised that it was quite distinct from *P. gracilipes*, especially in
the dormant condition when it forms a beautiful resting bud, the
size of a pigeon's egg, of overlapping mealy scales which completely
enclose the flower and leaf primordia. Once, with great difficulty,
Sherriff dug some of these resting buds from hard, frozen, icy
ground where the night temperature was 36 degrees of frost, and
found that the young leaves and flowers were ready to burst from
the bud. As the flowering season approaches the developing leaves
and flowers force the bud scales apart and at flowering time a
number of lovely delft-blue flowers nestle in the centre of a rosette
of finely-toothed leaves. With the fading of the flowers, the fruiting
capsules quickly develop and are carried above the leaves and into
the air on a stout stem. Such structure and behaviour is to be found
not in *P. gracilipes* but in *P. whitei* which Sir Claude White first
found on the Pile La, Tongsa, during a tour of Bhutan in 1905 and
and which R. E. Cooper collected several times in 1914-15. White's
and Cooper's plants had the petals finely and regularly crenulate
and the sepals often toothed and fimbriated.

Material similar to White's and Cooper's is to be found frequently in Ludlow and Sherriff's magnificent collections. But 1166
was different; and different also were plants which Ward gathered
on the Se La and on the Manda La in the Assam Himalaya in
1935. In these collections the petals were deeply tri-dentate, and
the sepals quite entire. On the basis of these characters, and with
the agreement of the collectors, 1166 and Ward's plants were
described as a new species under the name of *P. bhutanica*.

And thus it was until March 1947 when, at Tongkyuk Dzong in
the Pome district of SE Tibet, at 9,000 feet, Ludlow and Sherriff
came across a group of plants of the blue primula which they collected under 12299 and about which they wrote the following field
note: 'Specimens under this number should not be separated. All
specimens were collected from one spot and represent one species.
But it will be seen that the corolla (petal) lobes of some are finely
crenulately toothed whereas the lobes of others are coarsely tri-
dentate.' Thus the collectors had found the proof that so-called *P.*

Tsona

bhutanica and *P. whitei* represent one variable species and that the name *P. bhutanica* had to be reduced to the limbo of synonymy.

In 1936 Ludlow and Sherriff introduced *P. whitei* into cultivation in Britain by sending living plants by air mail, since when it has proved to be one of the most beautiful of early spring-flowering primulas.

From Trashigong to Tsona

The expedition left Sakden and returned to Trashigong on 18 March, moving on north-eastwards, the following day, for the valley of the Nyam Jang Chu, rather concerned about the numerous packages — they had collected a dozen bags of rice at Trashigong — and the transport which would be required to cross the Pö La. After descending steeply for a thousand feet they reached a great chain suspension bridge over the Dangme Chu or Manas river. The bridge was some 200 feet long, 6 feet broad, and hung 50 feet above the river. To allow the passage of animals such as ponies and cattle the footway and sides were of bamboo matting. The iron links of the chains, which were attached to wooden blocks built into masonry piers at either end, were a foot long. Though there are several such chain suspension bridges in Tibet this was the only one the travellers had seen in Bhutan and no one, not even the dzongpen, was able to say who had forged these great iron links. The dzongpen had accompanied them to the bridge and on bidding them goodbye warned them to be careful in their dealings with the Lobas in Tsari, who, he maintained, were a treacherous lot of people.

Marching through hot dry country where whole hillsides had recently been burned and where, even now, in a dozen different directions fires were raging, they reached Changpu on 21 March and a few hundred yards beyond the village a dry watercourse was pointed out to them as the Tibetan frontier. This was astonishing. Why had so insignificant a geographical landmark been used as a boundary when, not much more than a mile ahead, lay the great Nyam Jang Chu valley which had been discovered by Bailey and Morshead in 1913? As with the Manas river bridge, no one was able to enlighten them on this matter either. Not surprisingly for a river which has cut its way through the main Himalayan range the Nyam Jang Chu valley is a magnificent one. At no place is it very wide and on either bank the densely forested mountain slopes tower precipitously upwards to great heights. Occasionally, as at Pangchen and Lepo, the river meanders silently through grassy meads, but for the most part it is a swift torrent. It pierces the main range a few miles above Trimo where it is confined to a deep and narrow gorge. Despite the precipitous nature of the valley, as far as Trimo there were numerous villages inhabited by Mönbas who were under the jurisdiction of Tawang.

On the whole, the flowers, so far, had been rather disappointing, although at Gyipu (7,400 feet) and at Shakti (7,250 feet) *Cardio-crinum giganteum* had been growing in great abandon in all the damp places, literally covering the ground with, at this time of year, a miniature forest of 1–2 foot stems. At Pangchen (7,200 feet), where *Magnolia campbellii*, in magnificent flower, formed a vivid white streak along the hillside, the expedition had to halt for three days for both Ludlow and Sherriff were ill with fever and sore throats and had to stay in bed. But the other collectors were not idle and discovered a pale blue gentian with deeper blue spots, *Gentiana glabriuscula* (1264), not previously recorded; and a 30-foot rhododendron of tree-like habit, *Rhododendron epappilatum* (1260) whose pink flowers were heavily spotted with magenta on the three upper petals.

Approaching Trimo on 9 April, the magnolias at 8,000 feet were still a most superb sight and the hillsides were stained pinkish-rose with the flowers of rhododendrons; clear pale pink *Rhododendron virgatum* (1279), discovered almost simultaneously by Hooker in Sikkim and by Booth in Bhutan; a rose form of *R. kendrickii* (1280); the deep rose form of *R. arboreum* (1281) up to 25 feet tall; the bright rose *R. argipeplum* (1282), and the deep carmine *R. thomsonii* (1283); all were very common in the rhododendron and bamboo jungle between 8,000 and 11,000 feet. Primulas, of course, were not numerous, for the height of the primula season was still several weeks ahead. Even so, in damp grassy ground, or beside watercourses, at 10,000–11,000 feet, patches up to 2 feet across of the delft-blue *Primula whitei* (1291) and of the pink *P. gracilipes* (1292) were a most memorable sight.

At Trimo (10,500 feet), where serow and barking deer fed in the barley fields close to the camp, they halted for two days, arranging transport for the crossing of the Pö La into Tibet proper. Their anxieties concerning the pass proved to be unfounded. In six hours they climbed the 4,400 feet to the summit and saw the great Tibetan plateau stretching before them. Soon they joined the Tawang road which they had traversed in 1934, and reached Tsona late in the afternoon of 12 April. They forsook their 1934 camp near Tsona village and made camp near the steaming thermal springs near the Dzong.

Although an icy blast swept the bleak uplands and snow covered

22. The peaks of Takpa Siri

the ground, Tsona was awakening from its winter sleep and spring was in the air. 'Brahminy duck waddled about on the flat roofs of Tibetan houses searching for nesting sites. Bar-headed geese were paired and Brown-headed gulls, fresh from the Indian plains, flew lazily over the semi-frozen lakes. Snow-cock chuckled from the low cliffs behind our camp where a pair of Tibetan ravens were already feeding their young. Immense flocks of grandalas fed on the margins of the marshes.' [4]

To the holy peaks of Takpa Siri (*Map 7*)

After three days in filthy, wind-swept Tsona, the expedition, now a caravan of forty frisky yaks, struck eastwards across the bleak Tibetan plateau on 16 April and on the third day crossed the Nyala La (17,000 feet). The range on which the pass is situated is the watershed between the Manas and Subansiri valleys, and though the day was hazy a snow range to the south-east was clearly visible, as well as a peak which might have been the greatly venerated Takpa-Siri. From the pass they descended into the dry and barren valley of the Loro Karpo Chu to Trashi Tongme where, to Ludlow's great delight, in the thorny buckthorn thickets along the riverbank they found Prince Henri's Laughing thrush (*Trochalopterum henrici*). As with Beavan's bullfinch, this too had been considered a rare bird and, in fact, in April 1936 was represented in the National Collection in Britain only by a female collected by Bailey at Showa in SE Tibet in 1913, and by two other skins taken by Wollaston, at Kharta, in the upper Arum valley, in 1920. However, immediately Ludlow, Sherriff and Lumsden crossed the Nyala La and Kamba La and entered the basins of the Subansiri and Tsangpo, they found the bird extremely common. In fact it was the dominant Laughing thrush in these two great valleys, occurring on every pass and in every side valley wherever there was scrub vegetation. Moreover it was as much at home in the dry zone of the Tsangpo valley at 15,000 feet as in the wet zone of the Subansiri at Migyitun, at 9,000 feet. On this expedition fifteen specimens, including both sexes, were prepared for the British Museum.

From Trashi Tongme, where the Loro Nagpo Chu comes in from the south to join the Karpo Chu, and where the expedition first struck Kingdon Ward's route of the previous year, they moved eastwards through the wide arid valley to Chayul Dzong (11,200 feet), a sterile wilderness of grit and gravel lying at the foot of the main range. The Chayul dzongpen, a pleasant young lama, was most hospitable and anxious to help in any way. Readily he gave permission for the party to descend the Chayul river as far as the village of Lung, and promised to send interpreters who understood

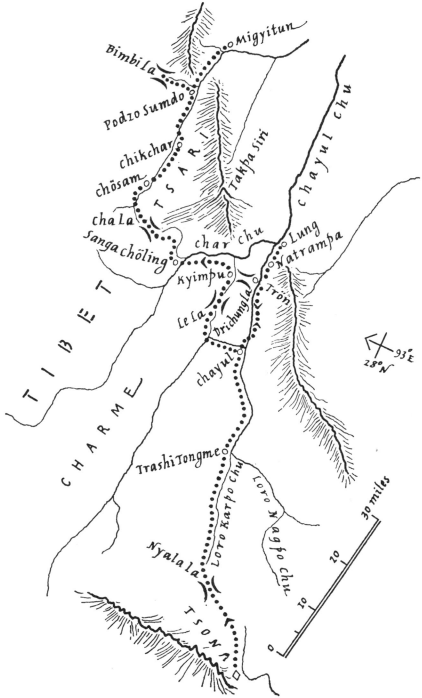

Map 7. To the Holy Peaks of Takpa Siri

Migyitun

Bimbi La

Podzo Sumdo

chikchar

chösam

chala

Sanga chöling

char chu

Kyimpu

Le La

Drichungla

Takpa Siri

Lung

Natrampa

Trön

chayul

Trashi Tongme

Loro Karpo chu

Loro Nagpo chu

Nyala la

TSONA

T I B E T

T S A R I

C H A R M E

Chayul chu

93°E
28°N

30 miles
20
10
10
0

87

the Loba dialects. With his blessing the expedition left Chayul on 24 April.

As they descended the Chayul river there were unmistakable signs of an approaching wetter zone; junipers and wild goose-berries were much in evidence whilst high up on the northern slopes of the main range conifers and rhododendrons were clearly distinguishable. And at the frontier village of Trön (10,200 feet), two marches below Chayul Dzong, the scenery changed in an intensely dramatic fashion. From the crest of a jagged spur which dropped steeply from the main range the travellers saw the Chayul river threading its way in one continuous cataract through a deep gorge where conifer forest and a dense undergrowth of shrubs clothed the hillslopes down to the river's edge. On 26 April they camped on a grassy flat at Natrampa (10,000 feet) and spent the following day exploring the surrounding country. And what a profitable day it was.

'We went along the Lung road to the first cascade about a mile downstream. There we found a most unusual red rhodo-dendron 1352 [*neriiflorum*], a beautiful mauve one 1354 [*cinnabarinum* var. *purpurellum*] and an unusual yellow one 1353 [*triflorum*]. Also a small pink one 1355 [*glaucophyllum*]. Then we followed up the stream, which came down in a series of waterfalls, as far as we could. Here, above us, was an impassable cliff with two streams coming down fine falls, close together. This was an excellent place as we found numerous rhododendrons, some of which, 1357 [*fulvum*] and 1358 [*arizelum*] were particularly fine. The latter has a most striking leaf covered with a thick bright rusty-red indumentum. Also there was a great deal of *Diapensia himalaica*, 1363, carpeting the damp rocks close to the waterfalls. . . . The vegetation, besides rhododendrons, consists mostly of fir and pine, with a good deal of bamboo and some fine larch, 1367 [*Larix mastersiana*]. I marked trees 1352, 1353, 1354, 1357, with tin labels.'[5]

At this time Sherriff had not developed the same keen interest in rhododendrons as in primulas, although by the end of the expedition he was fully committed to them, otherwise he would not have described *R. neriiflorum* and *triflorum* as 'unusual'. Both have been in cultivation for many years although the true *R. triflorum* appears to be none too common in gardens. Though not an unusual plant *R. neriiflorum* is certainly an interesting one from a geographical point of view for it forges a link between the floras of Western China and the Himalaya. So do *R. fulvum*, *R. arizelum* and the low-growing, glistening yellow-flowered *R. megeratum* (1361) which was also gathered on this day at Natrampa; all are found in Yunnan as well as in the Himalaya.

If *Rhododendron neriiflorum* wasn't very unusual, several other rhododendrons at Natrampa, all growing in the rhododendron and bamboo jungle at 11,500 feet, most certainly were; in fact they had never been collected before and were to be described as new to science. Ludlow and Sherriff 1359, with deep crimson, magenta-blotched bells was named *R. populare*, and 1360 and 1365, both with flowers of a bright lemon yellow, were named *R. dekatanum* and *R. amandum* respectively. Rather more exciting, and certainly rather more beautiful, were 1364 and 1354. The former, which was appropriately called *R. tsariense,* was 3–4 foot tall, had fine 2 inch leaves heavily plastered on the lower surface with a cinnamon-brown indumentum and carried an abundance of small trusses of pale bluish-pink flowers which in the bud were a rich carmine. In the middle of November seeds of *R. tsariense* were procured from the Nyingsang La under the number 2858 and from these seeds it was introduced into gardens in Britain. The seedlings have been rather variable in habit — some very dwarf, others up to 5 feet tall — as well as in flower colour, from white to pink. One seedling, under the name of *R. tsariense* 'Yum Yum', with exquisitely pink-flushed flowers was to receive the Award of Merit from the Royal Horticultural Society in 1964.

Ludlow and Sherriff 1354, which particularly impressed them, was not a new species but a new colour form of the well-known *R. cinnabarinum* and was named *purpurellum*; growing to a height of 12 feet, the slender branches were bending under the weight of the bright pinkish-mauve bells. This was not the last time Ludlow and Sherriff were to see this very distinct plant which is now well established in cultivation and received the Award of Merit in 1951.

Seven miles from Natrampa, down the Chayul Chu, was the deserted Loba village of Lung (9,200 feet), where another river, the Char Chu, tumbled headlong down a gorge. Here the expedition camped for two days in a forest of the giant *Cardiocrinum* and discovered more new plants, especially rhododendrons. 'We found eight more
    rhododendrons some of which are particularly fine. The prettiest I thought were 1383 [*fulvum*] with a nicely spotted upper half of the blush-pink flower; a fine big cluster of pale pink flowers, each on a long pedicel and a magenta patch at the base, 1385 [*hodgsonii*]; an almost equally big one, rather mauve-pink when fully open, 1386 [*puderosum*]; a beautiful deep crimson tree 1387 [*erosum*] which looks like a barbatum series, and the most delicate of the lot 1390 [*sherriffii*], a bell-

shaped flower, with umbels of 3–4 flowers, of a very deep carmine colour.'[6] The other three rhododendrons were the white or pink *vellereum* (1392); a form of the pale yellow *lanatum* which has been given the name *luciferum* (1389) because the thick tawny wool on the lower surface of the leaves is made by the Lobas into a wick which is widely used in oil lamps and goes by the name of Bané; and the crimson, dark-spotted, *populare* (1391) once again. Rhododendrons *puderosum, erosum* and *sherriffii* were all new species and are now in cultivation, the latter having received the Award of Merit in 1966, grown from seeds of L & S 2751 collected from this same locality in October 1936.

Apart from rhododendrons one of the most abundant plants was a yellow paeony which grew everywhere (1376); it was a vigorous plant, as much as 8 feet tall, and carried large bright orange-yellow flowers. Obviously it was closely akin to *Paeonia lutea*, which has been well known as a Chinese plant since Delavay discovered it in Yunnan in 1883, but was much taller-growing and did not hide its flowers in the foliage as *P. lutea* tends to. It was later named *P. lutea* var. *ludlowii* and is now firmly established in cultivation through several sendings of Ludlow and Sherriff seeds.

Just as the vegetation of the Chayul valley showed the influence of Western Chinese forms, so did the avifauna. All the five pheasants which were caught between Lung and Natrampa differed from those of the Manas basin and the distribution of four of them extended into Yunnan; Sclater's monal (*Lophophorus sclateri*), Temminck's tragopan (*Tragopan temminckii*), the Yunnan Blood pheasant (*Ithaginis cruentus kuseri*) and the Tibetan pheasant (*Tetraophasis szechenyii*).

During the halt at Lung a party of semi-barbaric Daflas ascended the valley carrying huge bundles of madder. Most were clothed in the skins of takin, barking deer and monkeys and a few had black shoulder capes which, though looking at first sight like bear skins, were actually made of palm fibre. Many wore close-fitting bamboo skull-caps, shaped into a spout at one end, and from this they both ate and drank. The cap was kept in place by a wooden or brass skewer which pierced a knot of hair hanging over the forehead. Into the headgear lammergeiers' feathers were sometimes stuck. All carried bamboo bows over 4 feet long, iron shod at one end, which they used as a khud stick when on the march, and their arrows were smeared to the barb with poisonous aconite. Yearly these Daflas cross the main range into Tibet carrying heavy loads of madder, rice and cane which they barter for salt, cloth and swords.

The expedition's next objective was the monastery of Sanga Chöling, no more than twenty miles from Lung. The quickest way thither lay up the gorge of the Char Chu but as the bridge over the river at Raprang had been reported unsafe they did not venture to use that route. Neither could they use Kingdon Ward's route over the Drichung La for the pass was still very deep in snow. They had, perforce, to return to Chayul Dzong and to cross into the Char Chu valley by the Le La (17,100 feet) and did not reach Sanga Chöling until 12 May.

The Le La is a knife-edge pass, or almost so, and they scampered down the eastern side until they were out of the snow. The primula season was fast approaching and, although several species were flowering, much the most spectacular was, strangely enough, the pigmy *Primula pumilio.* No more than an inch tall, it was perfectly lovely, up to 16,000 feet carpeting the ground with pink; it is another Himalayan link with China for it was discovered as long ago as 1880 in Western Kansu. At above 15,000 feet the rhododendrons began and from then on to Kyimpu (12,500 feet) the hillsides were covered with them. 'I don't think I have ever seen them so thick. The north face is covered with abies [*delavayi* (1572)] but there are a number of pale lemon-yellow rhodos too [*lanatum* var. *luciferum* (1557)], also any number of pink ones which all seem the same though they show different forms. . . . This area must be wonderful later on when everything comes out. The rhododendrons themselves must be a sight worth seeing.'[7] On the edge of the rhododendron forest the dwarf *R. anthopogon* (1565) was in great profusion and showing considerable variation through many shades of pink to pure white. Variable too, and of tree-like habit, was another rhododendron in the fir forest. Sometimes (1564) the flowers were of varying degrees of pink, blotched and spotted with magenta, and at others (1566) cream with a pronounced basal magenta blotch. They were both forms of a new species later designated as *R. dignabile.* Another species which impressed Sherriff was *R. aganniphum* (1567) with trusses of pale pink funnel-shaped flowers. This species again illustrates the influence of Western China on the flora of SE Tibet for numerous collections made by Forrest and by Rock have shown that it is a common rhododendron in Yunnan and Szechwan.

The most important bird find on the Le La was a new race of Tibetan babax which was named *Babax lanceolatus lumsdeni* in appreciation of the great assistance Dr Lumsden gave to the expedition and of the interest he showed in the avifauna.

Tsari

Sanga Chöling (10,700 feet) is situated in a gorge lying mid-way between the valleys of the Tsari Chu and Chayul Chu. As Ludlow and Sherriff intended to botanise intensively in both valleys, the monastery, a striking building with golden spires, offered obvious advantages as a base. Having deposited its surplus baggage there, in the care of the two kind ladies who were none other than the wife and daughter of the late Drukpa Rimpoche the holy incarnation of the monastery, the expedition left, on 14 May, for the Tsari valley, using the monastery mules for transport. These sensed that they were bound for the lush pastures of Tsari, whither they were sent each spring to graze, and moved northwards at a cracking pace. They made light of the snow-free Cha La (16,600 feet), the crossing of which was an easy one; on the other hand, only a few miles away, the Takar La was deep in snow and impassable. From the pass they descended into the Tsari valley to Chösam, surrounded with thick rhododendron and juniper scrub, and then reached the magnificent Senguti plain (13,300 feet) through which many branches of the Tsari Chu peacefully meandered. For some reason the whole of the Tsari valley is extremely wet and the mountain slopes on both sides are thickly forested with virgin silver-fir, juniper and rhododendron, in which the magnificent Sikkim stag enjoys sanctuary. Why the Tsari valley should be so wet it is difficult to understand for the headwaters of all other branches of the Subansiri are dry, until they cut through the main Himalayan axis which, in the case of the Tsari Chu, occurs two stages below Chösam.

Kingdon Ward had passed this way the previous year and at Chösam had gathered the lovely *Adonis brevistyla* which had been discovered by the Abbé Delavay in the mountains north-east of the Tali lake in NW Yunnan in 1886. Ward had also seen the plant in 1924, in the Tsangpo valley, at Tsela Dzong. In Ward's 1935 locality Ludlow and Sherriff now gathered fine specimens, the rather delicate-looking white flowers, with pale-blue or purplish-blue stripes on the reverse of the petals, giving colour to the floor of the juniper and rhododendron forest. The collectors gathered it under 1600 and it was under this number that plants were raised in a number of gardens in Britain, those raised by Mr T. Hay receiving the Award of Merit from the Royal Horticultural Society when exhibited in April 1940 under the name of *Adonis davidii*.

Tsari is holy ground where no life may be taken; bird-shooting

23. Dafla tribesman at Lung

therefore was out of the question. From their camp at the western end of the plain they looked on to the Takpa Siri group of peaks which, though of no great height, are regarded as very holy and their circumambulation an act of very great merit which attracts pilgrims from every part of Tibet. The inhabitants of the various valleys subsist almost entirely on the money they can earn by providing transport for the pilgrims as well as by begging in outlying districts. There are two pilgrimages. One, the *Kingkor*, is short; it is completed in a week or ten days, and occurs annually. The route leads down the Tsari valley from Chösam to Chikchar and then across seven steep passes, between 15,000 and 16,000 feet, before Chösam is regained. Until Sherriff made this circuit in 1936 the only European to have done so was F.M.Bailey. The other pilgrimage, the *Ringkor*, is a much longer one occupying a month and taking place every twelfth year. In 1936 no European had completed the *Ringkor* which also starts at Chösam, whence the route descends the Tsari Chu to the frontier village of Migyitun and then follows the river downwards into tribal territory until it joins the Chayul Chu. It then ascends the Chayul river as far as its junction with the Yume Chu and follows the latter river to Yume, whence Chösam is reached by the crossing of the Rip La.

In this marvellously beautiful country the expedition spent eight very busy days botanising, chiefly around Senguti, Yarap (12,400 feet) and the Chikchar valley, and Podzo Sumdo (11,000 feet), and remarkably rewarding days they were. Most of the rhododendrons which had impressed them in the Chayul valley were here also, some, such as the bright pink *R. puderosum*, in even greater abundance. But there were others which were quite new to them, two indeed which were new to plant science; *R. miniatum* (1627), a tree or shrub up to 15 feet high with deep rose rather fleshy bells, and the much lower-growing deep crimson *R. lopsangianum* (1651) which was named in honour of Nga-Wang Lopsang Tup-Den Gyatso, the late Dalai Lama of Tibet. *R. campylocarpum* (1628), one of the best of all yellow rhododendrons, which has been in cultivation since the early 1850s and has been one of the parents of many fine hybrids, was much in evidence, as indeed were two species which once again demonstrated the link between the floras of the Himalaya and of Western China; another yellow-flowered species *R. caloxanthum* (1656), and *R. calostrotum* (1649) the flowers saucer-shaped and grape-purple or a bright rosy-purple in colour.

And now the primula season had burst upon them. 'I have never seen more of any primula [the royal-purple golden-eyed

calδeriana] except *P. δikkimenδiδ* when going down towards
Mago [in 1934]. In places the grassy hillsides were covered
for areas of nearly 100 yards square. But the prettiest sight
was when it was in masses in glades in parts of the forest not
far above Yarap. There it really was most beautiful. Some of
the plants were magnificent ones. I counted on one particular
plant with one stem no less than 48 blooms. Of these nearly
half had petals in place of stamens.'[8]

Primula calδeriana of course was familiar to Ludlow and Sherriff,
albeit under the name of *P. roylei*, for they had found it growing
abundantly on the Me La in 1933. Indeed it was also well known to
botanists for it had first been collected by Hooker in Sikkim as long
ago as 1849. But at Yarap, and at Podzo Sumdo, Ludlow and
Sherriff found a closely allied primrose which neither they nor
botanists had seen before. On open hillsides beside the snow, leaves
and flowers were bursting from stout winter resting buds whilst
sometimes the flowering stems had grown a foot high and were
carrying clusters of deep rich purple or blue-purple, golden-eyed,
flowers. This primula, which lacked the strange unpleasant smell of
P. calδeriana, was common in the Tsari valley from 11,000 to 14,000
feet, and, after causing the Edinburgh Botanic Garden primula
authorities a good deal of thought and worry, was appropriately
named *P. tδarienδiδ* (1621, 1650). In 1937 and 1938 Sherriff was to
send home living plants by air. In some gardens these did remarkably
well, some of those grown by Major and Mrs Knox Finlay in
Perthshire gaining the Award of Merit from the Royal Horticultural
Society in 1953, and Mrs Finlay the Cultural Commendation. In
Mrs Finlay's garden, as well as in other Scottish gardens, this fine
species appears to be well established.

In places, at the edge of the snows and sometimes in the snows,
the companion of *P. tδarienδiδ* was the dwarf *P. vernicoδa* (1614)
which, with its flowers completely immersed in the rosettes of leaves,
carpeted the ground. First discovered by Kingdon Ward in 1911,
near the western front of Yunnan where it adjoins Tibet, it has
since been found not only in SE Tibet but also in Central Bhutan,
Upper Burma and the Assam Himalaya.

In 1933, on the Me La, Ludlow and Sherriff had collected *P.
δickieana* with yellow flowers. Now, at Podzo Sumdo, they saw
something of the remarkable variability of this species, when, at
12,500 feet, on steep very wet grassy slopes and on wet mossy
rocks, they gathered a blue-violet, orange-eyed, form of it (1654).

Also at Podzo Sumdo there was the exquisite *Paraquilegia
granδiflora* (1632), great aged clumps of it a foot across splashing

the cliff faces with blue-violet. Material under 1632(a) was sent to Britain and for a time this most desirable of plants was in cultivation. In 1949 Sherriff flew further material to Britain from Bhutan and several gardeners nurtured splendidly flowered plants, one such from Mrs Knox Finlay receiving the Award of Merit from the Royal Horticultural Society in 1951.

On 13 May the expedition moved down the valley to the frontier village of Migyitun which, although politically speaking is in Tibet, is outside the sacred area and thus is freed from the interdiction regarding the taking of life, freed even from the interdiction regarding the growing of crops for there was considerable cultivation of wheat, potatoes and turnips in, and around, the village. Here they stayed for over a week exploring the country in all directions and delighting in the richness of the flora which was more or less a repetition of that encountered during the journey from Chösam. Both *Primula tsariensis* and *P. calderiana* sheeted the slopes swept by snow avalanches and were among the first species to flower; they and the little white *P. vernicosa* which, as early as 30 May, was in fruit. At 9,500 feet the banks of some of the streams were a bright purplish-rose from the flowers of *P. yargongensis* (1662) the common Yunnan and Szechwan plant, which was growing with its roots in water; whilst 2,000 feet higher these same streams were aglow — and very fragrant — with the lemon-yellow, purplish-pink and claret forms of *P. alpicola* (1685, 1689). Also very fragrant was a beautiful golden-yellow primula with an abundance of white meal on the leaves, flower-stem and flowers. It proved to be one of two new species found at Migyitum in 1936 and was named *P. jucunda* (1732). The collections Ludlow and Sherriff made of the other new species (1708, 2118, 2796) were either in very immature flower or in advanced fruit and couldn't be accurately identified at the time. However, plants raised from seeds of 2796 flowered in 1938, at the Royal Botanic Garden, Edinburgh, and the elliptic hairy leaves, the compact head of 3–5 pendent widely funnel-shaped bluish-purple flowers on a stem no more than 4 or 5 inches high, convinced everyone that they represented a new species which was named *P. sandemaniana*. Migyitum appears to be its only known locality. It is a matter for regret that all the plants raised in Edinburgh from Ludlow and Sherriff's seeds only survived for some four or five years.

At the beginning of June, with a mixed lot of transport including

24. Sanga Chöling
25. *Primula tsariensis*

coolies, ponies and dzos, the expedition retraced its steps as far as Podzo Sumdo and then turned north up the pretty Bimbi La valley for a few days of plant hunting in the region of the pass (15,700 feet). The valley was well-wooded and in places very spectacular, the cliffs on the left bank towering vertically and magnificently upwards. Dense rhododendron, bamboo and silver-fir forest led, at between 14,500–15,000 feet, to rather bare open hillsides in places stained a brilliant purplish-pink with thousands of *Androsace adenocephala* (1765) and elsewhere a glorious rich blue-violet from vast numbers of *Primula macrophylla* (1777). At 14,000 feet there was another striking primula (1791) with hanging fragrant bells pale madder-pink within and almost white with meal without: 'a form of *sikkimensis* primula, perhaps *waltonii* or just a form'.[9] It was a *sikkimensis*-type primula, certainly, but a new one which Sherriff found several times again, which was later named *P. ioessa* and successfully introduced into cultivation, receiving the Award of Merit from the Royal Horticultural Society when exhibited by Mr R. B. Cooke of Corbridge in 1957.

Above 15,000 feet there was much rock and boulder, and some snow, and growing in wet cracks in rocks Sherriff found a lilac-flowered primula, heavily dusted with meal, which was new to him: 'a beauty with a very fine scent. I think it must be near to *P. rotundifolia*, with its colour and scent.'[10] As Sherriff suspected, it *was* a relative of *P. rotundifolia*; it was *P. caveana* (1768) first collected in 1909, in Sikkim. And again in rocky places, especially on cliff ledges, there was still another primula. 'At last I found specimens of a primula I have been watching for for ages, in bloom. Only two seem fully out. The calyx is large, the leaves small and one expects a very small flower. But it is really a beauty and the colour very pretty (2113). I hope it proves to be new.'[11] The lower part of the flower-tube was white and this gradually merged into the bright blue-violet of the petals; the head of 3–6 flowers was very elegant. But the primula wasn't a new one. Kingdon Ward had discovered it on the Temo La in 1924 and it had been named *P. cawdoriana* in honour of Ward's travelling companion of the time, Lord Cawdor. Moreover Ward had successfully introduced it into cultivation and it had received the Award of Merit from the Royal Horticultural Society in 1926 when exhibited by Mr A. K. Bulley.

Tsari was certainly proving a paradise for flowers and was more

26. The Bimbi La
27. *Primula calderiana* and *P. sikkimensis*
    with *Androsace adenocephala*

than fulfilling the expedition's expectations. But the flowering season was now almost at its height and Ludlow and Sherriff were tortured with the thought of the prizes, elsewhere, which might be eluding their grasp; it was tempting to want to be in a dozen places at once. If they separated they could at least be in two places at once, and this they decided to do. Sherriff would return to Chikchar, work the *Kingkor,* and then cross the Drichung La into the Chayul valley and collect on the main range above Lung and Natrampa. Ludlow, with Lumsden, would move east into Pachakshiri and then visit the Tsangpo valley. On 31 July they would all foregather again at Sanga Chöling.

With Ludlow to the Tsangpo (Map 8)

They parted on 12 June, Ludlow and Lumsden, with Danon, Ramzana and Pintso, crossing the Bimbi La and discovering a new species of willow-warbler (*Phylloscopus tibetanus*) in the process, descending through rhododendron, fir, willow, juniper and berberis jungle into the drainage basin of the Tsangpo, and halting at Tsemachi (13,700 feet) and Sumbatse (12,100 feet) before reaching Kyimdong Dzong (10,600 feet) 'a miserable tower-like building lying on an elevated fan at the junction of the stream which comes from the Lang La. There are very few houses and very little cultivation. . . . The dzongpen called in the evening. He is a lama but was partial to Cointreau and seemed willing to assist us. He knew nothing about the Pachakshiri country but seemed to have no objection to our going there.'[12] Neither did he seem to object to the great flocks of the Chinese paroquet (*Psittacula derbyana*) feeding in the barley fields and damaging his crops. This bird is a summer visitor only, arriving during the first half of May and departing about the middle of September.

From Kyimdong Dzong they turned east up the Palung Chu valley, heavily forested with fir, larch and birch. At Taktsa (13,000 feet) the deliciously-scented *Primula alpicola* was showing a wonderful range of colour — white, yellow, pale and deep violet and reddish-amber. The Lang La (15,800 feet) was crossed, a number of rare parnassius butterflies were caught, and a curious new species of primula was gathered; *P. advena* (1848), a close ally of the Chinese *P. szechuanica*, with greenish-yellow flowers, the lobes of which are completely reflexed along the tube. From the Lang La the descent was made to Molo in the Ne Chu valley, the north-facing slopes of which were clothed with fir forest and the southern mostly with willow and juniper scrub and the occasional belt of forest. Several times the expedition had found the existing map inaccurate — and so it was here. Molo (10,300 feet) lay at the con-

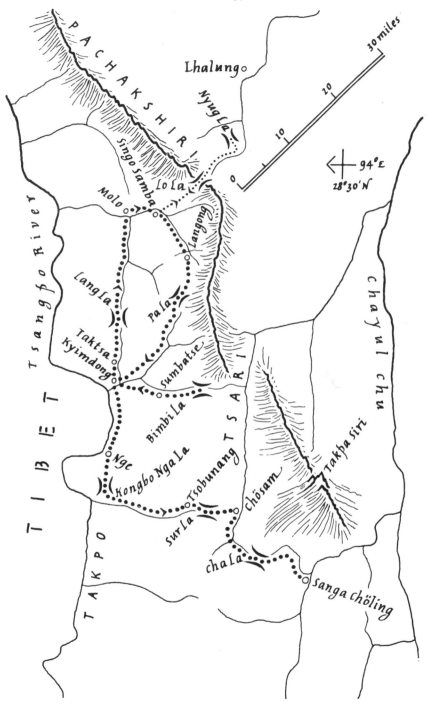

Map 8. *With Ludlow to the Tsangpo*

PACHAKSHIRI

Lhalung○

Nyug La

30 miles

20

10

0

←94°E
28°30'N

Tsangpo River

Singo Samba

Lo La

Molo○

Langong

Langla

Pa La

Taktsa
Kyimdong○

Sumbatse

TSARI

Chayul chu

TIBET

Bimbi La

Nge○

Kongbo Nga La

Tsobunang

Chösam

Takpa siri

Surla

Chala

TAKPO

Sanga chöling○

fluence of the river they had been descending, the Ne Chu or Lang Chu, and a large river, draining from the south, called the Langong Chu. The latter contained so great a volume of water that Ludlow suspected that it must drain a more extensive area of the Himalayan range than was shown on the map. And so it proved.

So far they had travelled in Kingdon Ward's tracks but at Molo they left his route and, still with a mixed assortment of coolies, ponies and oxen, turned south, up the Langong Chu towards Pachakshiri, a district no European had visited. Their first halt was twelve miles from Molo at the encampment of Singo Samba (11,400 feet) where a small stream came in on the right bank of the Langong Chu from the Lo La and where a bridge crossed the main river just above the junction of the stream. The bridge was a bad one; one large flattened tree-trunk about one and a half feet wide and another smaller one beside it; there were no hand rails and these two tree-trunks were raised on piers 20 feet or so above the river. The trunks quivered horribly as the expedition crossed, and although Lumsden made light of it Ludlow found the experience rather a sickening one. He was also sickened by the thought of having to make an enforced week's halt at Singo Samba whilst the coolies returned to Molo to attend some sort of annual fair; sickened because the camp was terribly confined, because the river was in spate and made a continuous and deafening roar, because the biting midges were a continual torment and, most important, because he was anxious to make his way to the unexplored Pachak-shiri district. However there was one recompense. The Lo La (13,300 feet) was an easy climb from the encampment and Ludlow was determined to spend the enforced halt in exploring both the pass and the Lo La Chu valley.

The first hint of the plant treasures to be found came on the first day of their halt, 26 June, when Danon found the beautiful *sik-kimensis* primula with the pale violet drooping flowers which they had recently found on the Bimbi La, *P. ioessa* (1868). But this was nothing compared to what was in store. Ludlow's first visit to the Lo La was a memorable one: '[it] gave me one of the greatest thrills of my life. On its northern slopes, in a region of incessant rainfall, grew the most amazing variety of plants I had ever seen. Day after day we scoured the hillsides, and always we returned with a bulging press and floral treasures new to our collection.'[13] And not only new to their collections but to botanical science as well! One such was the most lovely primula Ludlow had ever seen and which he had

28. The bridge at Singo Samba

the honour of naming after his mother, *P. elizabethae* (1886); 'an amazing [yellow] flower. Height only about 8–10 inches but with a flower stalk bearing 1–3 of the hugest flowers I have ever seen in a primula [up to 2 inches across]. It had a pale sulphur-yellow farinaceous eye and a bulbous stem [the scaly remains of the fat winter resting bud at the base of the plant].'[14] This marvellous plant is related to two other superb yellow-flowered species; *P. falcifolia* which Ward first found on the Doshong La in SE Tibet in 1924, and *P. agleniana* which Forrest discovered on the Mekong-Salwin divide in NW Yunnan in 1905.

Another new species almost equally lovely was *Primula laeta* (1910) with flowers of the darkest damson or of rich deep maroon, sometimes with a yellow eye. Still another was the minute rose-purple *P. subularia* (1912) forming on wet rocks mat-like growths, which were often intermixed with mosses, hepatics and a dwarf polygonum; its linear entire leaves distinguished it from all its dwarf allies. And a fourth new species was a close ally of the common Chinese *P. dryadifolia*; it formed thick compact cushions on rocks and boulders at 14,500 feet, the cushions being clothed with the withered remains of old leaves and studded with purplish-pink cream-eyed flowers, each with a pompon of hairs in the throat; it was named *P. tsongpenii* (1911) after the Lepcha collector, Tsong-pen. Another primula Ludlow now met with for the first time in his travels, and a glorious sight it was, was *P. valentiniana* (1885), first discovered in the extreme NW corner of Yunnan but now known to be more common in SE Tibet. Its nodding flowers, not unlike those of *P. kingii*, formed a rich crimson carpet on the summit of the Lo La where the deep purple form of *P. dickeana* (1878), seen for the first time at Podzo Sumdo, was equally abundant.

In the Lo La Chu valley, at between 13,500 and 14,000 feet, Ludlow was delighted to find growing cheek by jowl two supremely beautiful species of omphalogramma, and even more delighted when he realised, at a later date, that one of them had never before been found; it was named *Omphalogramma brachysiphon* (1887). Ludlow mistook it for a primula, and indeed, with its more or less cup-shaped regular velvety black-purple flower it is strongly reminiscent of a large solitary-flowered primula. The other species was the fairly well-known imperial-purple *O. elwesianum* (1891) discovered in Sikkim as long ago as 1878. And at 13,500 feet Ludlow found an abundance of 'two beauties', also growing together amongst grass on open rocky hillsides, and he couldn't decide whether they were lilies, or fritillarias, or nomocharis. They were in fact two low-growing lilies; *Lilium souliei* (1879) with shining flowers of a deep madder-pink, a native of China as well as of SE Tibet and once included in the genus *Nomocharis*; and the pale to deep lilac-purple *L. nanum* (1880) which when first found in 1845 in an unspecified locality in the Himalaya was mistaken for a fritillaria.

From the alpine zone of the Lo La, from 13,500 feet upwards, Ludlow took a rich harvest of prostrate or low-growing rhododendrons most of which must have been covered with 4 or 5 feet of snow in the winter. That which impressed him most was a form of *R. campylogynum* (1882), no more than 2 feet high, the small bell-shaped flowers of which had the colour and bloom of a muscatel raisin, 'an astonishing colour'. He also enthused about the prostrate *R. forrestii* var. *repens* (1883) with its large fleshy crimson bells, and about the precocious *R. mekongense* (1890, 1896) whose flat fleshy flowers were of golden-yellow, sometimes with a tinge of pink. All three rhododendrons are links between the floras of the Himalaya and Yunnan. One dwarf species, no more than a foot tall, was something quite distinct and quite new; the large saucer-shaped flowers, borne singly or in pairs, were a beautiful shade of primrose-yellow and spotted with reddish-brown at the base within. Very appropriately it was named *R. ludlowii* (1895) and under this number was introduced into cultivation. Of course not all the rhododendrons were dwarf. The form of *R. cerasinum* (1873) with the lower part of the flowers white and the edges of the lobes a bright rose, which Kingdon Ward had discovered in Upper Burma in 1926 and had called 'Cherry Brandy', was a handsome shrub up to 10 feet tall in the conifer forest at 12,000 feet. Of similar stature and in the same situation and elevation was the pinkish-red *R. keysii* (1904), different colour forms of which he and Sherriff had collected in 1933 and 1934. And another reminder of these earlier expeditions and inhabiting rocky hillslopes on the north face of the pass at 13,000, were the pale lemon-yellow *R. campylocarpum* (1893) and a form of *R. cinnabarinum* (1894) with clusters of orange-pink, pendulous bells.

As in the Chayul valley, so here, the avifauna, as well as the flora, showed the influence of China on SE Tibet. By far the most interesting bird on the Lo La was a tame, silent Three-toed woodpecker (*Picoides tridactylus funebris*) which is a characteristic bird of the forests of SW China.

On 2 July the coolies were due back from the Molo fair but as they arrived late the expedition was delayed for several hours before leaving Singo Samba and thus was forced to make camp 500 feet below the summit of the Lo La. However, they crossed the pass early the next morning, and in merciless rain and on the appalling track of a boulder-strewn watercourse they descended to a camp on the south face of the Lo La and in a forest clearing called Shinja (10,700 feet). The next day, along the banks of the Chudi Chu, the track was no better. They had to watch every step for great boulders, and fallen giants of the forest obstructed the path every-where, and their eyes dared not leave the ground. Often they sank to their knees in quagmires and were for ever climbing over rock faces and boulders by means of notched logs. Collecting under such conditions was utterly impossible and they did well to average a mile an hour, for six hours.

So far they had been travelling almost due south from the Lo La. Now, they turned east and ascended the Nyug La (11,000 feet), a minor pass on a spur running south from the main range. The name of the pass means 'the bamboo pass', and indeed, from the camp they had made at 9,000 feet in the forest, bamboo could clearly be seen on the summit. Here, Ludlow reluctantly decided to halt and to go no further into Pachakshiri. He dispatched Pintso, the interpreter, to Lhalung (6,300 feet), twelve miles distant, to buy rice and other provisions and to find coolies, whilst he and Lumsden hunted for birds.

Ludlow understood that both Temminck's tragopan and Sclater's monal were common in Pachakshiri but a week's hard work only produced a female tragopan and two juveniles. However he did refind the long-lost Hoary barwing (*Sibia* or *Actinodura nipalensis daflaensis*) first discovered by Godwin-Austen in the Dafla Hills in 1875 and then completely lost sight of until Ludlow now secured both sexes from the Nyug La on 7 July. Another find of great interest was the N W Yunnan form of the Long-tailed wren-babbler, *Spelaeornis souliei souliei*.

Not until 10 July were Ludlow and Lumsden able to leave Pachakshiri and its incessant rainfall and to retrace their footsteps over the difficult track to Singo Samba gathering on the way the elegant purplish-pink *Notholirion bulbuliferum* (1914) and a new species of lousewort, *Pedicularis fletcheri* (1915), which covered the grassy meads with its pretty white and purple flowers.

From Singo Samba they decided to return to Kyimdong Dzong by a previously unexplored route over the Pa La. Thus they fol-lowed the Langong Chu up to Langong (12,100 feet) near which the valley widened into magnificent grassy meadows gloriously scented, and brilliantly coloured, with all the many colour forms of *Primula alpicola*. Langong was a village of twenty to thirty

29. *Primula elizabethae* with *Lilium souliei*

houses; the inhabitants cultivated nothing except a few small fields of barley but they did breed very splendid horses and cattle which they housed in the winter in well-constructed wooden huts, each with its hay-loft. From Langong Ludlow and his party continued their march up the broad main valley before ascending 4,000 feet in four miles and by a series of steep zig-zags to the Pa La (15,900 feet) where several primulas were very much in evidence; a white form of *P. dickieana* (1922), the deep blue *P. macrophylla* (1917), the purple-pink *P. yargongensis* (1918), and a deep violet-flowered plant (1924) Ludlow hadn't seen before. This last wasn't new but it had only once been collected previously, by Kingdon Ward in 1926 on the Burma-Tibet frontier; it was *P. chamaethauma*, its violet flowers, with their conspicuous orange-yellow eye, developing simultaneously with the leaves from the fat winter resting buds.

Having crossed the Pa La, they entered the valley of the Ka Chu and reached Kyimdong Dzong on 21 July.

In ten days' time they were all due back at Sanga Chöling so after a day's halt at Kyimdong Dzong they followed the Kyimdong Chu downwards until it joined the mighty Tsangpo, at this point a majestic river with great whirlpools and eddies and containing an enormous volume of water, muddy from the recent rains. The valley was hot and dry and thus a most welcome change from the eternal wetness of Pachakshiri. They marched up the right bank of the Tsangpo and camped at Nge (10,500 feet) where flocks of paroquets were feeding on the crops now being harvested. Above Nge the Tsangpo enters a gorge and the road left the river and climbed steeply to the summit of the Kongbo Nga La (14,570 feet), a pass on a spur running down to the Tsangpo and finely wooded, almost to the summit, with larch and birch with an undergrowth of rhododendrons and other shrubs. From this pass they turned south up the unexplored Laphu Chu valley and on 26 July arrived at Tsobunang (13,500 feet) and at the first of a chain of four lakes a few miles below the summit of the Sur La (16,000 feet). 'They were crystal clear and their setting amidst the fir forest was one of the most beautiful sights of its kind I have ever seen in the Himalayas.'[15] Ludlow's camp at the head of the third lake, on a small grassy islet with a splendid clump of the fragrant yellow *sikkimensis* primula within a few feet of his tent door, was also one of the most beautiful in all his Himalayan wanderings.

A steady ascent of 2,500 feet in three miles, during which they gathered gentians, led to the summit of the Sur La, whence they saw Chösam immediately below them and no more than a mile away. From Chösam they crossed the Cha La (16,600 feet) and reached Sanga Chöling, where Sherriff was awaiting them, on 3o July. Eagerly they all exchanged notes. 'Lumsden and I rather prided ourselves on our collection of flowers but when we compared our results with Sherriff's much of the conceit was taken out of us. We had done well; but Sherriff had done much better.' [16]

With Sherriff around the Chayul (Map 9)

When Sherriff, with Tsongpen and Tendup, left the others on 12 June he returned to Chikchar to make the *Kingkor* pilgrimage. Before leaving Chikchar he and Tsongpen once again visited the lovely valley just to the south which they had botanised in mid-May. It was now a magnificent primula garden. '*Primula roylei* [*calderiana*] still holds on but the most numerous was the beautiful blue-purple one first collected here last time [*tsariensis* (1621)]. It is in masses, the whole hillside . . . . being covered with it. *P. atrodentata* is also there, but over now. The little white primula, 1614 [*vernicosa*], where the snow melts, comes into flower for its short season. *P. macrophylla* at about 15,5oo feet is fairly common and the lovely little blue primula 2154 [*cawdoriana*] was found in full flower on the cliffs on the r[ight] of the avalanche. It is a beauty, with such minute leaves and huge flowers. 2141, a white *sikkimensis*, is common [*ioessa* var. *subpinnatifida*], but the pick of the whole lot is a new one to us, 2153, which is a bright claret flower and is common on the bare slopes down which the avalanche sweeps every year. The ordinary *sikkimensis*, too, is fairly common. There are therefore nine primulas all flowering in this one quite small valley. We are still a little early for other flowers.' [17] No. 2153 was a new species, *P. odontica*, a small gem of a plant no more than 5 inches high, with 3–1o more or less pendulous cup-shaped fragrant flowers the claret colour of which glowed brilliantly when caught by the sun. Apart from this habitat at 14,5oo–15,5oo feet, its only other locality appears to be the Takar La and thus apparently it is confined to Tsari.

After a brief halt at Lapu (15,ooo feet), where a two-foot rhododendron with tight heads of the palest of pink narrowly-tubular flowers covered the hillsides on the north and east and proved to be a new species, *Rhododendron laudandum* (216o), and where the pale-lilac fragrant *Primula caveana* (2162), heavily dusted with white meal, grew profusely under overhanging cliffs and in the cracks of rocks, Sherriff climbed to the knife-edge Drolma La (16,1oo feet) and descended on the south side to Mipa (15,3oo feet) and to still another glorious primula garden. 'Mipa is on a plain, with water-falls coming down on all sides and a huge heap of avalanche snow beside my camp. The hillsides above camp are literally covered with primulas, in many places there is just moss with a primula shooting up every *inch*. *P. roylei* [*calderiana*], *macrophylla*, the purple one rather like *roylei* [*tsariensis*] .... are most common. Tendup came in with the minute one found at Lapu [the purplish-pink *P. rhodochroa* (2137, 2177) growing in moss on large rocks or on cliff faces] and the beautiful blue one which is just coming out everywhere [*cawdoriana* (2173)] and also the one Danon got at Natrampa and we kept in a box until it flowered [*dickieana* (2174, 2178), a reddish-purple form]. Tsongpen found a yellow one, 2175, which may or may not be new to us, I cannot remember [it was the new species, *P. jucunda*, which they had discovered at Migyitun at the end of May]. This place is chock a block with primulas but we have most of them by now. There is little else out.' [18]

En route for the Tama La (14,500 feet) on 20 June Sherriff was rather excited by a few plants of a pale golden-yellow meconopsis (2188) growing on a cliff face close to *Primula dryadifolia* (2181). '[It] looks like *M. horridula*. I wonder if it can be the *Mec. argemonantha* of Bailey. The latter is reported as being found in Mipak of the Tawang district. This is Mipa, but hardly the Tawang Dist. and the height is 15,800 not 13,800 feet. But I have my hopes about it.' The meconopsis was *not M. argemonantha* but a yellow form of *M. horridula* never previously seen and was later named *M. horridula* var. *lutea*. Rather remarkably, the pure white *M. argemonantha*, at that time very imperfectly known, turned up the next day on the Tama La. Sherriff found only two plants of it (2190), about a foot tall, growing on the ledge of a cliff not far away from *Primula flabellifera* (2184) a dwarf species with fan-shaped leaves and deep blue-violet flowers which Kingdon Ward had discovered only the previous year near Migyitun. The white meconopsis had first been gathered by Bailey in 1913 in the form of fragmentary specimens from which it had not been possible to assess its true relationship within the genus. From that time it hadn't been seen again until July 1935 when Kingdon Ward secured an excellent series of specimens in the Tsari valley at between 11,000 and 13,000 feet. Sherriff now also secured further fine specimens, both in flower (2531) and in fruit (2792) at 13,000–15,000 feet, all in the Tsari district, where it appears to grow in greatest abundance. From this ample material it was possible definitely to assign the meconopsis to its true position in the genus near to *M. florindae*, another SE Tibetan species.

Having crossed the Tama La Sherriff now turned west and at

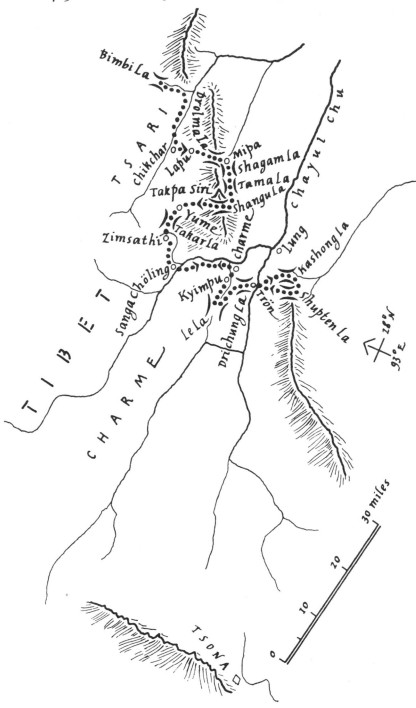

Map 9. With Sherriff around the Chayul

Bimbi La

TSARI

Drimala

chikchar

Lapu

Takpa siri

Mipa

Shagamla

Tamala

Shangula

Zimsathi

Yume

Takarla

Sangachöling

Kyimpu

charme

Lung

Kashongla

LeLa

Drichungla

Trön

Shupten la

Chayul chu

TIBET

CHARME

TSONA

28°N

93°E

30 miles

20

10

0

Taktsang (13,500 feet) met an officer in the Tibetan army, from Lhasa, who was also on the pilgrimage. 'He asked Tendup all sorts of questions about me. One was, what were all these flowers for ? Tendup told him I took them home and put them up as decoration on the wall.'[19] From the Tama La to Taktsang (13,500 feet), then to Tomtsang (12,600 feet), to Simoneri (13,500 feet) via the Karkyu La, and thence to Yume (11,800 feet), there was an abundance of flowers, especially primulas, the whole way, 'plenty of old friends'. There were also a few new friends, new species, including the wine-red white-throated lousewort, *Pedicularis filiculiformis* (2197), and a white and purplish-green gentian, *Gentiana leucantha* (2209).

Sherriff was fascinated with the names of the passes and rest-houses. 'Taktsang (Tiger's Lair) and Tomtsang (Bear's Lair) explain themselves. The coolies say that a year or two ago no one could pass these places, for the tigers and bears, till a lama spoke to the tigers and bears and put things right! Yesterday's pass, the Sha-ngu La, means "the Shao weeps" or "*even* the Shao weeps". Certainly it's a beast of a climb. Karkyu La means "Put your staffs down". I suppose it means that the pilgrimage is over as far as passes are concerned and everyone may drop their khud sticks there. I must have climbed a full 4,000 feet every day since the 14th and have got fed up with it. Tomorrow we merely go down 1½ miles to Yume where I will halt a day, I think.'[20]

Sherriff's next objective was Sanga Chöling which meant crossing the Takar La (16,700 feet) and then heading due south. There was still much snow on both sides of the pass, old avalanche tracks which are there the whole year round, which made the going difficult for the ponies especially on the descent to Zimsathi (14,000 feet). Once again there were flowers everywhere, especially primulas, including masses of the new red *P. odontica* and a little of the curious yellowish-green *P. advena* (2242) which Ludlow had discovered the previous week on the Lang La and which Sherriff now described as 'the most fragrant primula I have seen'. Nothing was more striking than a 12-inch tall androsace growing in great clumps both on open hillside and amongst scrub, at between 14,000 and 15,000 feet. The flowers were beautifully bright, the deep rose-pink of the lower surface of the petals contrasting well with the purplish-rose of the upper. It was none other than a form of the well-known *Androsace strigillosa* (2235).

Approaching Sanga Chöling on 29 June, Sherriff collected 'a queer lily-like affair, 2252, I don't know what it is.' The gloxinia-like flowers stood up boldly from the deeply divided leaves; they were lemon-yellow, heavily spotted and lined with red-brown and, with age, turned a deep red within. Sherriff had collected *Incarvillea lutea*, a Chinese plant which Forrest and Rock frequently collected there, and of which Forrest, never given to overstatement, wrote: 'Of the countless and beautiful alpines which inhabit the mountains of NW Yunnan, few stand out more prominently.' Here was still another reminder of the relationship between the floras of the Himalaya and Western China. When he reached Sanga Chöling on 29 June Sherriff was deeply worried; he had collected so many plants that he had practically run out of drying paper. Fortunately, he managed to procure some loose sheets during his enforced halt waiting for transport. When transport *was* available on 2 July, he marched south for eleven miles to Charme (10,200 feet) and then west for another eight miles to Kyimpu (12,800 feet) which the expedition had visited in the second week of May and where he now intended to stay for another few days of exploration on the Le La (17,150 feet).

Some of the prettiest plants on the Le La were the louseworts, making vivid splashes of colour in the wet grassy meadows and on the rocky hillsides between 13,000 and 15,000 feet; splashes of the yellow of *Pedicularis megalochila* (2290), of the deepest grape-purple of a form of *P. bella* (2293) and of the brilliant red-purple of *P. siphonantha* (2295). Of course there were also masses of the usual primroses as well as of one which Sherriff had seen only once before, on the Shagam La, where he had collected the new yellow form of *Meconopsis horridula*. It was the deliciously-scented pale blue-violet *Primula hyacinthina* (2294), so very like *P. bellidifolia* but usually with white farina on the lower surface of the leaves, and it was literally all over the hillsides from 12,500–16,000 feet. Still another primula, which he now saw for the first time, was *P. consocia* (2284). Indeed, only one other European had seen it before him and that was Kingdon Ward who had discovered it the previous year, only some eight miles away to the south-east, on the Drichung La (15,000 feet), the pass which Sherriff now had to cross on his way to Trön and the Chayul valley. And on this pass he discovered a new poppy which Ludlow thought was the gem of

overleaf
30. *Primula caveana*
31. *Rhododendron ludlowii*

32. *Meconopsis argemonantha* (above)
33. *Meconopsis horridula* var. *lutea*

the whole 1936 collection — 'rose-pink like the first flush of dawn on the snows'. Sherriff was rather more restrained in his enthusiasm; 'Came across a nice meconopsis, 2309, which I do not know, and cannot make out from Taylor's book. The flowers are a pretty pinkish wine-red, one on a scape and I don't think more than two to a plant. Leaves basal and cauline, all very thickly covered with bristles.'[21] Dr George Taylor (now Sir George) appropriately named this beautiful poppy which often grows no more than 8 inches high and carries 4 inch flowers with up to 8 petals, *Meconopsis sherriffii* (2309). Though seeds were collected in October Sherriff had to wait for several years before seeing good flowering specimens in British gardens.

Sherriff spent the next two weeks exploring the Chayul valley and each day yielded new plant treasures as well as many 'old friends'. A day on the Kashong La was especially memorable. 'A good morning till about 9 am. After that gradually becoming fouler and fouler. This evening it is just coming down in streams. Otherwise a day full of event. I went up the Kashong La, a climb of 1,000—1,500 feet, steep but otherwise easy. Blue skies this side and thick mist the other. On the way up saw some gentians coming on, but not open. Also the little white primula we got first between the Tama La and Taktsang [*muscoides* (2194)]. . . . There is a really fine yellow primula with enormous flowers 2370 [*hilaris*]. I measured some flowers at $1\frac{5}{8}$ inches across. They have a darker orange-yellow centre. We also got more of the minutissimae prim. 2350 [*barbatula*] which is very pretty indeed seen in masses. Gentians abounded, but hardly one fully open, which is no wonder [because of the foul wet weather]. The finest is 2357 [*phyllocalyx*] with a very big flower. The pass [Kashong La] lies east from here, and E. of the pass down the far side about $1\frac{1}{2}$ miles, and 1,500—2,000 feet below, is a large lake, a beautiful colour, which empties its water into the Chayul Chu. . . . . Over the Kashong La one is in an amphitheatre, with the large lake in the centre, and precipitous hills close all round. The path leads pretty steeply down over snow and shale scree for a mile, then round R-handed to the south, crosses the second valley and up over another pass. Then down to the Lopa country. We met some Lopas coming up to trade with Trön and they said their village was very low down and it took six days to reach here. From Karutra there is a short pilgrimage which we did today, over the Kashong La, round to the right and back by the Shupten La, which is about 3 miles south of here. The lake is called Lagyap Thungtso — 'thung' being one of the very long trumpets they blow in monasteries. The locals say that they can see one in the lake when it is clear. In the dip on this side of the lake we found other gentians in plenty and two more new primulas, both very pretty but nearly over, 2359 [*rimicola*] and 2373 [*chamaedoron*], the latter with huge 1½ inch flowers and only standing the same height from the ground. On this side of the Shupten La, where it is also pretty wet, we saw masses of the white *sikkimensis* primula we took on the Takar La [a white form of *P. ioessa* (2240)] and also, to my surprise, *Prim. cawdoriana*, and huge clumps of paraquilegia .... I have decided to stay here a week more. There is always this Kashong La to climb and almost anywhere one goes means climbing up 4,000 feet if not more. The ideal camp would be somewhere close to the lake, about 1,000 feet above it. A most interesting day and from the weather point of view a most foul one.' [22]

Primula hilaris (2370), *P. chamaedoron* (2373), *P. rimicola* (2359) and *P. barbatula* (2350) were all new species and the former three have never been collected again. *P. hilaris* is closely akin to the other new yellow-flowered species Ludlow and Sherriff had discovered at Migyitun at the end of May, *P. jucunda* (1732), as well as to the better known deep purple *P. griffithii*, one of the first primulas they had ever collected; all three form a fattish winter resting bud from which leaves and flowers develop coetaneously. *P. chamaedoron* with its usually solitary, large, violet, orange-eyed flower at some stages of growth almost as big as the rest of the plant, finds its nearest ally in *P. chamaethauma* which, a few days later (19 July), Ludlow found on the Pa La (1924). *P. rimicola*, seeking sheltered crevices on cliff faces, is a dwarf affair, no more than an inch tall, with, for the size of the plant, large solitary purplish-pink flowers half an inch across. *P. barbatula*, which Ludlow and Sherriff were to see on two other occasions, in 1938, is another dwarf whose flowers vary a little in colour from pinkish-purple to mauve or blue-violet.

Approaching Sanga Chöling Sherriff spent an interesting time on the Le La and picked a strange, though not very exciting, fritillaria. Growing in wet scree at 16,500 feet, its general impression was one of greyness. The leaves were a dull greenish-grey whilst the sombre grey of the outside of the flower was tinged with brown. Most of the colour was in the inside of the corolla where the very rough surface was of so deep a purple as to appear almost black. It was quite different from anything in the genus *Fritillaria* and was described as a new species with the name *F. fusca* (2459).

Sherriff arrived at Sanga Chöling on 29 July. He was in time to greet Ludlow and his party the following day. They were all very tired for the last six weeks had been strenuous ones. Tired, but elated at the success which had attended their plan of dividing their forces during the height of the flowering season. Ludlow very succinctly described Sherriff's collections; they were so good that 'they took the conceit out of us'. And Sherriff commented on Ludlow's:

'[he has] about 150 flowers including some lovely primulas. And among the primulas are 8 or 10 different to the collection. So we should now have nearer 50 species than 40. On the whole it has been fully worth while our separating. The best ground for collecting seems to have been the Lo La. South of the main range there was nothing. Round the Lo La there were a great many flowers and, rather surprisingly, a great number of rhododendrons. He must have brought back 15 other rhododendrons some of which will, I think, extend the known range of Chinese varieties a long way further west. The pick of the bunch seems to be what I think K.W. [Kingdon Ward] calls the Scarlet Runner [*forrestii* var. *repens*] and another raisin-coloured one [*campylogynum*] with bloom and all which was most noticeable.' [23]

They decided to rest for a little while at Sanga Chöling and in the meantime were busy packing up all the flowers and birds and once more rearranging their stores. They also sent off a mail, the Trön headman taking it as far as Chayul and being instructed to try to collect seeds of the pink poppy from the Drichung La in four—six weeks' time.

Mostly Gentians (*Map 10*)

On 6 August, with 23 pony loads, they marched leisurely westwards up the broad dry valley of the Char Chu passing several villages with their terraced cultivation of barley and peas before reaching Shirap (14,000 feet), a collection of a few dilapidated houses scattered about the wide open valley above the tree zone, with rolling hills on all sides. The Traken La (17,200 feet) now had to be crossed and they camped at 15,000 feet in a pleasant spot where there were lots of Tibetan partridges with broods of young as well as of colias and parnassius butterflies, like the travellers, rejoicing in the warm sunshine. Here, on 10 August, Ludlow celebrated his 51st birthday by gathering from the pass, at between 16,000 and 16,500 feet, the huge purple-flowered and very dwarf

34. On the Kashong La

Delphinium labrangense (1997); Kingdon Ward's pale mauve-flowered 'flannel-leaved aster', *Wardaster lanuginosa* (2001), here some 500 miles to the west of Ward's original locality in sw Szechwan; and a lovely little gentian which, in the sun, opened wide its star-shaped flowers, the lobes a bright azure-blue, the tube pure white, and which proved to be a new species, *Gentiana marquandii* (2006). And in the evening they all further celebrated with a 'huge birthday dinner of hare soup, roast partridge, paté de fois gras, apricots and cream and mushrooms on toast. Perhaps it was no wonder we slept badly.'[24]

They crossed the Traken La and worked eastwards through unexplored country to the head of the Tsari valley, on the way crossing two high passes neither of which was marked on the existing map, the Sokpo La (16,900 feet) and the Mihrang La (17,300 feet), and then paid another visit to the lovely lakes of Tsobunang by way of the Sur La. For the most part this was dry elevated country and though much of the flowering season was now past the gentians were fast nearing their peak. Ludlow and Sherriff made many collections of these lovely late-flowering plants the various species of which varied greatly in size, as well as considerably in colour, and occupied a wide altitudinal range from 10,000 feet to over 16,000 feet. Ludlow voted *G. waltonii* (2028) the finest gentian he had ever seen; on rocky hillsides, from 10,000 to 14,500 feet it sometimes grew in astonishing profusion, as near Sanga Chöling; with its semi-erect or decumbent stems often 15 inches or more long and carrying heads of 6–12 sea-blue narrowly bell-shaped flowers each at least $1\frac{1}{2}$ inches long it was a magnificent sight. So was *G. przewalskii* (2015, 2033, 2056, 2099) whose distribution extends into China; of similar habit, but inhabiting open scree, or the zone where the scree joined the grassy slopes, at the higher elevation of 15,000–16,500 feet, every erect 9-inch shoot carried a loose head of cream or white green-spotted trumpets each fully 2 inches long and nearly an inch across. At these higher elevations· white-flowered species, or at any rate species with a good deal of white in the flower, were fairly numerous. One such was *G. otophora* (2092) whose flowers are rather remarkable. The lobes open out at right angles to the tube. Inside, the corolla is white with· numerous minute blue spots on the margins of the lobes and on the plicae between the lobes. Outside, the plicae and one half of each of the lobes are white, with a few blue spots and striations, the other half of the lobe being striated with slaty-blue and purple which extends on to the greenish-white tube. Another basically white-flowered gentian, growing in open scree, proved to be a new species, *G.*

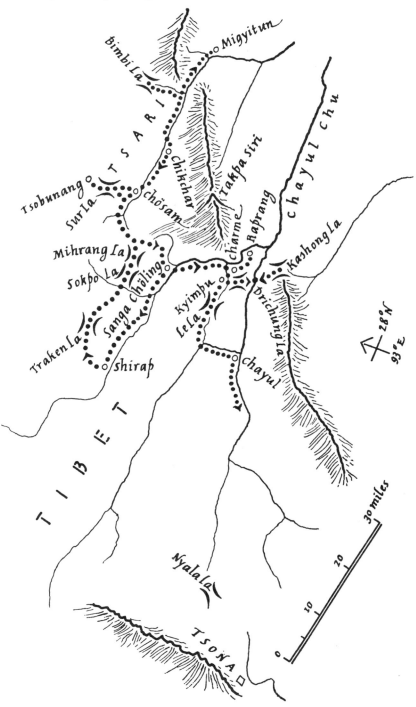

Map 10. *Departure from Tibet*

urnula (2704). The striations, varying in colour from pale slaty-blue to blue-purple, pass from the tips of the lobes to the middle of the otherwise white tube. Another new species, *G. namlaensis*, which grew in little tufts on the grassy hillside and among dwarf rhododendrons, was extremely variable in colour. Specimens taken on the Sur La (2054) were of a beautiful bright delft-blue, the lobes being spotted, and the tube striated, with deeper pigment. Other specimens collected on the Takar La (2063) were creamy-white, spotted and splashed with dark blue.

Of very different habit, especially on the Sur La, were several gentians which, lying flat on the grassy hillside, formed small compact rosettes of leaves in the centre of which stood the upright flowers. Two such were the cornflower-blue *G. infelix* (2053) and the violet-blue *G. tsarongensis* (2055).

The expedition stayed at the beautiful Tsobunang camp until 21 August and then, once again, descended the Tsari valley to Migyitun. At Chösam a new species of aconite was gathered, *Aconitum longipedicellatum* (2100), a most striking plant with its long lax inflorescence of blue-violet flowers on slender arching pedicels. Throughout the march from Chösam to Chikchar *Cyananthus lobatus* was everywhere, sometimes, as near Yarap and at Chikchar, staining whole areas of the steep hillside with purple-blue. Since they were last in this valley, during the latter half of May, everything had grown enormously and the grass, where it hadn't been eaten, was a couple of feet high. They reached Migyitun on 26 August, in high anticipation of fresh vegetables from the seeds they had sown in May. But alas; 'Migyitun is a foul spot. The vegetable seeds . . . . germinated, but the locals have no idea of how to grow them and they were choked with weeds. There are however a number of young carrots and turnips and some lettuce and spinach.' [25] In this very hot and enervating place and pestered with midges, flies and mosquitoes, they spent a week, harvesting some of the early seeds and collecting more plants, including further material of the lovely pinkish-mauve *Primula ioessa* (2514); another new gentian, *Gentiana nalaensis* (2527), the 1–3-foot stems trailing purplish wine-red, white-ribbed flowers over the steep hillside; and, once again, the long lost white poppy *Meconopsis argemonantha* (2531) of which, on the Pang La, Sherriff marked at least a dozen plants for future seed collecting.

Departure from Tibet

'We are now on our way home, having left Migyitun for the last time, and without many regrets' wrote Sherriff on 5 September. They were in fact spending several days on the Bimbi La where, above the trees, the great expanse of grassy hillside was carpeted with a mauve-violet cyananthus (2557) which was at first mistaken for *Cyananthus sherriffii* but which proved to be still another new species, aptly named *C. wardii*, which Ward had discovered on the Nambu La. Having dispatched two of the collectors to the Lo La in the Pachakshiri district to collect seeds, on 15 September the expedition began its return journey to Sanga Chöling to prepare for its departure from Tibet — and already Sherriff was formulating his plans for the future. 'I have pretty well decided to try a 3-months' trip next summer, May, June and July, to the Black Mountain in Bhutan, if permission is given from the Maharaja. And for the year after I am trying to find out all I can about the Pemako area from the south. It would save a huge round by Gyantse and I'm pretty sure the Bhutanese in Pemako would take me anywhere if only I can get in from Sadiya.'[26]

Before leaving Sanga Chöling, Ludlow, Sherriff and Lumsden were invited to lunch with the Chandzö or treasurer of the monastery and very entertaining the party was. 'Perhaps the course we appreciated most was *gyathu*. It is a kind of spaghetti mixed with minced meat and rice broth, served in pretty China bowls. To this one adds, as fancy dictates, morsels of chopped vegetables, chillies, prawns, bamboo shoots, etc. *Chang* — slightly alcoholic barley water — was of course available, and "Rosy Cheeks" [the daughter of the late Drukpa Rimpoche] waited on us and saw that our cups were full. The most interesting member of the party was the Chandzö. He had travelled far and wide and had many interesting tales of Lhasa, Western Tibet and encounters with Lobas. Then, somehow or other, the conversation turned on doctors. The Chandzö held the British medical profession in high esteem and insisted on its superiority over the Tibetan profession. I agreed and laughingly remarked that if one lost a limb, or an eye, or a nose or ear, all one had to do was pay a visit to the doctor and he gave one a new one. Lumsden blushed, and a look of incredulity spread over the features of our Tibetan friends. Then suddenly we saw the Chandzö fumbling with his mouth and before we could realise what was happening he brought his right hand down on the table with a bang, opened it, and displayed to our astonished gaze a gleaming set of artificial teeth. "The Sahib speaks truth", he said. "Look at that!" We looked, and, keeping straight faces with difficulty, politely admired the denture.'[27]

On 27 September they left for Charme. Here they halted for a couple of days and explored the Char Chu down to the Loba village of Raprang where the river cuts its way through the main Himalayan range and hurls itself down a tremendous gorge and where the change in scenery is startlingly abrupt. Above the village the hillsides are very bare, whilst only four miles below are the dense forests of Lung. As they stood by the old rickety bridge in brilliant sunshine they could see the rain falling in torrents two miles down the valley, and, fascinated, they watched the dark clouds surging up the valley only to vanish, as if by magic, as they met the dry air above Raprang. In July Sherriff had experienced an even more abrupt climatic change on the knife-edge pass of the Kashong La. On the summit and down the southern slopes, the rain had been incessant, yet on the north slope, 200 yards below the pass, butterflies had rejoiced in the warm sun.

Near to Charme the vinca-blue flat-flowered *Codonopsis vinciflora* (2714) was climbing through roses, barberries and a host of other shrubs and a most pleasant sight it was, even though some plants were now in fruit. Originally it had been found by Potanin a few miles to the north-west of Tatsienlu in Szechwan in 1893. Since then others have gathered it in the type locality and Kingdon Ward located it in the Tsangpo gorge on his 1924–25 expedition. Ludlow and Sherriff now gathered seeds and tubers, as in fact they were to do in 1938 and again in 1947, and plants were flowered in the Royal Botanic Garden at Edinburgh in 1937, since which time it has always been in cultivation, although rather sparingly so.

During these past weeks, in spite of the tremendous richness of the flora and the excitement of finding so many plants new to them, the ornithological aspects of the expedition hadn't been neglected. Though novelties had been few it was clear that the collections would provide a good deal of intensely interesting information on the distribution of various species. Moreover, at Kyimpu on 4 October, Ludlow had a great stroke of luck which thrilled him as much as had any of his plant discoveries; he obtained nine specimens of the Himalayan crossbill (*Loxia curvirostra himalayensis*). According to the textbooks of the day it was regarded as a rare bird and, though Ludlow had seen it on the Tian Shan mountains in 1930, this was the first time, in all his wanderings, that he had found it in the Himalaya. Here, at Kyimpu, it was in large numbers — at a modest estimate Ludlow saw at least 200 that day — feeding on the larch cones in the mixed larch, birch and fir forest. They were in large flocks, but apparently paired off within the flocks, and flew restlessly from one tree to another. Their strong flight and twittering notes reminded Ludlow of the Himalayan greenfinch. In no other locality was it seen again on this expedition. To judge from the collections of Forrest and of Rock it is more common in the forests of s w China.

The finding of the Himalayan crossbill wasn't Ludlow's only thrill at Kyimpu. There was a gentian to rival in beauty *G. waltonii.* 'Gentiana *sino-ornata* is in all its glory on a spur behind our camp. One particular bunch I saw filled me with ecstasy. It was situated on a ruined wall midst moss and short grass and within the circumference of a circle $1\frac{1}{2}$ feet in diameter there were over 100 blooms.' [28] This marvellous and well-known plant is, of course, a Chinese one, which Forrest discovered in N W Yunnan in 1904 and introduced into cultivation in 1910.

On 4 October the expedition once again divided its forces in order to harvest seeds over as wide an area as possible. Already two collectors had been sent to Pachakshiri. Now Sherriff returned to the Chayul valley via the Drichung La to collect seeds of his pink poppy as well as of other plants on the Kashong La and on the passes on the main range above Lung. Ludlow and Lumsden, on the other hand, returned to Tsona by way of the Nyala La, and thence by the Kechen La to Tawang and on into Eastern Bhutan. Not until 24 November were they all reunited, at Diwangiri, after ten months of intensive exploration which had yielded numerous valuable bird skins, nearly two thousand gatherings of magnificently pressed plants, two crates of living plants, and innumerable packets of seeds representing over three hundred collections.

When, in February, Ludlow and Sherriff had been disturbed by the news that Kingdon Ward had travelled in the districts of Chayul, Charme and Tsari only the previous year they had consoled themselves by the thought that probably he hadn't exhausted the possibilities of the country. They couldn't have been more correct in their judgment of the situation. Although Ward had forestalled them in discovering one or two new species they had found at least sixty plants new to science. Of the sixty or so different primulas they had gathered, fourteen (including *Omphalogramma brachysiphon*) were previously unknown; and of the seventy or so rhododendrons, fifteen were likewise. In addition there were the two exciting new poppies — the pink *Meconopsis sherriffii* and the yellow form of *M. horridula* —, the fine yellow paeony bearing Ludlow's name, several new gentians, saxifrages, aconites and other things.

New plants apart, the collections had greatly extended the known distribution of many previously recorded species many of which were shown to be forging links between the floras of the Himalaya and of Western China. Botanically, therefore, the expedition had been an outstanding one, one of the most outstanding of the 20th century in fact, and by their collections of seeds and of living plants Ludlow and Sherriff had done all *they* could (the rest was in the hands of the gardeners) to make it horticulturally successful. And, most important of all, this ten-month trip had stimulated the two friends to make their plans for future explorations in Bhutan and Tibet.

35. The Himalaya from Dungshinggang

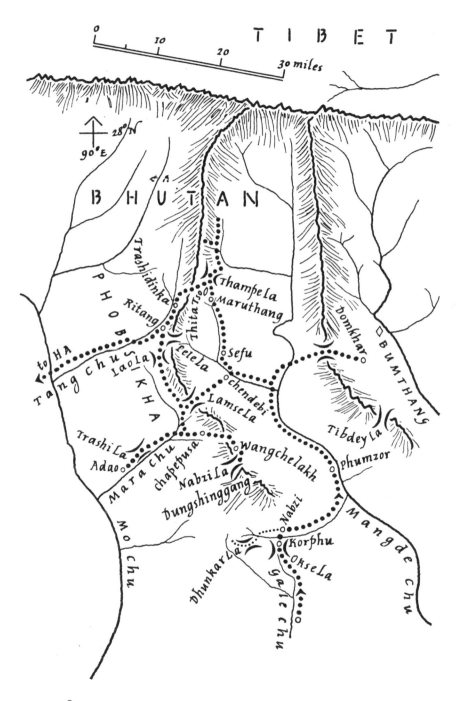

Map 11. The Journey to Phobsikha and Dungshinggang

TIBET

0 10 20 30 miles

28° N
90° E

BHUTAN

Trashidinka
PHOBS
Ritang
Thampela
Maruthang
Thita Tso
to HA
Tang Chu
Laola
Pelela
Sefu
domkhar
BUMTHANG
KHA
Chendebi
Lamsela
Tibdey La
Trashi La
Mara Chu
Chapepusa
Wangchelakh
Phumzor
Adao
Nabii La
Dungshinggang
Nabzi
Mo Chu
Dhunkarla
Korphu
Oksela
Mangde Chu
Galechu

Chapter Four

To The Black Mountain of
CENTRAL BHUTAN

o

In 1936 Sherriff had decided that, the following year, he would explore the region of the peaks in Central Bhutan known as the Black Mountain (16,130 feet) and now, in the last week of April 1937, he was in Southern Bhutan, in the Gale Chu valley, poised for a four-month journey. Though his friend Ludlow had other commitments in Kashmir and was unable to accompany him he had, for companions, the loyal and trusted Tsongpen, Tendup and Pintso.

On 27 April, Sherriff and his companions marched north, on elephant paths, through the densely forested Gale Chu valley passing temporary shelters used by the herdsmen in charge of the Maharaja's cattle, which came to the valley for the winter and now were on their way to Bumthang. The elephant tracks ascended steeply from 1,500–5,000 feet, the valley was desperately hot and sticky, and the leeches, 'a nasty looking kind, brown and big', as well as the ticks and dimdam flies were a constant plague and certainly would have prevented any botanising in the forests even had these not been far too thick to have been explored for plants. Occasionally a fallen bloom on the track would tell of flowers many feet overhead in the tall trees. For instance a few creamy-white petals which Sherriff picked from the ground, as well as a strong pervading fragrance, revealed huge 100-foot specimens of the magnolia-like *Michelia doltsopa* (2953). Whilst gazing in admiration at one of these splendid trees Sherriff spotted what seemed to him a rhododendron growing epiphytically on one of the michelia branches a full 70 feet from the ground, in, he thought, a completely inaccessible place. But Tsongpen thought otherwise and to prove his point climbed the tree and secured specimens of a rhododendron of poor scraggy habit with almost naked 8–12-foot branches but bearing terminal clusters of 2–4 huge flowers pure white or sometimes faintly tinged with pink and most magnificently scented; *Rhododendron edgeworthii* (2952). From similar habitats and of similar

129

habit Tsongpen also collected the rich cream, red-striped flowers of *R. rhabdotum* (2940, 2944).

After crossing the Okse La and passing the Bhutanese village of Korphu, they descended into the valley of the Jirgang Chu and to the monastic village of Nabzi (4,600 feet) with cultivated fields of buckwheat. In the valley west of Nabzi, in the region of the Dhunkar La (8,000 feet), Sherriff celebrated his 39th birthday, on 3 May, by drinking two cups of tea at breakfast and at teatime. Under the circumstances this was the greatest possible luxury for although the yearly rainfall must have been in the region of 200 inches there was precious little water at this time and such as there was was pretty undrinkable.

But if water was scarce rhododendrons were by no means so and from 8,000–10,000 feet were an astonishing spectacle, growing abundantly and to a great size. *R. grande* (2977) which had recently dropped its yellow flowers, *R. falconeri* (2983), *R. hodgsonii* (2987) and *R. eximium* (2989) all averaged between 40 and 50 feet and one particular specimen of *R. grande*, at 8,000 feet, was at least 70 feet high, with a girth, 5 feet from the ground, of 9 feet. Between 9,000 and 10,000 feet *R. falconeri* literally covered the hillside with its great flower-trusses and handsome bright-brown-backed leaves, some as much as 18 inches long and 9 inches broad. At the lower elevation the bell-shaped flowers were of a beautiful lemon-yellow with a large basal magenta blotch, and darkened in colour, with age. But at the higher elevation the colour was paler and the basal blotch smaller. At 10,000 feet *R. falconeri* gradually gave place to a dull dark wine-red form of *R. eximium* which was, above 10,000 feet in the bamboo zone, the dominant rhododendron along with the bright pale magenta *R. hodgsonii*.

Occasionally Sherriff caught a glimpse of the Black Mountain peaks through these trees, though never with a clear enough view for him to photograph. But if he couldn't photograph the peaks he got many a story of them from the locals. 'There is one old man who used to come up here twenty years ago to shoot and trap musk deer. When His Highness heard of it he had him beaten, so he has not been here since. There are said to be three kinds of lake near the top — one golden, one silver and one emerald. They say, also, that rock salt is found on the mountain but that the Tibetans prayed the Bhutanese not to use it, as they would have nothing to trade if they did. So now it is not used. This is hearsay, but three years ago a mad lama went up the Black

36. A *chorten* at Chendebi

Mountain and came back with a big piece of rock salt. So there must be something in the story.' [1]

The locals were worried about the proposed trip to the Mountain. 'Let the sahib go anywhere else and we will gladly take him. But if he goes up Dungshinggang [Black Mountain] it will bring disaster to us in the form of hail which will ruin our crops. I'm sorry they thought of that. I must go, but this is the time of hail-storms. They are certain to come and I will be blamed for all the damage they do. Pintso has told them that I did the pilgrimage of Tsari, and perhaps that will help.' [2] It did.

The word Dungshinggang, according to Pintso, meant 'the fir-range' (Dungshing = fir = abies; gang = range) and the locals believed there was a god of the mountain called Dungshing to whom they had built a place where they often worshipped.

Sherriff decided to approach the mountain from the north, from Chendebi in the Longte Chu valley, a four-day march from Nabzi first in a north-easterly direction up the Jirgang Chu and then north-west along the Mangde Chu valley; the map had to be corrected for a good deal of the journey. The march up the hot dry Mangde Chu was uneventful. Thick forests of *Pinus longifolia* clothed both sides of the valley and much of it had been ruined through the disastrous custom of grass burning. Near some of the villages there was considerable cultivation of buckwheat which would be ripe by the middle of May and then would be replaced by rice.

A week or two earlier the deep rich red form of *Rhododendron arboreum* must have inflamed the valley. But now, 11 May, as they approached Chendebi it was almost past and at Chendebi (8,400 feet) was replaced by the lax-trussed, pure white, fragrant *R. griffithianum* (3026), one of the finest of all rhododendrons, well known in cultivation, having received the First Class Certificate from the Royal Horticultural Society as long ago as 1886, and one of the parents of many splendid First Class Certificate and Award of Merit hybrids. The rhododendron forest was not entirely given over to the white flowers of *R. griffithianum*, for the slender boughs of some 10-foot shrubs were hanging with the weight of deep red trusses (3025). This was the plant which Farrer had discovered in Upper Burma and which is now regarded as a form of the well-known *R. neriiflorum*, and is named *phaedropum*. Sherriff's finding of it at Chendebi constituted a new record for Bhutan.

Sherriff recalled his visit to Chendebi in July 1933. 'We never saw more than 100 yards or so in any direction owing to the thick mist. It rained all the time we were here and dinner in our bamboo village was comic. Rain came in through the roof everywhere, and a young torrent found its way to the dining room. We all sat on the table with our feet on the chairs and ate like that. I also remember a plate of soup being brought in with a leech on the edge of it.'[3]

This year Sherriff's visit to Chendebi was memorable from quite a different point of view. On 12 May, through the small portable battery radio which he intended to present to His Highness the Maharaja, he heard the whole of the Coronation ceremony in London and was much moved by the taking of the Oath by the King.

Mainly primulas

Sherriff decided to use Chendebi as a sort of centre from which he would march in several directions. First, he would move south-westerly and on 14 May left for the Phobsikha district in the Mara Chu valley. Within a mile of leaving Chendebi he had what he called his 'first thrill'. 'On the right hand side of the path there is a cliff and just short of that, on the very steep hillside, we found a new primula (3052) — new to me at any rate. It is a small one, rather like *Primula tenuiloba* at first sight, but not so blue as that, and having a good deal of farina on the leaves. It was growing only at 8,000 feet, and that is the only place we saw it. There was a lot there too, growing mostly in large clumps, from 6 inches to 1½ feet across, in moss on very steep slopes and cliff faces.'[4] It wasn't new but it was *almost* so, having been collected only once previously, and that rather inadequately, by Smith and Cave, in August 1909, near the entrance to the Zemu valley in Western Sikkim. This original finding, too late in the year for satisfactory flowers, justified description as a new species only by reason of the fact that it was readily distinguished from all other dwarf primulas by the possession of long leafless flagellate stolons, each terminated by a small leafy bud; accordingly it was named *P. flagellaris*. Now, through Sherriff's ample material it was possible to amplify the original description especially in regard to the flowers; they are single or in pairs, funnel-shaped, about half an inch across and thus large for the size of the plant, blue-violet with a touch of red and with a white eye. On this expedition Sherriff extended the plant's range from Sikkim to Bhutan where he made seven different collections which showed that the species has a long flowering season from early May until July, or even early August, as well as a wide altitudinal range from 8,400 feet as at Chendebi to 15,000 feet as at Ritang. By August young plants had developed at the ends of the stolons and had become firmly rooted.

Six miles beyond Chendebi they crossed the Lamse La (12,400

feet) where, in the thick rhododendron and abies forest, a rich reddish-pink form of *Rhododendron hodgsonii* (3047) was very abundant, along with the pale lemon-yellow *R. campylocarpum* (3048), the bright red *R. smithii* (3049) and a salmon-pink form of *R. cinnabarinum* (3050). Once over the pass they descended into a broad open valley, with excellent grazing on the hills, with a good deal of cultivation of buckwheat and wheat around the several villages and with quail feeding in the green wheat crops. This was the Phobsikha district and *Primula whitei* (3053) was in such profusion as to colour large areas of the steep damp slopes of the rhododendron forest with delft-blue or pale blue-violet. 'Every time I see that primula I think it is the prettiest one I have ever seen and always I feel elated however many times I see it.' [5]

Almost equally abundant, at 12,000—13,000 feet on the steep slopes of the abies forest, as well as on mossy rock ledges, was another primula which Sherriff collected under the numbers 3060 and 3097. From a big resting bud developed usually heart-shaped leaves which sometimes were thickly coated below with cream farina, as well as a flower-stem, 4—15 inches high, bearing a whorl, occasionally two superposed whorls, of 3—6 bright yellow, or orange, or glistening greenish-yellow flowers. The Edinburgh primula authorities equated Sherriff's specimens with those which Kingdon Ward had collected in the Mönyul district of SE Tibet in 1935 and which had been described as a new species with the name of *P. barnardoana*. But Sherriff was never happy with this determination and, from his experience in the field, believed that so-called *P. barnardoana* was nothing more than a form of *P. elongata* which Hooker had first collected in 1849 in the Zemu valley of the Sikkim Himalaya and which Cooper and others had also gathered several times both in Sikkim and in Bhutan.

Still another primula about the identification of which Sherriff was dubious, for a short time at any rate, was one he had gathered already in flower early in May, in the Jirgang Chu valley where it was pretty common in the rhododendron and bamboo forest, often beside streams, from 9,500—10,500 feet (2984, 2991), and which he now collected in fruit at the end of May, at 8,000 feet in the Mara Chu valley (3162). It was a rosette plant, purplish-pink flowers with a bright orange eye margined with white nestling amongst finely toothed leaves. In flower the primula had much the facies of *P. gracilipes*, the first plant he and Ludlow had collected in 1933, and

37. *Primula flagellaris*
38. Camp in Phobsikha

yet he felt that, somehow or other, it was different. It was not until he saw the remarkable behaviour of the Jirgang Chu plant at fruiting time that he realised *how* different. The withered flowers and young fruits had been carried into the air on a 6–8 inch stem, or scape, at the apex of which a young plant had developed and in some cases had taken root when the scape had collapsed and fallen to the ground. Such behaviour is quite unknown in *P. gracilipes.* But such behaviour is characteristic of *P. bracteosa* which William Griffith had discovered in Bhutan in 1838, which R. E. Cooper had rediscovered in 1915 and which Sherriff had now found once again.

All over the Phobsikha forests, from 10,000–12,000 feet, grew an ally of primula, and an even closer one of soldanella and of omphalogramma, the monotypic *Bryocarpum himalaicum* (3054); from a winter resting bud develop a few thick-textured oval leaves and a 4–6-inch scape bearing a single pendulous, narrowly tubular, 7-lobed, yellow flower. Sherriff succeeded in gathering seeds some of which were sown by Lt.-Col. Shaw MacKenzie in his Ross-shire garden in Scotland in November 1938. The seeds germinated and plants were brought to the flowering stage in April 1940. Since that time this most interesting plant has been in cultivation, though always rather sparingly so.

On 18 May Sherriff and his party left Phobsikha and marched in the direction of Dungshinggang. All day they were in thick mist and saw nothing of the surrounding country and very little in the way of flowers. They camped first at Chapepusa and then at Wangchelakh (13,500 feet), where yaks were brought for grazing on the grassy hillslopes, where the rich maroon *Primula calderiana* (3076), smelling strongly of fish, grew thicker than anywhere Sherriff had seen it except at Tsari, and where *P. atrodentata* (3078) and *P. macrophylla* (3079) clothed the hillside up to 14,000 feet with rich deep blue-violet. The Pünsum or 'three brothers' of Dungshinggang showed up well in the clear sky of the early morning of 20 May — and all three peaks were pretty deep in snow. The party moved through the abies forest to a camp in a clearing on the north side of Dungshinggang hopeful of approaching close to the three peaks from the Nabzi La which also held a good deal of snow. The pass afforded a magnificent view. 'Had we been an hour earlier it really would have been as fine a view as any I have ever seen. All round to the north were snow mountains — the main Himalayan range. Owing to clouds the furthest west we could see was Chomolhari. Other big peaks showed up due north and then came Kula Kangri and another mass to the east of them which might have been Sangtopelri, near the Me La.

But there again clouds were just coming up and there was no time to study them. To the south were the peaks of the Black Mountain. Only one of the Pünsum could be seen, the others being behind it, but there were several subsidiary peaks close by.' [6] From what could be seen of the Black Mountain it held so much snow that there was little point in attempting to climb any of the peaks for another two weeks at least. In the meantime Sherriff would return to Phobsikha.

He began his return journey on 22 May and soon met with 'a most lovely anemone [*obtusiloba* (3094)]. It flowers both blue-violet and white, but looks as if it might come true to seed. I have never seen such a sight of anemones. For 100 yards or more the grassy hillside was thick with them, all jumbled together, a patch of blue, then joined on to a patch of white, and so on, a really beautiful sight. They were growing so profusely where the ground is probably pretty rich with yak manure, just close to a *"goat"* (=encampment for yaks or sheep) on the open grassy hillside. *Primula whitei* is again in masses, and bigger flowered here than I have ever seen it, and very profuse in fir and juniper forest. It seems to like the fallen leaves of *Abies spectabilis*.' [7]

Quick on the heels of this pleasant surprise came another. 'Soon after leaving camp I met a sepoy and a train of coolies from His Highness who . . . . had heard that I was up this way. Both his letter and the Maharani's were very friendly. "Since our first meeting we became such great friends that we are now like members of one family". Both sent presents; from H.H. (1) one load butter (2) one bag parched rice (3) two cases native brewery (4) one box native biscuits. And from the Maharani "a trifling present of native made *chadar* (=hand-woven cloth), one bag parched rice and a case of native brewery." The "native brewery" I had hoped would be chang (beer) but it is arak (spirit) which I can't touch. Everything else will be very good though, especially the 80 pounds of butter.' [8] This was not the first generous gift of food Sherriff had received from His Highness and his family, and it was not to be the last. In later years wherever Sherriff and his friends travelled in Bhutan they were to live very largely at the Maharaja's expense.

overleaf
39. *Bryocarpum himalaicum*

40. *Anemone obtusiloba* (above)
41. *Primula chasmophila*

And in the evening a third surprise. '6.0 p.m. It is a wonderful country, this. I have been cursing the rain, sleet, hail and mist most of the day. Ten minutes ago it was raining hard and we were in mist. Then suddenly it cleared and I looked out on a most wonderful scene away down the valley. The sky was blue-green with sunset-lit clouds in it. Below them an absolutely clear sky. In the distance a few peaks showing over a mass of valleys filled to varying degrees with clouds, some dull and some lit by the evening sun. And all around the blackbirds are singing. A most lovely evening. Most evenings lately I have been hearing a woodcock flying round for 20 minutes or so. This evening there must have been 8 or 10 of them. They all started together just after 6.30 and finished at 7.15. They flew round in circles fairly high . . . . Each one called for a few seconds, then swooped down "drumming", to resume its call a few seconds later. There were not more than 5 or 10 seconds between one or other of the birds drumming. All stopped almost simultaneously at 7.15 p.m.' [9]

Sherriff spent the next week in the Mara Chu valley, and spent it in considerable discomfort for the dimdam flies were appallingly bad and the leeches were increasing in numbers every day. Hoping to find new plants he made one excursion downstream in the direction of the Trashi La ridge (7,900 feet) which has been cut through by the Mara Chu which then falls precipitously about 3,000 feet in 1½ miles. Beyond the Trashi La lay Adao, a collection of three villages on both sides of the river, the right bank of which, at this point, was completely covered with a very fine forest of *Pinus longifolia*. At this time of year Adao was almost deserted for the villagers had moved up to Phobsikha and only periodically returned to their homes to attend to their crops. A little downstream from Adao the river once again flowed dramatically in a series of terrific falls, dropping at least 2,000 feet in little more than a mile. Spectacular though the country was it held little in the way of new plants. *Cardiocrinum giganteum* (3155) was in fine fragrant flower and *Primula bracteosa* (3162) had developed its long fruiting scapes. *Rhododendron edgeworthii* (3132) and *R. rhabdotum* (3136) were both very common on oak and other trees, some young plants of the latter, barely a foot tall, carrying a couple of flowers. But the rhododendrons which most interested Sherriff were a couple of deliciously fragrant ones with 3-inch long, tubular-funnel-shaped white flowers, at first delicately tinged with pink. Clearly they were very closely akin. One was *R. polyandrum* (3164), the mature flowers yellow-flushed in the throat, and in one locality it was fairly abundant growing amongst *R. arboreum* and other trees and shrubs. The other was *R. maddenii* (3147) and there was very little of it.

Sherriff now decided to make a two-week journey to the north and left Phobsikha on 2 June. For two days he camped on the Lao La (11,000 feet) where, in abies and bamboo forests *Primula whitei* (3185), in immature fruit, was hardly recognisable, its leaves having grown to three or four times the size they are at flowering time; where *P. geraniifolia* (3173), which Sherriff hadn't collected before, was in full rich deep pink flower, as many as fifteen blooms to the head; and where the pink *Androsace geraniifolia* (3175) was luxuriating in any wet situation and especially by the margins of streams. After descending steeply through dense forest on the north side of the Lao La, marching through Ritang (8,175 feet) which he had visited in July 1933 and where, now, great clumps of *Cardiocrinum giganteum* were in splendid flower, and after spending a night at the village of Trashidinka, Sherriff made camp on the Pele La Range (11,055 feet) east of the Tang Chu, there to stay until 10 June. And there to make careful and detailed observations on some of the primulas he gathered; observations which were to help enormously in solving some of the taxonomic problems of this group of plants. From this point of view Sherriff's 1937 journey, as well as the diary of this journey, are of particular interest and value.

For instance, on 7 June he collected '*Primula griffithii* sp. No. 3205 which will cause the Professor some thought I'm sure'. The Professor was Professor Sir William Wright Smith, Regius Keeper of the Royal Botanic Garden, Edinburgh, and the authority on the taxonomy of the genus primula. About 3205 Sherriff wrote this note: 'This seems most likely to be a variety of *P. griffithii*
[which he had first collected in the Chumbi valley in 1933]. I have not yet come across *P. griffithii* [on the 1937 journey] which is almost always found in forest, under abies or rhododendron at 10–11,000 feet. No. 3205 comes from 14,000–15,000 feet. It resembles in many ways my No. 1621 from Chikchar, Tsari, taken in May 1936. The habitat is the same, open steep grassy slopes where snow has lately melted. It grows in masses; it is almost precocious and it has little or no farina. Probably to this can be added, it never flowers in whorls and is very large flowered. Where it most resembles 1621 is in flowering when only 1–2 inches high, when the leaves have hardly started to open, and in its habitat — well separated by 2,000 feet from the forest.' After examining the dried herbarium specimens of 1621, the Professor had named them *P. griffithii* — and Sherriff hadn't altogether agreed with him. Later,

with the observations of Sherriff before him, the Professor changed his mind and described 1621, 3205 — and several other gatherings — as the new species *P. tsariensis*.

And another example. On gathering 3227, *P. sikkimensis* var. *hopeana*, on 9 June, Sherriff made this observation. 'If I am right in determining this primula as *P. hopeana* then it seems a very early flower. Cooper got it on 20 June and 1 July, whereas last year, in Tsari, we were finding it in August and quite late in August (and in September). Its habitat seems much the same as in Tibet. It was certainly seen there, more where the snow had prevented it from coming up earlier, but this flower approximates much closer to *sikkimensis* than our collections in Tsari.' The Tsari collections had had white to creamy-white flowers which were sometimes tinged with purple or violet and the Professor had first called them *P. hopeana*, and later on a variety or subspecies of *P. sikkimensis* with the name *subpinnatifida*. Sherriff's observations prompted him to make a further examination of the Tsari collections as a result of which he related them to *P. ioessa* and renamed them *P. ioessa* var. *subpinnatifida*.

Before leaving the Tang Chu area, Sherriff collected the white form of *Primula vernicosa* (3240) — the first record of this species for Bhutan; the small, ivy-leaved, pale or deep rose-pink *P. listeri* (3192), smelling strongly of *Geranium robertianum* and inhabiting only one small area in the bamboo and abies forest at 12,000 feet — another first record for Bhutan; and on mossy cliff faces at 14,000 feet, and a most lovely sight, clumps of *P. tenella* (3249) barely two inches high and studded with large one-inch flowers of a rich blue-violet, sometimes with a touch of red, and always with a prominent white eye. Cooper had gathered this splendid dwarf for the first time in Bhutan in 1914.

It was now the middle of June and Sherriff returned to the rather barren and windy Phobsikha valley preparatory to marching once again towards the Black Mountain on 16 June. But Pintso could not march: 'Pintso has fever and I am leaving him behind tomorrow to come on with the coolies on 21st if he is fit. He is not very bad, but has pains, some fever and "*chukkas*" (=giddiness) .... He will remain with a bottle of quinine and of aspirin.' [10] By 21 June, when Pintso had recovered sufficiently to join him at the camp near to Wangchelakh, Sherriff too was ill: he 'had got some kind of fever which makes me feel pretty queer and very weak; also a throat which feels completely raw and is most painful. So I only stayed out for 3 or 4 hours this morning. We found nothing. I have thought Tsongpen was a little worn too and this evening he has just come in to say he also has fever, got it at the same time as I did, but he says he thought his inability to climb was just due to staleness and nothing else.'[11]

Though still feeling 'pretty rotten' and with a painful swollen neck making swallowing difficult Sherriff marched on, and on 23 June made camp near Dungshinggang, just south of the Nabzi La. 'We have a small collection of flowers — only two — although Tendup and I were out from 6.0 a.m. to 3.30 p.m. But I never mind how few when the list of primulas is added to. Today we got No. 3301, which we had seen last time just coming up, now in full flower, but rather scarce. It is a pretty primula, colour of *P. macrophylla* exactly, and growing in many ways like *P. cawdoriana*. It is a brute to get a photo of, as it won't keep still in the slightest breeze. I presume it is *P. umbratilis*, but I cannot imagine it in "forest 12,000 feet" as Cooper records. It seems essentially a high cliff-ledge primula, growing among grass. I am much better today, but still have a head and weak knees. Tsongpen started well but was done by the time we reached camp . . . . We are camped by a little lake which is . . . . very warm for this height, the temp. being 64°F. There are a lot of small lakes here. On the way up we had organised searches for blood pheasants' nests, but never found one though we were pretty sure they were near us and had eggs. The locals say they breed later on Dungshinggang and that is certainly correct. None have their young out yet.'[12]

Primula no. 3301 was not Cooper's *P. umbratilis*. But it *was* Cooper's deep violet-flowered *P. chasmophila*, growing no more than three inches high, which he had collected in 1915 on the Black Mountain, its only known locality. Cooper's specimens had been in advanced fruit and too imperfect for identification. But plants raised from Cooper's seeds had been identified as a new species and figured in the *Botanical Magazine* in 1919 (Plate 8791), under the name of *P. chasmophila*. Sherriff's specimens were thus the first flowering ones to be collected in the wild. The plants raised from Cooper's seeds were quickly lost to cultivation, and so, unfortunately, were those grown from Sherriff's 1937 seeds.

Though the search for the eggs of the Blood pheasant (*Ithaginis cruentus*) was unsuccessful, Sherriff was enabled to make some most interesting observations on these birds. 'I believe that Blood pheasants practise both polyandry and polygamy. Certainly on two occasions I have seen two males to one female. In the first case (Lao La camp) we stopped an hour beside them, when chicks were out and running about. Both males kept near the place the whole time. Again today [23 June], although we never saw eggs or nest, there were two males to the one female and all kept close to the original spot, though six of us were searching the forest. Three locals have taken the eggs of Blood pheasants and eaten them. One says he took six from one nest last year. These three also say they have often seen two females to the one male and maintain that it may be either way about.' [13]

On 24 June Sherriff determined on an early start at 4.30 a.m. in the hope of attaining the summit of one of the three Black Mountain peaks. 'We spent our time going up and down the most awful steep hills. But we really found very little . . . . We found no new primulas on a long day over very likely ground but got some more of, and saw lots of, 3301 [*Primula chasmophila*]. And a most beautiful primula it is too. We also got a full collection of the little *nivalis* primula collected first under 3271, now taken under No. 3310 . . . . I believe W. Smith put a similar primula which we got last year on the Bimbi La as *P. macrophylla* forma. I cannot agree that this one is the same as *P. macrophylla*. The latter grows and flowers in profusion all around 3310 and is recognisable at once. It is always much bigger and always the same colour, and always has the deeper blue-violet eye. The tube also is dark coloured, not nearly white as this is. The habitat is much the same but 3310 grows on almost bare cliff faces and in wetter places. [As a result of these field observations, and the examination of fruiting material, 3310 and the Bimbi La collections of 1936 — 1778, 2112, 2561 — were named *P. macrophylla* var. *macrocarpa*.] Over the first hill we came to what they call the Door to Dungshinggang, a huge hole through the hill about 70 feet by 20 feet. Away below us on the rhododendron clad hillside we could see the shadow of our hill and this huge hole showing up in it. Down there were about 100 sheep brought up by the Nepalis from Chirang direction. We went on, up and down the most precipitous places till the "three brothers" peaks of Dungshinggang were very close, but we did not attempt them. One would need more time than we had to spare. The Yum Tso (Yu Tso), or amethyst lake, was very well named. We saw a lot of lakes of all colours usually about 50 yards long . . . . The three peaks of Dungshinggang are very steep and the rocks rotten but I think they should be possible to get up, though difficult. I had hoped we might try, but camp would have to be a good deal nearer than this, especially if one were to get up before the mist covered everything.' [14]

Two days later Sherriff was back again at his base in Phobsikha having noted, on the return journey, a fine mass of *Primula chasmophila* 'which will do for seed later on if I can send these people back here in September', as well as an interesting variation in the flower colour of what 'the Professor' called *Primula sikkimensis* var. *hopeana* which at 14,000 grew in profusion by the margins of streams in open grassy meadows. Something like 97 per cent of the great colony (3316) had pendulous bells of creamy-white whilst the remaining 3 per cent (3317) had the flower-tube and the base of the petal lobes stained with bright cerise. As a rule plants with the red-tubed flowers were in clumps of three or four. Sherriff marked four groups of 3317 for future seed collecting; he was anxious to see if the colour variation was maintained in the offspring and possibly wondered at the relationship of these plants to those from Tsari with creamy-white flowers tinged violet or purple in the tube which had been named *P. ioessa* var. *subpinnatifida!*

After halting for two days Sherriff moved on to Chendebi for a trip to the north which would occupy the first half of July. For four miles he marched up the Longte Chu having as his constant companion for most of the way *Lilium nepalense* (3339) which was a grand spectacle on the steep rocky hillsides, its rich creamy-yellow perianth segments reflexed to the extent of revealing the rich deep red star within. From the valley of the Longte Chu he moved into that of the Rinchen Chu and to the village of Sefu (9,000 feet) where the wheat crops were being threshed; thence onwards northwards to Maruthang (11,800 feet) where the continuation of the Rinchen Chu is known as the Maru Chu and where an abundance of primulas splashed much of the landscape, especially on marshy ground, with white and yellow; the white flowers of *P. involucrata* (3355) and the yellow ones of that form (3353) of *P. alpicola* known as '*luna*' which Kingdon Ward had discovered in SE Tibet in 1924.

From Maruthang, on 3 July, Sherriff sent Tsongpen and Tendup to their camp of 6–10 June on the Pele La 'to pick up a few things there' whilst he and Pintso explored the Maru Chu valley, and some of its side valleys, to the north. Sherriff found little to interest him. 'We saw a good deal of the primula 3353 [*alpicola* '*luna*'], a good deal of *Meconopsis bella* (entire leaf form, 3361), but little else of interest that was new. *Primula sapphirina*, *pusilla* and *hopeana* are everywhere at the right height. A pretty beastly wet day.' [15] The following day was equally wet and equally disappointing — until the return of Tsongpen and Tendup in the evening 'with a good haul of things, but no seeds. Far and away the best were two primulas, or two forms of one primula, 3366, 3367, both magnificent flowers . . . . I measured one of the specimens of 3367 at $1\frac{3}{4}$ inches full, across. No. 3366 was in masses, by the hundred yards; 3367 in rather a small patch, but closely packed. These are the best things collected so far. . . . A most cheering evening to an otherwise foul day of pouring rain and strong wind. Everything is pretty sodden and flowers take a horribly long time to dry. The drying paper seems as wet as ever only an hour or two after changing.'[16]

Sherriff's work for the day was not finished for he continued to ponder on the nature and the identification of the primulas 3366 and 3367 — and his thoughts on both of them he entered into his diary.

'3366. Prim. sp. Appears at first to be a white form of P. 2373 ? [*chamaedoron*] which was collected last year on the Kashong La. Could it be a white form of No. 3364 = 3205 ? [*tsariensis*]. It is growing about 1,000 feet higher than 3364, but on open grassy hillside, similar to 3364. It was found in a very big patch, 200–300 yards square and where found like this was entirely the white form, with the exception of a very few pale yellow forms. However a few white ones were found among 3367.'

'3367. Prim. sp. Thought at first to be the same as 2373 [*chamaedoron*] taken on the Kashong La, Tibet, last year. It seems to be very near that primula. But may it be No. 3364 = 3205 [*tsariensis*] which I have called a form of P. *griffithii* ? If it is the same as 3205 then 3205 is not P. *griffithii*. And yet I think that 3205 is the same as 1621 and others collected last year in Tsari. 1621 was named P. *griffithii* by the Professor I think, though I did not altogether agree with him.'

Such observations as these were of immense value to Professor Wright Smith who was monographing the genus. At first he had named 1621 P. *griffithii*, but after hearing Sherriff's comments had changed his views and had described it, with 3205, 3364, and other gatherings, as the new species, P. *tsariensis*. He was now able to discount Sherriff's first idea of equating 3366 and 3367 with P. *chamaedoron* and named 3367 P. *tsariensis*, and 3366 a beautiful white form of this wonderful species.

Sherriff's enthusiasm for primulas was now quite remarkable, especially when he thought he had discovered something new; and how the monographer blessed him! On the evening of 5 July, at his camp on the Omta Tso (14,200 feet), four miles north of

42. *Primula strumosa*

Maruthang, he described his feelings during the day. 'I spent the day in alternate cursing and rejoicing. Cursing the weather which was really as foul as anything could be, and rejoicing in the flowers, especially primulas; rejoicing in their beauty or new-ness, and cursing because photography on a day like this is really very trying. Most exposures for colour were 10–15 minutes, in pouring rain, with a wind, and under two umbrellas and various people trying to keep wind off the flowers and rain off the camera. However it was a good day; any day must be good when one gets primulas like Nos. 3383 and 3384. The former just covers the hillside on the [slopes leading down to the] Thita Tso [14,200 feet]. The latter was hard to get, only on the most sheer cliff faces, where even Tendup could hardly reach. In fact we only got enough to press. There is lots more, but it will be a job to get seeds, unless the cliff is quite dry. Here at camp *P. hopeana* is in masses.' [17]

Sherriff described 3383 as 'a magnificent primula, covering hundreds of yards of steep open grassy hillsides on the slopes leading down to the Thita Tso; corolla a pale golden yellow with a large rich golden eye; flowers up to 1⅝ inches across; tube pale yellow; calyx dull red-green; as a rule no farina.' He was quite certain that he had discovered a new species, but the monographer thought otherwise, regarding it as a variation of *P. strumosa* (to which he gave the name var. *perlata*), which Cooper had discovered in Bhutan, at Champa Pumthang, in 1915 and which Sherriff and Ludlow had collected at Sakden in 1934. Sherriff sent plants of 3383 to Britain by air mail and although they flowered they did not produce the startlingly large flowers as those found in the wild.

'Is this a white form of *Primula umbratilis*?' was the question Sherriff asked when examining 3384. It was.

After passing the Omta and Thita lakes and after crossing the Thampe La (16,000 feet), where, at 15,500 feet, *Primula pusilla* and *P. sikkimensis* var. *hopeana* covered the hillsides with blue and white respectively and *Rhododendron anthopogon* (3400) and the dwarf *Diapensia himalaica* (3392) with pink, and where from an inch-wide crack in a rock face he took a specimen of *Meconopsis bella* (3395) with a tap-root exactly 56 inches long! Sherriff marched north to Changsethang (14,200 feet) on 8 July and there camped until the morning of the 11th in a country full of great stones and boulders, perhaps the result of an earthquake. 'The hills all around are crumbling to bits, very steep, and don't look much good for flowers. However we brought our primula total up to 39 with a dwarf one . . . . very little of which we saw, No. 3413.' [18]

This was the purplish-pink, yellow-eyed *P. concinna*, no more than an inch tall, which only Cooper had previously gathered in Bhutan — Hooker having discovered it in Sikkim in 1848.

The previous day, 9 July, Sherriff had gathered another of Hooker's Sikkim discoveries of the middle of last century and by so doing had recorded it for the first time in Bhutan — a white form of the minute *P. muscoides* (3407); it was inhabiting moss-covered rocks, surrounded by a dwarf thick mass of old withered leaves.

On the return journey to Chendebi Sherriff had the most wonderful time. Below the cliffs on the north side of the Thita Tso 'there were just masses of flowers: Primulas *hopeana, pusilla, calderiana, atrodentata, sapphirina*, 3383 [*strumosa* var. *perlata*] and *glabra: Meconopsis bella, horridula, paniculata*: geraniums, saxifrages, salvias and many more. It was all very pretty and the big cushions of androsace make it look much nicer than ever.' [19] And 13 July, on the march to Chore (14,200 feet), was: 'one of the nicest days I have ever had. Except for one shower it was fine until 3.0 p m when we came in . . . . We first saw a lot of 3383 and it certainly is a beauty. When I stopped to admire that, I was standing on a primula . . . . 3438 [*uniflora* coming into flower]. There also was *Gentiana phyllocalyx* (3439) in plenty. I have never seen so many alpines out together as on this march. In places the hillsides and cliffs were just covered with them and the variety was great. . . . . We came to a little grassy hollow and here we found the most extraordinary collection of coloured primulas. There must have been the most awful intermarriage going on. There was *P. calderiana*, quite true and apart. Then there were all shades of colours from mixtures of 3366 [white form of *tsariensis*], of 3367 [blue-violet *tsariensis*], and of 3383 [golden-yellow *strumosa* var. *perlata*]. I counted seven variations in colour and all were mixed together. We came on to camp under a huge sheer cliff which has many flowers at the bottom, and half way up a fair amount of the beautiful white [*Primula*] *umbratilis* No. 3384. I tried to reach it from top and bottom but could not get near. It is a pity: I should like some more of it. However we have a fair number of new flowers today and it has all been great fun.'

Sherriff made beautiful and ample collections of 3432 — pale yellow, golden-yellow-eyed; of 3433 — pale blue with a golden eye; of 3434 — white, with golden-yellow eye; of 3435 — deep violet to purple-violet with a golden eye; 3436 — a rich yellow, golden-eyed form of *P. strumosa*; and of 3437 — the velvety purple,

orange-eyed *P. calderiana*. And on these gatherings he thus commented. 'I do not profess to lay down the law about these. The specimens taken were the more obvious variations in colour. About eight shades could easily have been found and they were growing right up against each other. Presumably they are due to hybridisation of 3366, 3367 and 3383, and yet these three primulas grow in their masses apart. Especially 3383: I have seen it by the thousand, with never a colour variation among the whole lot. Nos. 3366 and 3367 are in smaller masses but they also keep apart as a rule. One or two slight variations were seen among them, but very few. I also took *P. calderiana* from here, No. 3437, but *P. calderiana* has nothing to do with these colour variations. It can always be separated by its extremely unpleasant smell, noticeable 100 yards off, and by the fact that it is farinaceous. Nos. 3432−3436 must presumably be an interesting example of hybridisation [between *strumosa* and *tsariensis*]. The area in which they were all found was not more than 50 yards by 50 yards and there were not a great many of the species in that area. Half a mile on the true yellow form 3383 [*strumosa* var. *perlata*] was very common but kept to itself. All these colour variations seem to tend to many more flowers in a head.'

It is tempting to wonder at the fate of these collections in the hands of the taxonomists had not Sherriff made his observations. How many of these natural hybrid seedlings might have been described as new species ? Years later, in 1949, in the Bumthang Chu valley of Bhutan, Sherriff met with an identical situation which caused him to change his mind regarding the parents involved in the hybridisation. *P. calderiana* after all, and not *P. tsariensis*, had been the species which had mated with *P. strumosa*.

On 16 July they were all back again in hot fly-ridden Chendebi. 'Our collecting for the year is now almost over. We will get more things of course but the high altitude work is finished and we can only hope for one or two more primulas and those certainly known ones. Among primulas the two most obvious misses are *P. dryadifolia* and *P. griffithii*. We may have a var. of *griffithii* [3205] but I would not put it down as that. We now have, according to me, but perhaps not according to Wright Smith, 40 species of primula. About a few I am doubtful. . . . . I am quite satisfied, more so, as on the Black Mt. we started so badly.' [20]

43 & 44. The Maharaja and Maharani of Bhutan

The guest of H.H. The Maharaja

Sherriff now had to move eastwards to visit His Highness the
Maharaja in Domkhar. Before leaving Chendebi he was able to
make himself reasonably presentable for the occasion by taking a
much needed bath and by having his clothes washed by Tendup.
He was also able to replenish the wireless, a gift for His Highness,
with new batteries which Tobgye had sent. On 23 July he reached
Domkhar and was met by the army: 'all dressed up in very smart
khaki drill. They played us in with three bands, a pipe band,
fifes and bugles. All were quite good and I much admired the
buglers who bugled hard even when climbing a 1 in 4 hillside.
We passed the new palace, a grand looking place, and came
on ½ mile beyond where a special house had been erected for
me. It is a grand place, with hall, reception room, bed-writing
room, bathroom and usual offices a little way away. The whole
thing is very pukha and all lined with cloth, walls and ceiling,
while carpets are on the floors. Naku [the younger brother
and right-hand-man of His Highness] met me and we talked
for an hour or more, but he has forgotten nearly all his Hindi,
and conversation was not too bright. Then I went and saw
H. H. with the Maharani and Jigmie who is a grand little boy
of 10 (8 according to our counting). H. H. has not changed at
all and he is just the same charming man as before. I just
stayed a couple of hours with him and then came away.' [21]
However, Sherriff saw much more of His Highness during the
following week. 'H. H. is very strict with all his people, servants,
Kazis (= aristocrats) and even Naku and the Maharani. None
is allowed before him unless called for. Naku is the Donyer
and brings people to see H. H. but he himself must not come
unless called. I noticed how he had always to stand behind
H. H. and this must have been rather awkward for him once
when I was explaining a camera to H. H. and kept turning
the camera in different directions. As H. H. turned so Naku
had to run behind him. I have had a little difficulty with language, H. H. talking Hindi and I answering in north country
Urdu .... The army now has its own lines, cookhouses and
everything, and they parade regularly most of the morning.
They are a smart lot and seem pretty keen too. I played football against them once, and had an amusing but not scientific
game. Another day I asked to see the 2·75″ gun. It is all here
but the ammunition is missing; it is all in Trongsa. The army
has its own signallers who practise daily, and a band of drums,
bagpipes, fifes and trumpets .... I presented H. H. with the

wireless and went to hear it one evening when it worked very well indeed. On 31st H. H. arranged to have sports, which included boxing, relay race, women's race, three-legged race and high jump. It finished up with a race for the women of the local villages and then H. H. gave a 'tea party' — tea and rice — to everyone present, about 400. I then found that he had been quietly . . . . giving money to winners in my name. On the 30th presents started arriving in numbers, and I really don't know what to do with it all. H. H. produced a sword, two daggers, pair of boxes, an ivory and silver cup and reams of cloth, and the others were in proportion.' [22]

Whilst Sherriff was being thus entertained Tsongpen and Tendup had marched south to the Tibdey La chiefly to hunt for the remarkable wine-red-flowered, white-hoary-bracted *Lobelia nubigena*, in structure and organisation so like the giant lobelias of the Mountains of the Moon in Central Africa, which Cooper had discovered on this pass in 1915. The two collectors had no difficulty in locating it. In the spot where Cooper had found it, and nowhere else, it was very common, growing to a height of 3 feet among dwarf rhododendrons above the limit of the abies forest (3489).

Tendup also left the main party for a second time, for a brief visit to Phumzor in the Mangde Chu valley, in search of seeds of the beautiful blue-violet *Iris decora* (3001). However he joined the rest of the party at Chendebi when they all returned there on 3 August bringing with him a good collection of the iris seeds which were ripe at 4,500 feet, unripe higher up, and had all been thrown lower down.

The end of the Journey

On 5 August Sherriff, Tsongpen and Tendup left Chendebi for the last time and marched north again to Maruthang, then west into the Tang Chu valley and thence to Trashidinka and Ritang. Pintso, on the other hand, returned to Dungshinggang before joining the others at Ritang. On the departure from Chendebi Sherriff gathered, on the grassy hillside at 8,000 feet, a 4-foot tall aster with pale lavender, or pale blue-violet, ray florets; it proved to be a new species and was suitably named *Aster sherriffianus* (3522). Four other species new to botanists were discovered during the next two days. On the march to Maruthang on 6 August a new saxifrage with pale orange-yellow flowers, heavily spotted with a richer colour, inhabited the thick mixed forest at 10,000 feet and was later described as *Saxifraga strigillosa* (3528). And at the higher elevation of 12,500 feet, on an old stony river-bed 'a pretty little gentian, No. 3531 did me down badly. It was fully open when I first saw it

so I ran for my camera, but the rain started as I opened the camera and the flowers shut up in about 20 seconds.' When closed the corolla was velvety-black with a tinge of green; when opened, the lobes were slightly reflexed and of a rich velvet-blue; it was named *Gentiana melanensis* (3531). Then, on 7 August, on the journey to Chizukang, a codonopsis with purple cup-shaped flowers and with the usual evil smell, and rather abundant on the hillsides at 14,500 feet, proved to be a new species allied to the familiar *C. ovata* and was named—but not until March 1972!—*Codonopsis bhutanica* (3539). The fourth new species was *Delphinium muscosum* (3537), a desirable plant with rich blue-violet, white-bearded flowers and lovely lace-like foliage, inhabiting wet stony scree at 15,500 feet. Sherriff encountered it in later years and introduced it into cultivation.

Unluckily for Sherriff, a plant he prayed might be new was not so. It was the little primula (3438) on which he had trampled when admiring the large-flowered *P. strumosa* on the slopes of the Thita Tso in the middle of July. From Maruthang, Tsongpen had returned to the Thita lake once again and had collected further material under 3536. 'This is a magnificent primula, much finer than it appeared to be going to become when first taken. Now in full flower, on a very small patch of open grassy hillside. It opens more fully than any other primula I know in this section [Soldanelloideae], being almost flat and looking straight out at one, not pendulous at all. The biggest flower taped $1\frac{5}{8}$ inches across.' Though not a new species, this 2−4 inch tall jewel, with the great solitary (occasionally two) blue-violet, saucer-shaped flowers, had not previously been found in Bhutan. It was *P. uniflora* which had first been collected in 1848, by Hooker in the Sikkim Himalaya.

This was virtually the end of the expedition. They left Ritang on 10 August and a week later reached Ha where Sherriff spent a pleasant nine days with his friend Tobgye, and where, on the Ha La and the Kyu La, and particularly between the two, he found, as in 1933 and as Bailey had found ten years earlier, *Meconopsis superba* (3573). By September he was back once again in Srinagar with six hundred and sixty gatherings of plants and a great deal of information on the primulas of Bhutan which was to be of immense value to those concerned with the taxonomy of this group of plants which had so completely captured his interest.

45. *Primula umbratilis* (above)
46. *Primula uniflora*

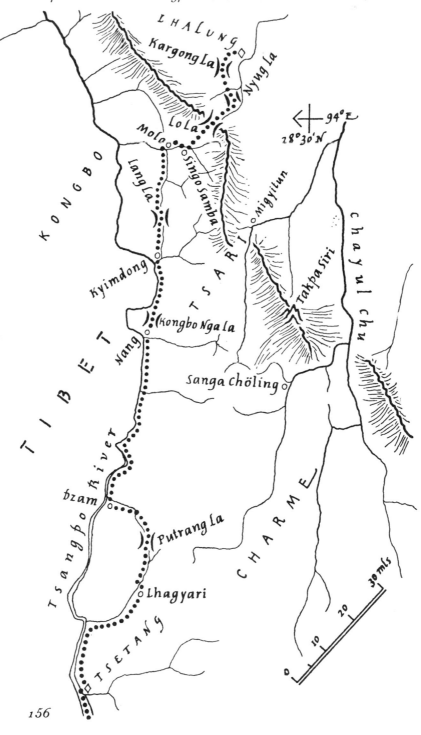

Map 12. Down the Tsangpo to Pachakshiri

LHALUNG

Kargong La

Nyug La

94° E

28° 30' N

Lo La

Molo

Singo samba

KONGBO

lang la

migyitun

TSARI

Takpa Siri

chayul chu

Kyimdong

Kongbo Nga la

Nang

Sanga chöling

Tsangpo River

TIBET

Dzam

Putrang la

CHARME

Lhagyari

30 mls

20

TSETANG

10

0

156

Chapter Five

Within the Drainage of the Tsangpo,
S.E.TIBET

o

In 1938 the two friends were again free to join forces to further their endeavours to carry biological exploration eastwards, from the western frontier of Bhutan to the gorge of the Tsangpo. In 1936 they had collected in the upper reaches of the Subansiri river, had travelled through the Tibetan districts of Chayul, Charme and Tsari, and had paid hurried visits into the Tsangpo valley. Now they planned to spend ten months moving further eastwards through the provinces of Takpo and Kongbo and along the main Himalaya to the neighbourhood of the great mountain Namcha Barwa (25,445 feet) which, with the rather less impressive, albeit more massive, Gyala Peri (23,460 feet) forms the portal of the tremendous gorge which the mighty Tsangpo cuts through the Himalaya, falling 6,000 feet in 150 miles, to emerge (as the Dihang) from the foothills of Assam. The Lo La was the most easterly pass they had reached in 1936. From here to Namcha Barwa the main range gradually converges to the Tsangpo over a distance of some 90 miles and it was in this area that they now proposed to concentrate their collecting effort (see Map 12).

Not all the ground they planned to work was virgin, though the main range between the Bimbi La in Tsari and the Doshong La near Namcha Barwa certainly was. For the rest, Bailey and Morshead in 1913, and Kingdon Ward and Cawdor in 1924, had traversed the Tsangpo valley but only the latter two had explored the Gyamda Chu. 'We felt pioneers and were thrilled at the thought. It was good to be living in an age when new lands and flowers and birds still awaited discovery.' [1]

The 1933 and 1934 expeditions had begun badly. So did this one. Dr George Taylor (now Sir George) of the Botanical Department of the British Museum had been given leave of absence to join the expedition and was due to meet Ludlow and Sherriff in Calcutta as near to 12 February as possible. But in Kashmir, and shortly before they left for Calcutta, Ludlow and Sherriff received a cable from Taylor telling them that he would not be able to keep the Calcutta appointment as he was due to enter hospital for surgical treatment. However, in Calcutta, on the evening of 12 February, Sherriff spent 80 rupees speaking to Taylor by telephone for six minutes in his London hospital and received the good news that he proposed to reach India in April and that, if a cook and instructions regarding the route Ludlow and Sherriff proposed to take could be left behind for him, he would join them at a specified meeting place. Molo, a small village in Kongbo province, which Taylor had never heard of, was decided on as the rendezvous 'round about the middle of May'.

To the Pachakshiri district

Before meeting Taylor, Ludlow and Sherriff proposed to explore the district of Pachakshiri, south of the main range, where, they hoped, plants would be in flower long before those on the colder northern slopes of the range. The route followed in 1936 no doubt would have been the shortest one to take, but as it was early in the year and they were fearful of snow in the Tsari district they decided on the more circuitous route which passed through Gyantse, Tsetang and the Tsangpo valley. Leaving Kalimpong on 22 February, they reached Chaksam on the Tsangpo, just opposite the Yab La and two stages from Lhasa, a month later. The Kalimpong-Lhasa road often has been described before and need not concern us now; wonderful views of the icy pyramid of sacred Chomolhari, nearly 24,000 feet high, and indeed of the main Himalayan range from Dochen, from before Phari to Dochen; and from Phari to Chaksam the, for the most part, dusty, icy, windy, desolate Tibetan plateau. The Tsangpo valley, at Chaksam on 19 March, was a delightful contrast to the plateau; pleasantly warm, with willows and poplars bursting their leaf buds, the grass turning green and iris leaves piercing the sandy soil. And an even more delightful contrast to the cold tedious trek across the Tibetan plateau was the leisurely and luxurious journey, by yak-skin coracles, down the Tsangpo from Kongka to Tsetang. Floating down the river, they had ample time for bird-watching; Bar-headed geese, Black-necked cranes and Brahminy duck were everywhere, two species of gull plied up and down the river and cormorants dived for fish. One cormorant was killed by a catapult. It had recently swallowed a fish over a pound in weight and was unable to rise. Ludlow and

47. Chomolhari from the Phari road
48. Yak-skin coracles on the Tsangpo

Sherriff ate the fish and their servants the cormorant. As Ludlow said, 'Chacun son goût!'

Tsetang (11,850 feet), where they halted for a couple of days, was a picturesque place, at any rate from a distance, with two red and brown gömpas and with the other houses white-washed, and with many fine and very old willows and poplars. But the incessant wind and the sandy dust which covered and got into everything was quite awful. There was one very interesting and very talkative inhabitant, a Ladakh trader named Atta Ullah Khan who had befriended Bailey and Morshead in 1913 and had cashed the former's cheque after he had been robbed of all his money during his journey up the Tsangpo. The old trader produced Bailey's letter of recommendation, as well as one written by Ward and Cawdor in 1924. He was very proud of these letters, as well as of the one he now received from Ludlow and Sherriff.

At Tsetang they dispensed with the coracles and proceeded, with donkey and bullock transport, to the rich and important dzong of Lhagyari (13,100 feet) which is perched on a cliff overlooking the river. On the journey they were greatly impressed by the way in which, below Tsetang, the Tsangpo cuts through a range of hills and descends in a series of rapids through a narrow gorge. A short distance beyond Lhagyari the road left the main valley and ascended that of the Changra Phu Chu to a pass called the Putrang La (16,500 feet). Probably no European had trod this way before; certainly Bailey had avoided it in 1913 and Ward and Cawdor in 1924. On the summit of the Putrang La, whence the snow peaks near Kyimdong Dzong (some seven marches distant) looked very splendid, a great surprise awaited them. Since leaving Phari, except for willows and poplars growing in areas of cultivation, hardly a tree had been seen. Now, on the eastern slopes of the pass there were patches of birch and juniper forest as well as dense thickets of *Rhododendron vellereum* (L S & T.[2] 3587); the rhododendron was not yet in flower of course, and at 15,000 feet was only 3 feet high, but lower down easily attained 12 feet, and its leaf indumentum varied from cinnamon to almost white, even on the same plant. And in this thick rhododendron undergrowth, the Eared (*Crossoptilon c. harmani*) and Tibetan (*Tetraophasis szechenyii*) pheasants lurked. Quite clearly the travellers were on the threshold of a much wetter region than any they had passed through since leaving the Chumbi valley a month ago.

From the Putrang La they descended to the level of the Tsangpo at Dzam (11,000 feet) where willows were in leaf and great pollarded trees, with trunks 5 feet in circumference, were in full bloom.

Thence along the right bank of the river for four days to Nang Dzong (10,700 feet) where they again left the river to avoid a gorge, and ascended the Kongbo Nga La (14,570 feet) which Ludlow, with Lumsden, had crossed in late July 1936. Half way up the pass amongst burnt larch trees, and greatly to their surprise, they saw 'a strange satanic-looking woodpecker' about the size of a jackdaw and jet-black save for a flaming crown and crest. It was the Great Black woodpecker (*Dryocopus martius khamensis*), which George Forrest had recorded in Yunnan, and which neither Ludlow nor Sherriff expected to find so far west as the 93rd meridian. Here was yet another example — and the Tibetan pheasant was another — of the close relationship between the avifaunas of SE Tibet and SW China. From the Kongbo Nga La they took the 1936 route to Molo but were held up at Kyimdong Dzong (10,600 feet) for five precious days waiting for transport for the crossing of the Lang La (15,800 feet), and not until 14 April did they reach Molo (10,300 feet) where the hills were thickly clad with abies, holly and oak and where the rose-pink flower clusters were opening on the 5–15 foot bushes of *Rhododendron vellereum* (3623) and *R. hirtipes* (3624) and on the aromatic, twiggy, 3 foot tall *R. kongboense* (3629).

The expedition was now within reach of Pachakshiri, though the main Himalayan range still barred the way and three passes had to be crossed, the Lo La (13,300 feet), the Nyug La (11,000 feet) and the Kargong La (8,800 feet), before Lhalung, the final goal, could be reached. Accordingly, at Molo, immediate steps were taken to cross the Lo La which Ludlow and Lumsden had visited at the end of June and the beginning of July in 1936. But the locals raised all manner of objections to the journey, the chief one being that there was so much snow on the pass at this time of the year as to render it impassable. Ludlow and Sherriff thought otherwise. Experience had taught them that most knife-edge passes — and the Lo La is one such — can be crossed with comparative ease at night when the snow is frozen. The Lo La proved to be no exception. Though snow lay at breast height for more than three miles on the north face and for two miles on the almost vertical south face, the expedition had crossed the pass before dawn on 24 April — after an enforced halt of six days at Molo — and by sunrise was well down the southern slopes and into the conifer zone and the district of Pachakshiri.

The crossing was a very different affair to Ludlow's in 1936. Now, rhododendrons and other shrubs which lined the path two years ago were completely buried under snow, as, of course, were all the primulas and other herbaceous plants. But Sherriff 'was shown the places where *Primula elizabethae, Omphalogramma minus* and *O. brachysiphon* and some of the rhododendrons were under the snow . . . . We dug down near the *P. elizabethae* place, but after going three feet found another two feet to go, so gave up. And yet it will be in full flower in 60 days. Other flowers are nearly over by then, though covered so deep now with snow.'[3]

Below the snow-line, however, plants were awakening from their winter rest; 'it is grand to see some flowers at last after such a long pause'.[4] 'Suddenly we saw a rhododendron in bloom. It was only a common "Grande" but the sight of it quickened our pulses and we plucked a huge truss of its rosy flowers and arranged it, almost reverently, in the press.'[5] The rhododendron in question probably wasn't a 'Grande' but a 'Falconeri', *Rhododendron hodgsonii* (3643), a common tree up to 30 feet high in the abies and rhododendron forest from 10,000–11,000 feet and with great trusses of bright pink to magenta flowers. But this was by no means the only rhododendron. There was the fleshy dark-crimson-belled *R. lopsangeanum* (3635) and, of much the same colour, the prostrate *R. forrestii* var. *repens* (3642), no more than six inches high, both of which had so much impressed Ludlow in 1936, the former having been found then for the very first time. There was a very beautiful and very distinct form of another of their new species of 1936, *R. tsariense* (3645), this time with pale yellow, red-spotted, cup-shaped flowers instead of the cream or pale-rose forms of two years ago. And there were two species which they hadn't gathered before; the glandular bristly, rich crimson-flowered *R. exasperatum* (3634) with its attractive bronze young shoots, and always on overhanging cliff faces in the rhododendron and abies forest, the 2–4 foot primrose-yellow *R. sulphureum* (3641, 3644) which, like the dwarf *R. forrestii* var. *repens*, was still another link between the floras of sw China and se Tibet.

Other plants offered striking evidence of the same relationship. There was a yellow-belled form of that beautiful creeping sub-shrub, *Diapensia himalaica* (3638), whose solitary terminal flower may be white, yellow, lilac or rose-purple, and which ranges from Burma through se Tibet into Yunnan. And there was *Primula lacerata* (3649); 'a most beautiful primula with the exact habit of

> *P. normaniana* but with a much smaller altitudinal range [9,000–12,000 feet] and not growing nearly so high. It grows mixed up with 3650 [*P. normaniana*]. In places covers the banks of paths and streams in masses. In the sunlight seems to keep its colour well [pink with bright orange-yellow eye, the

corolla-lobes very finely cut]. A much bigger flower than *P. normaniana*, with a brighter eye, and not quite so much pink in the flower. Strikingly beautiful.' [6] This was the first record of *P. lacerata* for SE Tibet. Previously it had been collected in NW Burma by Kingdon Ward in 1919 and by Farrer the following year; and although it has not yet been recorded from SW China, its closest allies are three species which are confined to Tonkin, Szechwan and Yunnan — *P chartacea*, *P. petelotii* and *P. veitchiana*.

Sherriff sent to Britain living plants of *P. lacerata*, which for a year or two grew vigorously and produced an abundance of flowers without however ever attaining the size of those of native specimens. That the species should have been lost to cultivation so quickly is rather surprising for it offers a ready means of vegetative propagation. The scape produces vegetative buds at the apex exactly as in the manner of the better-known *P. bracteosa* and when the scape is pegged down to the soil the vegetative buds produce roots in a few days. Thus although *P. lacerata* was slow to produce ripe capsules one would have thought that it would have been possible to have maintained it in cultivation by vegetative means.

Two other primula gatherings were not only new to SE Tibet but new to plant science also. One was *Primula geraldinae* (3640), a most attractive dwarf species with dark lilac, mauve, or rich wine-red flowers as large as the rest of the plant. A short stout stem, covered with the withered remains of old leaves, anchored the plant into the wet moss on almost inaccessible cliff faces at 11,000 feet, and the one-inch flowers, sprinkled with yellow farina around the eye, and either single or in pairs, almost hid the compact rosette of leaves which were brilliantly yellow-farinose below. The other new species was akin to *P. chamaethauma*, which had been collected in 1936 and was to be recollected on several occasions in the very near future, but had bright golden-yellow flowers, instead of blue-violet ones, nestling in the leaf rosettes. 'I think this may

> turn out to be a new species. It is a very pretty one when seen in mass, but singly is not striking. It was growing in a typically compact mass, all rooted together. The colour is a strikingly bright golden. We only saw one patch but a snow avalanche had come down beside it and very likely more lurked under the snow. Probably only in flower for a short period, like *P. vernicosa*.' [7] It was named *P. chionogenes* (3648) and an attempt

was made to introduce it into cultivation by sending live plants to Britain by air later in the year. These flowered two or three times at the Royal Botanic Garden, Edinburgh, but survived only for some six years.

The journey from the south side of the Lo La to Lhalung was a desperately bad one. 'Very nearly ⅓ of the "road" must be on wood. The other ⅔ are in mud and water. The path goes over huge boulders . . . . and across streams and in each case logs are laid to walk on. If up and down then they are notched.'[8] Fortunately the tedium of the march was relieved by the sight of some marvellous plants; trees 20 feet high of *Rhododendron grande* (3663) carrying huge trusses of pale lemon-yellow flowers; trees equally as large, with lovely smooth light brown bark and with fine dark crimson bells, of *R. hookeri* (3676); magnificent forms of the straggly, sometimes epiphytic, and always marvellously fragrant *R. lindleyi*, some with the great trumpets pure white save for the golden base (3665), others delicately flushed with rose; and the sticky, green-bristly, bright rose-flowered *R. rude* (3670). These were all gathered on the Nyug La, were first records for SE Tibet and this was the first time *R. rude* had been found outside Yunnan.

And it was the third time that a rhododendron (3750), clearly allied to the well-known dark red *Rhododendron thomsonii,* but with a loose truss of copper-red fleshy bells and with a glandular lower surface to the leaf which was sticky to the touch, had ever been found. It was flowering in the thick mixed forest beside a waterfall at 9,000 feet and there was very little of it. In October seeds (6567) were collected and from these seeds this rather strange and striking species was introduced into cultivation. Plants were raised and flowered and in the garden of Messrs Gibson, of Glenarn, Rhu, Dunbartonshire, for several years bore the name of *R. thomsonii* var. *pallidum.* It is now realised that this Lo La rhododendron is a new species and has been named *R. viscidifolium.*

Ludlow, with Lumsden, had thus far travelled in 1936. Now, he and Sherriff pressed onwards through the densest of rain forests along an atrocious track to Lhalung (6,300 feet) which is situated in the middle of an extensive plain about eight miles long and a mile broad. At its northern end three streams unite to form the Siyom river which flows down the middle of the valley. On both sides of the river large areas had been cleared of forest and were now covered with bracken and pasture-land. Cattle grazed on these open downs which were hemmed in on all sides by luxuriant and almost impenetrable rain forest. It was now 27 April and the expedition planned to stay in the Pachakshiri district, at Lhalung, for the next ten days.

49. *Primula lacerata* (above)
50. *Primula normaniana*

Pachakshiri is bounded on the south-east by the territory of the savage Palo Lobas and on the south-west by that of the equally savage Morang Lobas and both tribes had been a source of great anxiety to the Pachakshiribas. The Palo Lobas were only a day's march from Lhalung, and a few years previously the Tibetan Government had sent a detachment of troops to Lhalung to punish them for various offences. But the troops had done nothing and when they had returned to Molo the revengeful Lobas had attacked the innocent and unfortunate people of Lhalung. However, in 1938, the Pachakshiribas were not so much afraid of the Palo Lobas as of the Morang Lobas who, during the previous winter, had ambushed and killed several Lhalung men at Chudi, at the foot of the Nyug La.

Whilst the expedition was not anticipating trouble from these savage tribes it could not avoid trouble from other sources, chiefly from myriads of the biting *ðimðam* flies (a species of *Simulium*). The name Lhalung is derived from Lha (Gods) and Lungma (valley); a lama reincarnation visited the place and thinking it so beautiful named it the Valley of the Gods. Ludlow had other names for it! He seemed to be more sensitive to the bites of the flies than his companions. 'They raise an itching blood blister wherever they fasten on to you . . . . My hands are puffed up to a size that would rival any washerwoman's. In addition to leeches there are ticks and I extracted from my groin a very painful one today.'[9] Despite the heat Ludlow eventually had to wear gloves.

Floristically, as early in the year as this, Lhalung was a great disappointment. So much so that Sherriff left on a six-day trip up the ridge to the north, leaving Ludlow to hunt for his birds. But Sherriff's luck was out. 'This is, I'm afraid, a wild goose chase.

> On the ridge there is nothing but dense forest, so dense that every step has to be cut the whole way. We took a good six hours for what is certainly not more than two miles. It is next to impossible to leave the ridge as it is knife edge and very steep indeed both sides. However we came on until the coolies refused to go any further. Then strangely enough we failed to find water on either side. After an hour's search we found a little, and here we are in, without exception, the foulest camp [I have experienced]. The midges . . . . are too awful for words. They simply swarm in dense clouds everywhere and it is impossible to stand still even for a minute. I have a smoke fire in my tent which is infinitely preferable to the midges . . . . I'm afraid a new primula on my birthday tomorrow [3 May] is hardly likely.'[10] And on the morrow it 'rained all day with-

out stop. A perfectly filthy day. Tsongpen and I, with two men, cut our way up as far as we could along the ridge. There cliffs blocked our way and we could not go on, either side. So, soaked through and freezing cold, we returned to camp . . . . At the highest point we reached we found some rhododendrons in flower, but only those seen before. There was no sign of any primula or of anything else for that matter. Heavy rain and dense forest is too much to compete with, either for birds or flowers. So I am just sitting in camp hoping for the day to pass quickly, which of course it won't do. The camp is just mud now and everything we have is wet and filthy.' [11]

Early next morning as they were about to leave the mud, the wet and the filth they heard tragopan calling close to camp, and Tsongpen shot one in perfect plumage. It was a splendid gift to offer Ludlow who had found Lhalung disappointing from the avifauna point of view. It was not a case of the birds not being there but a case of the track being so bad and the forests and undergrowth so dense that it was impossible to work the place thoroughly. Ludlow had only been able to snatch a few birds here and there. He was delighted with his gift of the male tragopan (*Tragopan temminckii*) which appeared to change his luck as far as his quest for these beautiful game birds was concerned. In 1936, he and Sherriff had spent countless fruitless hours searching for specimens and the net result had been one mature female. Now, within the space of a week, they obtained all the specimens they required. In fact Temminck's tragopan appeared to be fairly abundant in Pachakshiri between 7,000 and 12,000 feet, inhabiting thick rhododendron and bamboo jungle in the densest of forests. Moreover the Pachakshiri people appeared to be able to snare this pheasant without difficulty. At any rate one evening one local produced five fresh skins but skins so badly damaged that Ludlow refused to accept them. Geographically the pheasant is of interest in that it links the avifauna of sw China — Yunnan — with that of se Tibet.

It was exactly the same story with another pheasant, Sclater's monal (*Lophophorus sclateri*). During the 1936 expedition not a single specimen had been obtained although Ludlow was convinced that on the Na La, above Migyitun, the three pheasants with the conspicuous white band on the tail which flew from out a rhododendron break at 12,000 feet had been this species. Now, on 14 May during the return journey to Molo, Ludlow secured a good series of specimens of the rare pheasant on the southern slopes of the Lo La at 11,500 feet; the birds were breeding in the dense rhododendron undergrowth of the silver-fir forest. Once again the natives seemed to have little difficulty in snaring this monal. They chose a small clearing in the forest where the birds were accustomed to feed and surrounded it with a brush-wood fence in which holes were left for the passage of the birds. Bamboo nooses were suspended over the holes and the birds were snared as they squeezed their way through. These pheasants apart, the most important capture at Lhalung was that of the Bar-wing (*Sibia* or *Actinodura nipalensis daflaensis*). It had been discovered by Godwin-Austen in the Dafla hills in 1875 and then completely lost sight of until Ludlow now procured specimens in Pachakshiri.

The Return to Molo

From Lhalung the expedition returned to Molo by the one and only route via the Lo La. It was now 16 May and there was still much snow on the pass. Ludlow's *Primula elizabethae* was still covered with three feet of snow which, though melting rapidly, probably wouldn't be away for the best part of another month. By mid-October it would be back again, leaving just four short months for all the lovely herbaceous plants to flower and set their seeds.

At Singo Samba during the journey to the Lo La in April Ludlow had funked crossing the river by the quivering tree-trunk bridge preferring to ford the river some 400 yards below the bridge. But now so high was the river that the bridge *had* to be crossed and Ludlow found the experience even more frightening than in 1936. And frightening, too, to some of the coolies who had to be blind-folded and carried across. Others crawled across on their hands and knees, whilst still others, 'the gay young lads of Molo', took it on the run. Molo was reached on 17 May at 2 p m and precisely at 2.30 Dr George Taylor walked into camp. Was there ever such immaculate staff work?

Taylor didn't simply walk into camp; he brought three hundred gatherings of plants with him and these were just as surprising to Ludlow and Sherriff as was his sudden arrival — there had been no news of him since Sherriff had spoken to him on the telephone on 12 February. Taylor had left Gangtok on 7 April and thus, hurrying to join the expedition, had collected three hundred specimens in less than a month and a half. Little wonder Ludlow and Sherriff were surprised for in five months in 1933 they had gathered some 500 specimens; in five months in 1934, about 600; in nine months in 1936 about 2,000; and in 1937, on his four-month trip, Sherriff had

51. *Rhododendron lindleyi*
52. *Rhododendron rude*

taken some 600 gatherings. They now realised that they had been rather too selective in their collecting. 'Taylor collects everything from mosses and fungi upwards to lilies' said Ludlow.

The three friends spent the following week at Molo, reading — and answering — the two months' mail Taylor had brought with him, drying and preparing specimens, developing films, discussing future plans and reorganising the stores and kit for the next journey. It was now the end of the third week in May and to cover as wide an area as possible during the height of the comparatively short flowering season it was clearly advantageous for the party to separate. Accordingly Sherriff decided to remain for a time in the Molo area and work the main Himalayan range from the head-waters of the Lilung Chu to Tsela Dzong, concentrating on the Lo La, Pa La, and Tsari Sama passes. Ludlow and Taylor on the other hand would explore the range from Tsela Dzong to Gyala and also visit Pemako and the passes in the vicinity of the Doshong La which Kingdon Ward had shown to be very rich in plants. They would reunite at Tsela Dzong on 31 July. Thus at 7 a m on 24 May Ludlow and Taylor marched north through the well wooded Lilung valley and Sherriff south-west to Langong down the lovely broad Langong Chu valley with open grassy meadows on the left bank and forests down to the river on the right.

With Sherriff on pilgrimage (*Map 13*)

Sherriff reached Langong (12,100 feet) on 27 May and for the next month used this village as his centre, either making daily expeditions from his camp there, or making new camps within a distance of ten miles. In dividing the expedition's resources to cover as much ground as possible there was one considerable disad-vantage; one part of the expedition was denied the expert bird-skinning techniques of Ramzana. He was now with Ludlow and Taylor and although Gulla, Ramzana's assistant, would be a great help to Sherriff, the latter decided to experiment with a few birds with the object of possibly dispensing with the process of skinning. On the march to Langong on 27 May he shot a young black woodpecker and, instead of skinning it, injected it with eight drops of a 1 to 20 solution of formalin, with a further two drops up the anus. However by 11 July the corpse was a mass of maggots and had to be thrown away. The formalin method of bird preserva-tion was not to be a substitute for the time-consuming skinning on this expedition — and Sherriff was disappointed. And also a little disappointed with the flowers. 'There is nothing in any of these valleys till one gets to 13,500 feet or so. Then primulas appear . . . . Everything here [Langong] seems to be late. Snow lies pretty

Map 13. Sherriff on Pilgrimage

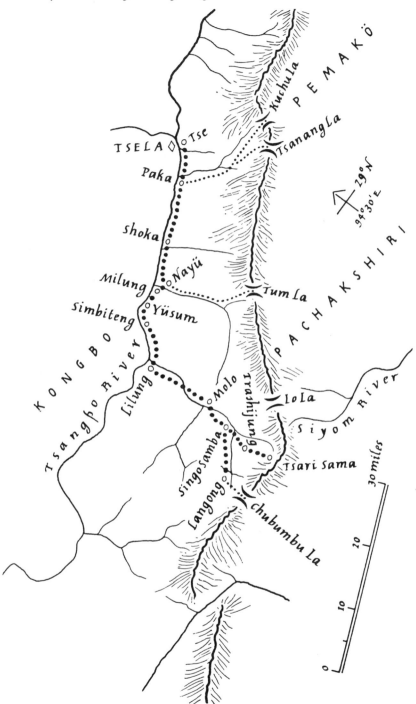

PEMAKÖ

Kuchula

Tsanangla

Tse

TSELA

Paka

29°N

94°30'E

Shoka

Nayü

Milung

Tum la

PACHAKSHIRI

Simbiteng

Yūsum

KONGBO

Tsangbo River

Lilung

Molo

Trashijung

lo la

Siyom River

Singosamba

Tsari Sama

Langong

Chubumbu la

30 miles

10

10

thick on the hills yet and nearly all passes are shut.'[12] In fact, above
14,000 feet, on the grassy hillsides from which the snow had just
melted, some primulas were already casting off their flowers, so
short is the flowering season. *Primula chamaethauma* (3893) and
P. calderiana (3894) were growing together;[13] some specimens,
plastered with farina, were bursting from their winter buds whilst
others, having lost most of their meal, were in immature fruit. And
in the Langong area on all the hills above 13,500 feet both north
and south of the main Himalayan range the glorious *P. tsariensis*
(3923) was everywhere, the winter bud scales, the scape, the
pedicels and the sepals all stained a dull red, and the corollas a
rich blue-violet to blue-purple and always yellow-eyed. Sherriff's
critical eye had noted that from Tsari to the Langong region *P.
tsariensis* was fairly constant in character, any deviations from the
norm taking the form of white flowers or of more robust habit.
However, on the Chubumbu La (13,900 feet) at just below 14,000
feet, he gathered what he took to be a hybrid between *P. tsariensis*
and the golden-yellow *P. chionogenes* (3983) which, on the south
side of the pass, was literally covering large areas of the open hillside preferably beneath cliff faces and down avalanche slopes. The
suspected hybrid (3984) occurred in clumps of 6–8 plants, always
among *P. chionogenes*, and the flowers varied in colour from a pale
purplish-yellow to a dull yellowish-purple, always with a yellow
eye. Until they had the benefit of Sherriff's observations these
plants were regarded as nothing more than minor colour forms of *P.
tsariensis* by the Edinburgh primula authorities.

It was on the Chubumbu La, between 13,000 and 14,000 feet,
that Sherriff, for the first time, saw the glory of Ludlow's *Primula
elizabethae*. 'We went down the most slippery and difficult grass
slope I have been on and we all took many tosses. On it, again in
masses, was *P. elizabethae* (3985) and I must now, rather reluctantly, say that it is the finest primula I know. It extended down 200
yards of this very steep slope. As Ludlow said, "a truly magnificent
primula".'[14] The avalanche slope yielded specimens in the winter
resting bud stage, specimens with the young leaves and flowers bursting
from the bud and specimens in full glorious bloom. Even the winter
buds were things of great beauty, dark crimson with a tinge of
yellow and velvety smooth to the touch. Many of the fragrant,
usually clear primrose-yellow, fully opened flowers, were faintly
stained with crimson, especially at the base of the tube, whilst the

53. *Primula chionogenes*
54. *Lilium souliei*

sepals, pedicels and flower bracts were all touched with the same rich dark colour. 'When I tried to photograph *P. elizabethae* in mass, the wind, mist and rain were so bad I had to give up after half an hour's waiting and hoping. The flowers would not keep still. We finished the day by glissading down the avalanche snow from the pass for nearly a mile.' [15] Some of the resting buds were flown home but by 1941 all the plants had died without flowering.

From Ludlow's experiences and findings in 1936 it was obvious to Sherriff that the Lo La would prove to be his most profitable hunting ground and accordingly he now planned to go there, stay for a week or so, and then undertake the short Tsari Sama pilgrimage. Tsari Sama means the 'New Tsari'. It is called either 'Sama' or 'Sarpa' and became a place of pilgrimage through a lama called Giaogama from the Tsübo Gompa in Lhasa. The pilgrimage simply involves journeying from Trashijung to Tsari Sama and back again. Accordingly Sherriff left Langong on 10 June, crossed to the right bank of the Langong Chu by a good cantilever bridge with one mid-stream pier, and reached Singo Samba at 9 a m on the 12th, to find that the bridge which had to be crossed if the Lo La was to be visited had been wilfully destroyed by some of the Langong people. As the river had now risen to a huge size and was impossible to cross the exploration of the Lo La was now out of the question and Sherriff was bitterly disappointed. However he would endeavour to make the best use of his misfortune by working the areas to the east and west of the pass. And he began, on 13 June, during the journey to Trashijung and thence to Tsari Sama, by making interesting observations on the primulas of the Sikkimensis section for the special benefit of the Edinburgh Primula authorities.

'L.S. & T. 5537, 5538, 5539, 5540. Compare all these together and with them 5541. [5537; corolla pale pinkish-mauve, paler towards the eye where there is some white farina, tube pale wine-red, fragrant: 5538; corolla bright rosy wine-red, white with farina inside, fragrant: 5539; corolla pure white, in bud tinged palest yellow, farina inside corolla, fragrant: 5540; in bud rosy brick-red, paling to pale rose-pink, flower white with farina inside, fragrant: 5541; corolla mauve or violet-mauve with tube usually more wine-red with a tinge of violet, inside almost white with farina, very fragrant.] I take 5541 to be the real *Primula ioessa*, although taken much earlier in the year than the type 2514 [27 August 1936]. It alone seems to be a true species, or at least to be constant in colour, shape etc. It grows slightly higher [in elevation] than the other numbers, but comes down to where they grow. Down the fairly steep streams, on the wet banks, it is plentiful and always constant. Then the streams come to an open boggy meadow. Here is also *P. alpicola* and here it is that an extraordinary mixture of colours is found, not in *P. alpicola* which is barely out yet, but in these *P. sikkimensis* varieties. [15] I have chosen four, quite distinct, but could take a dozen shades. Some, if not all, are most beautiful varying between the white of what looks like *P. hopeana* to the wine-red of *vinosa* [*waltonii*]. My own opinion is that *P. hopeana* and *P. ioessa* are true species, but after seeing this collection here, I should think that any others must be doubtful in the extreme.' [16] Sherriff's herbarium specimens of these numbers were named *P. ioessa* except for 5539 which was identified as *P. ioessa* var. *subpinnatifida*.

In a day or two Sherriff was to collect another primula of the Sikkimensis alliance which was quite new to him; 'its leaves are unlike any *sikkimensis* primula I know.' They were ovate or elliptic, round at the apex, heart-shaped at the base, deeply toothed at the margin, and were those of *Primula firmipes* (5575). Umbels of up to 8 nodding very pale yellow flowers were carried on scapes up to 12 inches high. A more fragile plant than *P. sikkimensis* and never growing in the spectacular masses of this species, it is, nevertheless, a beautiful and most graceful plant which Forrest discovered in the Tibetan province of Tsarong in 1919 and which has also been recorded from Upper Burma, the Assam Himalaya and NW Yunnan.

Sherriff was now on his brief Tsari Sama pilgrimage (15 June). 'I have seldom had such a bad day for weather. It was raining hard at 4.00 a.m. and has not yet stopped and there is a high wind with it all. We had hoped to cross to the south of the Himalayas but this pilgrimage apparently keeps to the north. At first there was nothing at all to be found. Then we got into good country, and here [Tsari Sama] it is very good and must be like the Lo La north side, only a little higher. Of flowers there is masses of *Primula calderiana* — a very fine form, *P. tsariensis*, *P. valentiniana* . . . . *P. elizabethae* and others, besides one new to me — *P. subularia* (5561) [new to Sherriff in the field; the dwarf new species with purplish-pink flowers, forming moss-like clumps on wet rocks, which Ludlow and Lumsden had discovered on the Lo La in 1936]. Rhododendrons too are good and of interesting ones seen today, there are *R. mekongense* (5555) and *R. campylogynum* (5560). Another nice one is one of the Glaucum [Glaucophyllum] series, No. 5565. The finest sight of all is certainly *Primula valentiniana* which is in masses. We can see when the mist rises a little, vast areas covered a deep wine-red. I went this afternoon to have a look at it and was amazed at how much there is. We reached here [Tsari Sama] about 12.00 noon, all soaked to the skin.' [17]

Sherriff was not quite correct. 'The finest sight of all' was certainly not *Primula valentiniana* which, though quite common (5569), was swamped by the more robust *P. kingii* (5570) a few plants of which Sherriff had first found in the Ha valley in Bhutan in 1933. Now, its velvety, rich wine-red flowers were deeply dyeing hundreds of yards of open wet meadows and patches of grassy hillsides from 13,000–14,000 feet. Apart from *P. valentiniana*, *P. elizabethae* was also growing among it.

Some of the Tsari Sama rhododendrons were also very fine. There was 5565, the purplish-pink or apple-blossom-pink, rather shallowly cup-shaped-flowered *R. charitopes*, which was one of Farrer's discoveries in Upper Burma, was known also from Yunnan and was now found for the first time in SE Tibet. There was an abundance of Ludlow's new species from the Lo La, *R. ludlowii* creeping along the moss-covered rocky soil of the hillsides at 13,000 feet its flattish primrose-yellow flowers being flushed and spotted with pink. Later in the year, in October, Sherriff collected seeds (6589, 6600) and introduced the lovely plant into cultivation. Another prostrate species, higher up the hillsides, was by no means common; it was a rather remarkable form of *R. forrestii* (5572) with fleshy slightly orange-pink bells instead of the usually brilliant scarlet ones. Obviously closely akin to this, and of much the same height — no more than a foot — covering rocks on the hillside with an admirable blending of dark green foliage and the palest of yellow flowers, was a new species which was appropriately later named *R. trilectorum* (5582) to commemorate the partnership on this expedition of Ludlow, Sherriff and Taylor. This was not the first time the three friends were to be honoured in the one name. A form of *R. wardii* with a conspicuous dark magenta blotch at the base of the pale lemon-yellow flowers, seeds of which were collected later in the year, received the Award of Merit from the Royal Horticultural Society when exhibited by Captain Collingwood Ingram; he gave his seedling the cultivar name of 'Elestee' [El (L) es (S) tee (T)]. These rhododendrons apart, some of the rock ledges at 14,000 feet were clothed with the pale yellow form of *R. tsariense* (5581), the lower sides of the leaves plastered with a thick fawn

55. *Meconopsis argemonantha* var. *lutea*

or pale greyish-brown felt, whilst in large groups on the stony hillsides were two of Forrest's Yunnan discoveries; *R. callimorphum* (5584), one of the most charming species in the entire genus with its trusses of half a dozen soft rose, often crimson-blotched, cup-shaped flowers, much darker rose when in bud; and a form of *R. erythrocalyx* (5568) which tended to hide among the oval leaves the pale pink flower-trusses.

Apart from the primulas and the rhododendrons two other plants particularly impressed Sherriff. One was a wine-red lousewort all over the wet grassy slopes and the other a prostrate fleshy cream-flowered daphne growing in one small area only. Both proved to be new species and were named *Pedicularis tayloriana* (5578) and *Daphne macrantha* (5585) respectively. In point of fact the daphne was infected with a mildew-like fungus which was also a species new to science — a new species of the genus *Coleroa*!

His brief pilgrimage over Sherriff returned to Molo — and in rather low spirits. 'It is sad to think that the longest day is past and I don't think I have got very much yet. It has been disappointing so far, and I hope for better things further east if only we can get where we want to go. There are three passes east of Lilung, the Nayü, Shoka and Lamdo Las, each of which I hope to visit, but I feel that we will be lucky indeed if all that comes off. Ludlow and Taylor will also, I expect, have found that this is not such a good area as that we visited in 1936. Or it may only be that everything east of Tsari is considerably later, due to more snow. Today was very cold indeed and I don't remember any days as cold as this in Tsari.'[18]

Sherriff left Molo on 29 June with the object of exploring the passes east of Lilung. He and his party proceeded north, following the route Ludlow and Taylor had taken five weeks earlier. And a pleasant route it was. Below Molo the steep slopes of the Lilung Chu were finely wooded with the Sikkim spruce, *Picea spinulosa* (5691), and with the Sikkim larch, *Larix griffithii*, the two dominant species. With their long pendulous sweeping branches they are both most beautiful trees. The spruce, discovered by Griffith in Bhutan about 1841, was especially splendid, frequently attaining a height of at least 100 feet, with many fine specimens up to 180 feet, and clothing all the side valleys leading down to the Tsangpo river. Approaching Lilung and the dry valley of the Tsangpo, seventeen miles and a two-day march from Molo, Sherriff passed through thick forests of the evergreen oak, *Quercus aquifolioides* (4466) and through an almost pure stand of *Pinus tabuliformis* (4469), every

Primula kingii

tree approaching 80 feet in height. This is the most common pine in the mountains of w Yunnan and w Szechwan. Almost daily now, examples of the relationship of the flora, and of the avifauna, of s w China and s e Tibet were claiming the attention of the travellers. And sometimes there were pleasant, and no doubt nostalgic, reminders of Britain; colonies of the native British *Adoxa moschatelina* (4415) with a profusion of minute green globular flowerheads, and shrubs of the Sea buckthorn *Hippophae rhamnoides*, carrying their clusters of orange berries and smothered by the thick fleshy yellow flowers and silken plumed fruits of *Clematis orientalis* (4472, 5703). The great white flowers of *Clematis montana* var. *grandiflora* (4458) were also much in evidence draping many of the undershrubs of the pine forest.

From Lilung (9,800 feet) a nine-mile march eastwards took the party to Simbiteng (9,900 feet) with conifer forest the whole way, with the Chinese *Ceratostigma minus* (5706) just beginning to open its blue blooms on the banks of the Tsangpo, and with *Iris decora* (4461) studding the turf with its brilliant blue-violet, golden-crested flowers. Thence for seven miles through a good deal of cultivation to Yüsum (9,700 feet) where walnuts were abundant, and then another seven miles to Milung (9,800 feet) only to find that the bridge over the river Nayü, which had to be crossed if Sherriff was to visit the Tum La, had been washed away. Apparently Sherriff wasn't having much luck with bridges on this expedition. Fortunately he had to stay at Milung only for one night, for next day, 4 July, the local gyimpu, or headman, came to his aid. 'It was
about 8.0 a.m. before a move was made this morning and then we found there was a "boat" consisting of three or four logs bound together which could go down the Tsangpo to the Nayü junction. It must have weighed tons and had about 2" freeboard. But off we went and there was only about 40 yards to cross once we reached the river Nayü. This "boat" only took four or five loads at a time, so it promised to be a lengthy business. But the gyimpu of Nayü arrived with a *kowa* (= coracle) and everything was very quickly taken across with that. He met me there and took me up on his horse to Nayü — the first time I had ridden for nearly four months. He is a nice man, most obliging and helpful and not like any of the locals we have come across for a long time. There is no trouble about going to the Tum La and staying there for 3 days. I "dined" in his house this evening and asked to see the Kongbo wooden teapot about which the Edinburgh Museum has asked so much. It is called Tibti, and is made of a wood called GIUGO [*Acer*

caesium (5740)], not of TSENG, which wood comes from the Lopa country only. Giugo is found here and I hope to have it pointed out tomorrow. The teapot is used for tea, but of course the tea is not heated in it. When lamas come to a house to do *puja* (=worship), tea is made for them in a big pot, and then poured into the Tibti for distribution . . . . Just as I was leaving the Tsangpo valley today it started to clear in the east and I caught a glimpse of a lovely sight which I hope I will see again. Straight down the valley, beyond Tsela Dzong, is a lovely snow-covered mountain locally called Gyala Tsutum (G. Peri of the map ?). It is said to have had its head cut off, to be carried to Samye. But when at Yüsum it was stopped in some way by a female (deity ?) and planted there. The little hill just south of Yüsum is now said to be this peak and pilgrims go round it.' [19]

From Nayü (9,800 feet) Sherriff marched south through the swampy Nayü Chu valley which, probably, at some time had been a huge lake. Most of the time the party waded in 6 inches to a foot of water the river seemingly unable to carry the water away fast enough. Water lilies were there in profusion especially the pure white, golden-stamened *Nymphaea tetragona* (5759). Taylor's favourite group of plants, the potamogetons, must also have been here in profusion although Sherriff neither commented on them nor collected them. However, on the Tum La (12,000 feet) he did gather one of Taylor's other speciality, a meconopsis. 'The mist was very thick and we could see nothing of the country. In fact it was so bad that we were completely lost for over an hour on our way back, this in spite of having a local with us who had often been this way. We luckily recognised one place we had passed and the Lopa was able to track our footsteps in the grass. It was a disappointing day on the whole . . . . The one bright spot was a meconopsis (5790) which reminds me very much of *Meconopsis argemonantha* but it has yellow flowers not white ones. We left enough for seeds if we can again find the place which will be difficult as we could recognise nothing to mark it by, today in the mist.' [20] Taylor later identified the plant as a new yellow-flowered form of *M. argemonantha* and gave it the varietal name of *lutea*.

The finding of the meconopsis was really not, as Sherriff had said, the only bright spot of the day. It was always a matter for rejoicing when he found a primula new to him and this he did on

56. A Dorje tribesman of Lamdo

this day of 8 July. Unfortunately it (5785) was not in flower. Even so, it was most elegant on the grass-covered cliff ledges at 12,500 feet the narrowly strap-shaped leaves, beautifully and regularly toothed, very dark green above and thickly coated with yellow farina below, surrounding 2 feet tall sturdy scapes carrying fat dull-crimson capsules. Sherriff found it again a few days later (5872), with glorious large rich purple flowers, up to 10 in the truss, and said of it 'I think it must take pride of place this year, with *Primula elizabethae*.'[21] He had found *Primula calliantha*, discovered in 1883 by the Abbé Delavay in the mountains near Tali in Yunnan. It has been found on many occasions since, for its beauty is such that no explorer can refrain from collecting it. Abundant locally, it is not widespread in the mountains of Western Yunnan and apart from the Tali range it occurs chiefly in the NW corner of the province with extensions into the adjoining parts of SE Tibet and Upper Burma. Twice in the past this marvellous plant had made fleeting appearances in cultivation in Britain; in 1908 seedlings were raised by Messrs Bees from Forrest's seeds, and plants flowered at the Royal Botanic Garden, Edinburgh, in 1925–26, again from Forrest's seeds, but in both instances the plants quickly died. Now, Sherriff sent home living plants by air mail but they did not take kindly to the British climate and to attempts to cultivate them and were soon lost.

Sherriff regarded his short trip to the Tum La as being no more successful than that to Langong, chiefly because the flowering season of the primulas seemed to have passed. As the neighbouring Shoka La was little higher than the Tum La he saw little point in going there, as he had planned, and decided to return to Nayü and follow the Tsangpo north-east to Tse where he was due to meet Ludlow and Taylor at the end of the month.

On 12 July he marched seven miles to Shoka (9,600 feet) taking with him a new volunteer to the expedition, a Khampa called Kesang who had made the journey with them to the Tum La; Kesang vowed that he would accompany Sherriff wherever he went and would do anything Sherriff asked of him. Throughout the short journey *Codonopsis vinciflora* (5817) was much in evidence trailing its flat periwinkle-blue flowers over the shrubs, even over a new species of barberry, *Berberis taylori* (5821) which, with its clusters of greenish-yellow flowers, was conspicuous on the banks of the Tsangpo. From Shoka, a further ten miles to Paka (9,600 feet) with many mulberry trees, the fruits scarcely ripe, on the way, and thence to Samar, eight miles distant, for an exploration of the Tsanang La (13,900 feet). As at several previous camps, at Samar there was trouble with the coolies, Kusho, who was in charge of the coolie transport, being a poor substitute for Pintso who was now with Ludlow and Taylor. At Samar there was very heavy rain which greatly raised the level of the river. Seven coolies were due to come from across the river, from Lamdo, and the others from Paka and Kangka. When the Lamdo coolies arrived there were only five of them; when asked where their missing companions were they calmly said that as three of them were fording the river, one, a bit of a lunatic, had been washed away and drowned; they seemed unable to account for the other missing one.

Coolie trouble was not the only one. The rains were incessant and now that Sherriff was in striking distance of the Tsanang La, where he was quite determined to go, the river was unfordable. However with Tsongpen's help he felled a large fir tree and with all the coolies pulling managed to span the river with it and thus use it as a bridge. This was not the only felling required for approaching the Tsanang La they had to cut their way through the thick abies forest to make a path. But once on the steep hillsides above the upper limit of the forest they were among the dwarf rhododendrons; the deciduous *Rhododendron trichocladum* (5844) usually no more than a foot high, the clusters of pale yellow flattish flowers appearing on the bare branches before the leaves unfurled — the first time the species had been found outside north-east Upper Burma; a low growing form of *R. charitopes* (5848) with flowers of pinkish wine-red; a deep crimson form of *R. campylogynum* (5847) sometimes less than a foot tall; and the saucer-shaped, pinkish-purple-flowered *R. calostrotum* (5855).

Growing with these dwarf rhododendrons, especially on cliff ledges, and usually no more than 6 inches tall was a lovely cassiope carrying innumerable white bells with brilliant red sepals. At first Sherriff believed he had found *Cassiope wardii* but closer investigation showed that his plant was a natural hybrid between *C. fastigiata* and *C. pectinata* (5846, 5846(a)). Sherriff also believed that he had found two new primulas — 'new to us anyway — after I had almost given up hope.'[22] But one was not new; it was *Primula calliantha* (5872) which he had found in fruit on the Tum La and of which he now wrote on seeing the royal-blue flowers for the first time, 'it must take pride of place with *P. elizabethae*.' The other one, as Sherriff thought, *was* new, and he accurately assessed its position in the genus. '[It] is, I think, another of the Dryadifolia section, or so it would appear from the roots and leaves and general habit. But it also has a marked pompon of hairs at the throat [of the flower], which I thought was peculiar to the Bella section. It is a pretty little primula (5865) and oddly enough grows close to *Primula jonardunii* [which was very common about 14,000 feet].' [23] Drifts of *P. valentiniana* (5866) were also in the same company. Like those of *P. jonardunii*, the bright purplish-pink flowers of 5865 were immersed in the dryas-like leaves which, unlike those of the other species, were completely lacking in farina. The new species, with great propriety, was named *P. tsongpenii*.

For three days and nights it had rained incessantly and still was pouring when, at 5.30 a m on 21 July, Sherriff and Tsongpen decided to climb a hill to the north of the Tsanang La valley. 'This morning I was not feeling fit, very weak and with a bit of fever. The rain was almost too much for me and for a while we hesitated. However I felt I must go on . . . . the climb proved much further and much steeper than I had expected. I could only go slowly and we did not reach the top till nearly mid-day, but we got there alright. Then the weather suddenly cleared up and we had a most unexpected perfect evening with the sun out full blaze and the clouds on the hills gradually clearing away. For a moment I could see all three of the valleys leading to the Lolung Leku La, the Tsanang La and the Kucha La. The latter looks good. The place we went to is known as Go-Nyi-Re — "there are two heads" — and lies more or less directly between here [near Tsanang Gompa] and Tsela Dzong. Being south of the main range it is a comparatively dry area, at any rate not as wet as the main range. Having made up my mind to do the main range, I have not touched this area yet, but having these two days to spare thought one of them should be used to explore it. So up we went. The first flower of interest was *Primula baileyana* (5887) fairly common in a limited habitat and not nearly as high as I would have expected. Then above that was one of the *bellidifolia* primulas (5888). Beside this was a small primula which was unfortunately over and of which we only saw three flowers. This looks somewhere near *P. atrodentata*, but I don't think can be this. I cannot place it. This was 5890. We are too late, and a most interesting primula found next, not in flower though, is P. (nivales sect. ?) 5889. The capsule looks nivalid, but the plant does not and I have no idea what it will turn out to be. At the top of the hill there is a pass with a little wall built up, and close by this, on the south side, is the little meconopsis which I found at the Tum La under No. 5790.

57. *Primula tsongpenii*

Here it is common and was taken under No. 5898. We should certainly get seed of this from Go Nyi Re. But the best find today was another meconopsis, taken under No. 5891, a fine big plant 2 ft. high with from 3–10 flowers of 3–4½ inches across. What this is I don't know. I gave Taylor his own Mec. book [*The Genus Meconopsis*, by George Taylor] as I did not expect to find any myself. We hope for a new species as I can't think of anything like this from SE Tibet. If this is so, this will certainly be a red letter day. We got quite a lot of other things too and I felt like cancelling the Kucha La trip. But that may as well be done now. There are but 10 days left till I am due at Tse and after that we will visit the drier areas. Except for seeds I am not going off to the main range again. In fact it would be poor policy. What a joy it was to see and feel the sun again and to feel dry. We were all soaked by mid-day but in a very short time were dry. I feel very weak this evening and have a headache.' [24]

Primula baileyana (5887) was growing in larch and rhododendron forest and above the tree zone at 12,500–13,500 feet, a beautiful species with roundish or kidney-shaped leaves thickly coated with white meal below and with white meal dusted over the rest of the plant, including the fragrant pale violet, white or pale lemon-eyed flowers. It is known only from SE Tibet where it was discovered by Kingdon Ward on the Nam La and Nambu La in 1924. No. 5888 was another of Ward's SE Tibet discoveries, in the valley of the Loro Chu, in 1935, the sweetly-scented *Primula hyacinthina*. Ludlow and Sherriff already had collected it, in 1936. It differs but little from the better known *P. bellidifolia* except for the white farina which usually covers the lower surface of the leaves. In fact Sherriff later believed that *P. hyacinthina* probably was no more than an eastern form of the other species and one which, due to climatic differences, has, as a rule, so much more farina. Still another Kingdon Ward discovery was 5890, *P. rhodochroa*, a charming dwarf affair no more than an inch tall with 1–4 purplish-pink, sometimes cherry-red, orange-eyed flowers half an inch in diameter and immersed in the dense tuft of narrow leaves thickly farinose below; Ward had discovered it on the Doshong La in 1924. The primula which Sherriff suspected as belonging to the Nivales section (5889) no doubt was of this alliance and probably was the purplish-violet *P. amabilis* which Forrest had first found in the province of Tsarong in SE Tibet in 1919.

The meconopsis which Sherriff now gathered under 5898, and which he had first taken from the Tum La under 5790, was the new

Meconopsis argemonantha var. *lutea*. Had he had with him Taylor's meconopsis monograph he would most certainly have identified his 5891 as *M. integrifolia* and have realised that he had not found a new species but one which had been discovered as long ago as 1872, by Przewalski in Kansu, whence it extends in the north to SE Tibet and appears to be most abundant in NW Yunnan, where Delavay discovered it on the Likiang range in 1889 and where Forrest and others have often since collected it, as well as in Szechwan where E. H. Wilson saw thousands and thousands of plants and whence he introduced it into British cultivation on behalf of Messrs Veitch in 1904. Since that time the Lampshade poppy, as it is commonly known, has always been in cultivation and has been the parent of several hybrids. Crossed with *M. betonicifolia* and *M. grandis*, *M. integrifolia* has produced, respectively, the hybrids x *M. sarsonsii* and x *M. beamishii*. We shall hear more of meconopsis later.

On 23 July, Sherriff, still with a tummy upset, rested in the lovely valley of the Kulu Phu Chu preparatory to three or four days of botanising on the Kuchu La (13,200 feet). The valley was really a great open swampy meadow much used by yak-herds who were no respecters of plants for a marvellous colony of *Notholirion bulbuliferum* (5923) had been half destroyed by them. Individual plants were nearly 5 feet tall, some stems carrying as many as thirty flowers each one dark purple-mauve and tipped with green.

Although the Kuchu La country looked perfectly wonderful for flowers there was really very little to whet Sherriff's appetite and he stayed at the pass for two days only. At 14,000−15,000 feet some of his old friends, and new, were present in some profusion on the open hillsides and on cliff faces and he gathered a very fine form of *Primula macrophylla* (5936) with an abundance of white meal, especially on the leaves, with almost black bracts and sepals, and with rich blue-violet corollas. *P. tsongpenii* (5931) and *P. dryadifolia* (5932) were also in plenty, as well as a species he hadn't seen before. This last was clearly akin to *P. bellidifolia* which Sherriff knew well, and to *P. hyacinthina*, but instead of having the usual funnel-shaped flower it had an almost globular one, dark velvety-purple in colour and delicately dusted with white meal. It was yet another of Kingdon Ward's discoveries, *P. concholoba* (5938) which Ward had collected on three occasions in 1926, in the Delei valley and at Seinghku Wang on the Assam-Burma-Tibet frontier. Sherriff's finding of it on the Kuchu La on 25 July was not the first record for SE Tibet for Ludlow and Taylor had forestalled him in the Lusha Chu on 9 June. At Sherriff's camp *Meconopsis betonicifolia* abounded and not far away there were other species; a poor coloured form of *M. simplicifolia*, *M. horridula*, and the pale satiny-blue *M. speciosa* (5940). And as the weather was now fine and sunny, gentians opened their flowers and gave further colour to the hillsides. Sherriff gathered three species and had he realised at the time that two of them were new to science he might have been less depressed with the Kuchu La flora. These were *Gentiana taylori* (5929), the corollas dark blue without and pale slaty-blue within, which Taylor had first found the previous week on the Doshong La; and *G. leucantha* (5934), the corolla lobes white and the tube slaty-green, which Sherriff failed to recognise as the same species he and Ludlow had collected on the Karkyu La in June 1936. The third species was the pale bluish-green *G. prainii* (5930).

On the evening of 27 July, at his camp in the Kulu Phu Chu valley, Sherriff wrote: 'I hope to finish this book of field notes up to 6000, which will give me up to 920 specimens this year. Taylor will likely have 1500 or so.' In fact when the three friends were reunited at Tse on 31 July Ludlow and Taylor brought with them close on 1100 specimens. 'Sherriff was disappointed with his results but he is, I think, a trifle pessimistic, and when his plants are worked out I imagine he will find he has done very well indeed. At any rate he has over 50 different primulas and as we have at least 20 he hasn't got, we have so far over 70 different species and varieties of the genus. This is really amazingly good for a strip of the Himalayas 60 miles long!'[25]

With Ludlow and Taylor (*Map 14*)

Ludlow and Taylor had left Molo on 24 May, had marched up the valley of the Lilung Chu and had reached Lilung the following day, heading for Tsela Dzong, five easy marches down the Tsangpo valley where the Tsangpo and the main Himalayan range have converged to such an extent that it is barely a days' march from the river to the summit of the various passes. In the bed of the river there were vast accumulations of sand, drifts extending in places to 500 feet above the level of the river, and being colonised by pines. Many swollen torrents rushing down from the main range had to he crossed, often with difficulty. One of these, the Nayü Chu, was in high flood and the bridge, which had collapsed, was being rebuilt by a party of workmen including a number of Lobas who were said to be slaves. It was also said that the price of a slave was a dagger for each limb, a sword for the head and a sack of flour for the body! The Loba women, resplendent in blue bead necklaces, were no

58. The Tsangpo at Tsela Dzong

more than 4½ feet tall and found the fording of the rivers, more than breast height to them, rather troublesome. But no matter the trouble they were always all smiles and giggles.

On 30 May Ludlow and Taylor's little party reached the small settlement of Tse (9,600 feet), opposite Tsela Dzong, where the Tsangpo is a placid river a mile wide with the huge Gyamda Chu coming in from the north and joining it in a multitude of branches. At Tse there were only two houses but the abundant cultivation around the village showed that in the past it must have been much more populated. Though the smallness of the population was attributed to the depredations of the Pobas who had raided the district some years ago and had caused the inhabitants to quit, Ludlow was convinced that the dwindling population was due to disease and a high death rate, goitre, leprosy and venereal and other diseases being rampant.

At Tse a halt had to be made for four days because respects had to be paid to the dzongpen at Tsela Dzong and his agreement to Ludlow and Taylor's future plans ensured. The travellers were ferried across the Tsangpo in a yak-skin boat, found the Dzong to be a poor and dirty building even though it was the most important one in Kongbo, and the dzongpen to be a pleasant young man of 28. He promised to give every assistance to the travellers whilst they were in his province but was adverse to their venturing into Pemako, where, he said, there was a smallpox epidemic. Ludlow and Taylor learned that this was perfectly true, but the real reason why he withheld permission was that there was trouble with the administration of Pemako and he was fearful that the advent of foreigners would complicate matters.

> The foreigners were pleased with their reception. 'We gave the dzongpen a mauser rifle and 150 cartridges with which he was mightily pleased, as he should be. Tibetans love firearms though they like them mainly for display. We thought it expedient to give the dzongpen this handsome present as we shall be spending a long period in the district over which he yields authority. His wife was much concerned about the freckles which had appeared on her face and asked us persistently how they could be banished. We promised face-cream and powder and decent soap and regretted we had not a supply of lemon juice which we told her was a certain cure.'[26]

Ludlow and Taylor decided to make Tse, rather than Tsela Dzong, their base as it was more conveniently situated for their work on the main range. And it proved unexpectedly interesting for flowers. Immediately above their camp was a narrow, steep-sided,

Map 14. With Ludlow and Taylor

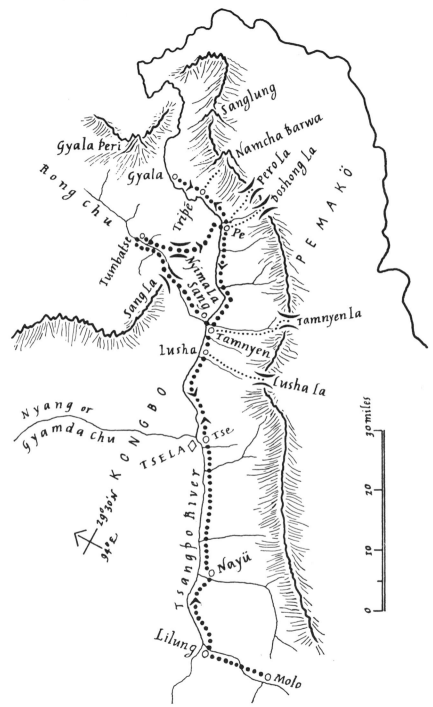

Sanglung

Namcha Barwa

Gyala Peri

Gyala

Pero La

Doshong La

Rong chu

P E M A K Ö

Tumbatse

Tribe

Pe

Nyima La

Sang La

Sang

Tamnyen La

Lusha

Tamnyen

Lusha La

Nyang or

Gyamda chu

K O N G B O

30 miles

Tse

TSELA

29°30'N

94°E

20

Tsangbo River

10

Nayü

10

Lilung

0

Molo

191

richly wooded valley where Taylor spent the whole of one day, reaching the fringe of the alpine zone and returning late in the evening with his presses bulging with flowers. Not far from camp, at 10,500 feet, in a small grassy clearing very prone to frequent flooding, he was delighted to find huge clumps of *Meconopsis betonicifolia* cohabiting with *Primula bellidifolia*. The poppy was finer by far than anything Taylor had seen in cultivation being 4 feet tall with many four-inch sky-blue golden-bossed flowers. Delighted with the poppy, yes; and utterly lost in wonder at the sight of *Paraquilegia grandiflora* (4608) clinging to the dry rocks — 'for sheer delicacy, poise and refinement [it] must be supreme.' Close by, in similar situations between 11,000 and 12,000 feet, grew the lovely *Primula baileyana* (4604, 4604a), whilst *P. calderiana* (4605) covered the mossy floor of the rhododendron forest, two of the main constituents of which were the pure yellow unblotched form of *Rhododendron wardii* (4620) and the white, purple-blotched form of *R. dignabile* (4621), the new species which Ludlow and Sherriff had discovered on the Le La, at Kyimpu, in 1936. And in his presses he held a new saxifrage, cushions of small encrusted leaves and white-starred flowers, which was later named *Saxifraga subternata* (4593).

Leaving Tse on 5 June for two days they journeyed eastwards down the right bank of the Tsangpo towards the village of Lusha (9,500 feet) which lay at the mouth of the large Lusha Chu valley coming in from the south. Here they camped beside the Lusha Chu amidst the sweetly-scented *Primula alpicola* in an admixture of colours but mostly of the *violacea* form (4665), and the handsome dark lustrous purple *Iris clarkei*. They were anxious to explore the Lusha La and accordingly on 8 June marched up the Lusha Chu valley. The march was typical of that up every other valley in this area leading to the main Himalayan range; a gradual ascent for three-quarters of the distance followed by a steep and abrupt ascent for the rest. For the first couple of miles the track led through the holly-oak zone of the dry Tsangpo valley into a belt of pine forest, and then as the rainfall increased, through the usual spruce, larch, poplar and birch forest with, here and there, extensive meadow flats and bogland. As soon as the track rose steeply the abies zone was entered and this gave way to dwarf-shrub moorland and the alpine zone of rock and snow.

They pitched camp amidst a cirque of rocky crags just below the

59. Lopas at Lusha
60. *Rhododendron pumilum*

timber-line at 12,500 feet, there to remain for a week exploring all accessible ground on both sides of the Lusha La (14,600 feet) in what Taylor described as 'a perpetual fury of rain', for the monsoon, heralded by a series of sharp thunderstorms, broke in earnest that evening (8 June) and it rained and rained and rained without respite for the entire week they spent on the pass. But neither rain nor anything else could dampen the enthusiasm of the plant hunters, especially of Taylor who, authority that he was on the genus meconopsis as represented by dried specimens in herbaria, could scarcely contain himself and was utterly entranced by the sight of the various species growing in the wild.

Just above camp, and given the protection from the depredations of yaks by various willows and by bushes of the spiny *Berberis macrosepala*, he found splendid colonies of *Meconopsis simplicifolia* (4789) in magnificent flower, as many as 8 or 10 two-foot bristly stems to a plant each carrying a large more or less nodding golden-centred flower either of the purest sky-blue or of a deeper blue. The colour variation must have interested him for it is generally supposed that in cultivation there are two more or less clearly defined strains of this species, a deep blue one which is usually polycarpic, and a sky-blue one which is usually monocarpic and dies after flowering. Higher up the valley slopes, at 13,000 feet, in the crevices of rocks with a dryish southern exposure were a few plants of the silky-blue-petalled *M. speciosa* (4792), the leaves with their strange straw-coloured, purple-based prickles. This desirable species is a representative of the flora of NW Yunnan where it was first discovered by Forrest in 1905.

In the rhododendron zone there were many species to catch the eye and to find their way into the plant presses, acres of dwarf plants no more than 2 feet tall forming a thick colourful and aromatic undergrowth; *Rhododendron forrestii* (4751), sometimes a prostrate creeper, sometimes more or less erect, with bright red or scarlet bells varying much in size; *R. calostrotum* (4711) with its flattish magenta-purple flowers; *R. paludosum* (4784), a very dominant carpeting shrub no more than 12 inches high with its usually solitary flowers in all shades of magenta to lilac-purple; these are all representatives of the Yunnan mountain flora above 12,000 feet. Codominant with *R. paludosum*, and of the same habit, was *R. fragariiflorum* (4785), its shallow pinkish-purple or purplish-crimson or even crushed-strawberry flowers in clusters of up to six, whilst the dwarf *R. anthopogon* (4781) clothed the north aspect of the hillside with sheets of all shades of pink and *R. pumilum* (4765) stained the avalanche slopes with pink and rose. These are fairly common Himalayan species. Amongst the taller growing species there was a pale form of *R. wardii* (4747), another Yunnan and Szechwan plant, the saucer-shaped flowers of ivory-white blotched at the base with claret; the white-flowered form of the new species, *R. dignabile* (4808), forming dense thickets on the hillsides; and still another new species conspicuous on the broad rock ledges, *R. tsariense* (4757), the leaves cinnamon-felted below and the reddish flower buds maturing to trusses of white, delicately pink-flushed and slightly red-spotted cups.

Other ericaceous shrubs, the cassiopes, in vast numbers at 12,500 feet, were at the peak of their flowering, and most memorable were they. *Cassiope selaginoides* var. *nana* not only formed compact carpets in the open spaces among the rhododendrons but crept on to the moorland and on to moss and soil-covered rocks, the myriads of cup-shaped nodding flowers, pure white except for the crimson base within and carried on reddish-brown flower stalks, being utterly enchanting, as were those, almost identical, of *C. wardii* (4734). This latter plant, a most striking and beautiful species, in large colonies on exposed gravel banks and rock-strewn slopes, grew almost a foot tall, its fastigiate branches of tightly ranked leaves being beautifully adorned with copious long silky hairs. Towards the end of October Taylor returned to harvest seeds of it, under 4734(a). Seedlings were raised by Mrs J. Renton of Perth and one was presented to Mr R. B. Cooke of Corbridge, Northumberland, in the spring of 1940. Mr Cooke exhibited his plant at the Royal Horticultural Society's Show in May 1949 and so greatly did it please the experts that it was awarded the Society's Cultural Commendation, the Award of Merit, as well as the Farrer Memorial Medal. Closely akin, but now assigned to the family Diapensiaceae, were two other marvellous alpines. Dense low mats of minute evergreen leaves, studded with large rich rose-pink flowers, of *Diapensia himalaica* (4754) cushioned many of the rock ledges and rocky and gravelly slopes of the alpine grassland, whilst *Diplarche multiflora* (4753), described as the nearest approach to a heath to be found in the Himalayas, its tight globose heads of pink flowers terminating densely leafy wiry stems, favoured gravel banks and avalanche slopes. Both are well-known Himalayan plants and both extend their territory into N W Yunnan.

Strangely enough, 'the two beauties' which Ludlow and Sherriff had found growing together in the Lo La Chu valley in 1936, were again cohabiting here amongst the dwarf rhododendrons and willows, the shining dark maroon-belled *Lilium souliei* (4783, a, b, c) and the dull mauvey-purple *L. nanum* (4782).

For Ludlow, the chief glory of the pass was 'a beautiful purple Nivalid [primula] which bears two huge purple flowers (4745). I have never seen this primula before. It grows in protected and shady situations. I only found three or four plants and they were on moss-covered stones on the outskirts of the abies forest near a stream but protected from the direct rays of the sun, and from wind by the fringe of the fir forest.' Thus wrote Ludlow on 8 June. The following day he found masses of it and clearly saw that, whatever it was, it was closely allied to his own *P. elizabethae.* Whenever they found a plant which they thought might be new and in which they were particularly interested, Ludlow and Sherriff, especially under the tutelage of Taylor, always endeavoured to present the botanists with as much information as possible in the form of field notes which, frequently, are more detailed than any others relating to plants collected in the Himalaya and in sw China. Ludlow and Taylor's field note on LS & T 4745 is typical of a great many such:

'4745. Lusha Chu, 12,500 feet. 9.6.38. Flowers 1–3. Scale leaves and under-surface of foliage leaves sulphury farinose. Scape reddish-brown towards apex and sparingly farinose. Calyx lobes reddish-purple, farinose. Corolla reddish-purple in included portion, upper part and segments rich violet-purple, efarinose on outside. Segments slightly paler within. Corolla-tube with a large sulphury-farinose eye. On rock ledges and in open marshy places and also in abies forest in wet situations and also common along streams with large saxifrage. Abundant on Lusha La (13,500 ft.) and south side of the pass on the moss grown boulder scree.'

It was the primula which Sherriff found on the Tsanang La, and of which he said that it must take pride of place with *P. elizabethae;* it was *P. calliantha.* The three friends did everything possible to introduce this marvellous plant into cultivation. They collected an abundance of seeds and Sherriff even sent home plants by air. Sadly, little success attended their efforts.

Two other primulas Ludlow now saw for the first time. Gritty ground by the edges of loose screes was colonised by another nivalid species, *Primula rigida* (4767), a dwarf plant 4 inches high with narrow rigid leaves very mealy below and with 5–10 lavender-blue or purple, yellow- or white-eyed flowers, which was discovered by Forrest in sw Szechwan in 1918 and had been found only once previously in se Tibet — by Kingdon Ward at Tümbatse

61. *Primula dickieana* (above)
62. *Primula jaffreyana*

in 1924. The other species, Kingdon Ward's *P. concholoba* (4724), relieved the shade of rocks in the rhododendron thickets with its tight heads of dark purple flowers. Sherriff was to find it three weeks later on the Kucha La.

For the rest, the charming dwarf *Primula genestieriana* (4758), discovered by Handel Mazzetti in Yunnan in 1916, brightened sodden gravelly soil beside snow-fed streams with its small compact rosettes of bright green leaves and up to two-inch stems of loose heads of violet-purple or mauve flowers; the similar but more compact-headed pinkish-purple to bluish-violet *P. glabra* (4803) inhabited wet grassy flushes; damp mossy rocks sparkled with the delightful, minute, pale or very dark pink *P. rhodochroa* (4736); *P. dryadifolia* (4744) covered rock crevices and ledges with its farinose leafy mats splashed with rose-crimson, and *P. chamaethauma* (4746) the wet avalanche slopes and open moorland with a profusion of blue-violet. The soggy avalanche slopes at 13,500 feet were the habitation of the creamy-yellow alpine form of *P. sikkimensis* called *pudibunda* (4749); both the white and purple-flowered forms of *P. vernicosa* (4756), as is their wont, displayed their precocious flowers beside the melting snow; and in the damp alpine meadows and on the grassy rock shelves an abundance of *P. dickieana* (4804) showed a range of colour from white through magenta to violet-purple.

Though an amazing diversity of plants found the perpetually wet atmosphere of the Lusha La and of other passes in this region so much to their liking, birds shunned the sodden moorland and dripping woodlands. The woodsnipe (*Capella* or *Gallinago nemoricola*) was an exception and it was often flushed by day and heard drumming by night sometimes at altitudes as high as 14,000 feet. Possibly these were the birds which had drummed around Sherriff's camp, at 13,000 feet, on the Black Mountain in Central Bhutan on 22 May 1937. Be this as it may Ludlow never dreamed of these birds occurring in the Eastern Himalayas in the breeding season at such high altitudes.

On 15 June the expedition returned to Lusha. 'Ramzana called us at 2.30 a.m. and says he mistook the light of the rising moon for dawn. It rained heavily throughout the night and throughout our march [of fifteen miles]. We got soaked. And when we arrived at the ford opposite Lusha we found the river in spate and nearly up to our waists. One woman fell with a flower box but on examining the pressed flowers inside they appeared to have suffered no harm.' [27] This was a stroke of luck and it is permissible, here, to anticipate an even greater one. Sherriff and

Ludlow always had the best of good fortune with their collections and never lost anything of real value. The most amazing example of their good luck occurred later in the year at Lilung, on 3 October, when they were homeward bound. A mule, carrying a load of pressed plants representing months of toil and labour, stumbled and fell whilst crossing the bridge over the river. The girth rope snapped and the two yakdans fell, one on each side of the mule. The mule unconcernedly rose, and, without touching either box, walked quietly on, leaving the boxes perched precariously on the edges of the bridge and overhanging the rushing torrent.

After a couple of days at Lusha drying their collections and clothing and enjoying the respite from the ceaseless rain of the high tops, Ludlow and Taylor continued down the right bank of the river to the small hamlet of Tamnyen (9,500 feet) there to spend a week in the Tamnyen Chu and in the exploration of the Tamnyen La (14,500 feet). The flora of this area was by no means as luxuriant as that of the Lusha Chu. Unfortunately the interminable rain was at least as abundant and caused the frustrated travellers to beat a hasty retreat to the sandy shores of the Tsangpo once again to dry both themselves. and their specimens. One of their best finds had been the reddish-glandular-haired, deep purple-violet-flowered *Omphalogramma minus* which they gathered under 4923 and of which they were to harvest seeds (4923a) in September.

Mainly meconopsis and primula

They now decided to leave the main range for a time and to cross to the other bank of the Tsangpo to the drier ranges behind the persistent monsoon screen, whence, in 1924–25, Kingdon Ward had reported such a rich flora. This time the yak-skin coracles, or kowas, were not available and the crossing had to be made in a weird-looking craft locally known as a tru: two 40-foot long conifer dug-outs lashed together with bamboo thongs. It was a version of the craft Sherriff was to use to cross the Nayü and was propelled by great oars with enormous blades, the oarsmen beginning each stroke in a standing position and ending it seated. The Tamnyen locals were of a mind that one tru would be sufficient for the 36 loads of baggage, the 36 coolies to carry the loads down to Sang whence Ludlow and Taylor were bound, the expedition members, and six oarsmen. But Ludlow thought otherwise and, much to the disgust of the locals, decided on two crossings. A few powerful strokes by the oarsmen carried the unwieldy craft into the fast current; it was then allowed to drift rapidly downstream and was guided to the other bank of the river near to Sang village where an enormous sandhill over 300 feet high was popularly supposed to

cover a large Dzong and much buried treasure.

Ludlow and Taylor made their camp at Sang (9,600 feet) on an alluvial flat and reconnoitred the area prior to exploring the Sang La (14,500 feet) which lies at the head of a valley leading to the Rong Chu above Tumbatse, the area which Kingdon Ward vividly and charmingly describes in his *The Riddle of the Tsangpo Gorges* and from which he made the bulk of his collections in 1924–25. The fact that Ward's exploration had not included the Sang La made them quietly optimistic of discovering new plant treasures there.

In the dry evergreen oak forests about 1,000 feet above Sang they were given their first surprise, the pink-petalled *Primula jaffreyana* (5000), 2–10 inches high, revelling on the exposed sun-baked banks of the forest. Later in the year they saw much more of this remarkable plant. In the dormant state it shrivels into a minute bud with absolutely no external sign of life, anchored by a few wrinkled and contracted roots, and surrounded by the withered leaves which powder to the touch. Remove the outer tight bud scales and within there is the tiniest green germ from which the next year's plant will develop. A number of these desiccated plants (6240) were collected, survived a two-month journey to Britain and produced lovely flowering specimens the following year. Most of these plants remained in cultivation for some fifteen years.

On the steep climb up to the alpine zone of the Sang La through mixed abies and rhododendron forests there were other primulas. On grassy banks or on bare black soil there was *Primula cawdoriana* (5020) in most magnificent flowering condition, up to a dozen pendent flowers to the head, each flower dusted with farina, greenish-white in the tube and with the lobulate petals dark violet-purple. On the loose earth of the avalanche slopes there was the mauve-petalled, green-eyed *P. baileyana* (5026), almost every part of the plant, except the upper surface of the leaves, thickly covered with beautiful white farina. And on the same slopes there was an abundance of a superb form of the common Yunnan *P. sinopurpurea* (5027) carrying several tiers of drooping violet-purple flowers on sturdy farinose scapes.

The spectacle of the primulas was as nothing compared to that which greeted the plant hunters as they emerged from the forest. 'Our hearts gave a leap at the prospect before us; the rolling moorland was a billowy sea of dwarf rhododendrons and other

63. The *tru* ferry at Tamnyen
64. *Meconopsis integrifolia*

shrubs. Spires of yellow poppy flowers pierced this matting and all about were colonies of the sky-blue *Meconopsis simplicifolia*. Our admiration of the scene was unbounded and we pitched our tents on a broad exposed ridge in the midst of this glorious alpine garden about 200 feet below the Sang La. We were quite oblivious to the sharp sleet showers as we traversed the hill slopes, eagerly cramming the choice plants into our presses and vascula. Failing light brought our labours on this memorable day [28 June] to a close and as we turned for camp in the gathering dusk we had superb views away to the south beyond the Tsangpo of the mighty snow range, muffled in an ever changing cloud mantle.'[28]

No other locality, except the neighbouring Nyima La, excelled, or indeed approached, the Sang La for variety and profusion of meconopsis. Allow Taylor, the meconopsis authority, to tell the story. 'The most conspicuous plant on the moorland was *Meconopsis integrifolia* [5043] whose fountain of yellow flowers rose elegantly through the carpet of *Rhododendron laudandum* [5042] and *Potentilla fruticosa* var. *grandiflora*. At this elevation, about 13,000 feet, the plants were up to four feet in height and very homogeneous in character. All had prominent, slender, cylindrical styles and the ovaries were densely covered with golden-brown adpressed hairs. *Meconopsis simplicifolia* grew in association but was not so prominent, as its flowers barely showed above the hummocky rhododendrons. Colonies of the species grew in small clearings. But the most exciting plant was one [5053, a, b, c], bearing pure white to pale-yellow flowers, which occurred sporadically in association with *M. integrifolia* and *M. simplicifolia*. At a glance this was recognised as Kingdon Ward's Ivory poppy, which was discovered in 1924 on the nearby Temo La. In my *Account of the Genus Meconopsis* I tentatively assigned this plant to x *M. harleyana*,[29] but little did I imagine then that I would have the opportunity of confirming my opinion in the field. The hybrid, in habit and flower colour, was easily picked out and we counted about twenty individuals in the area where the plants overlapped. The Ivory poppy was not found isolated from both parents. In the majority of the specimens the flowers were borne on basal scapes and the leaves were usually notched, characters which have been derived from *M. simplicifolia*. The flowers were up to five inches in diameter, and in texture, shape and colour showed the influence of *M. integrifolia* though occasionally the petals had a faint flush of mauve from *M. simplicifolia*. In its ovary characters, the plant was intermediate between the parents. The capsules of x *M. harleyana*, in contrast to the turgid state of those of the parents, were narrow and spindly and when opened they showed rows of abortive ovules. All the plants examined were monocarpic and by September had completely withered with the gaping capsules containing powdery undeveloped seed.

'In sheltered pockets of black soil on the block boulder scree at 15,500 feet another form of *M. integrifolia* was found. This plant was barely two feet in height, with up to twelve narrow petals, and with the style so contracted that it was concealed by the dark brown hairs of the ovary. Such states have been treated as species by some authors but study of some hundreds of herbarium and cultivated specimens and examination in the field show that the forms of this very polymorphic species merge and do not justify taxonomic separation.

'The other species seen on the Sang La were *M. horridula* [5061] —the racemose form just coming into flower on the scree slopes: *M. impedita* [5024] on earthy banks under *Rhododendron* and bearing intensely violet-purple satiny flowers on spiny basal scapes, and *M. speciosa*. I remember hearing the late George Forrest extolling the virtues of *M. speciosa* and he was full of regret that the species had never become established in gardens from his expeditions. Having seen the plant in its native habitat — in boulder scree or in crevices of dry rocks with a southern exposure — I can well understand Forrest's sentiments. The flowers are usually of a beautiful silky azure-blue, though I saw some plants with rich maroon petals. We collected a quantity of seeds of these forms but apparently no success has attended our attempted introduction.'[30]

Apart from meconopsis, the flora of the Sang La at this time of year was not particularly exciting, so on 1 July Ludlow and Taylor struck camp, crossed the Sang La and descended into the lush Rong Chu valley. Gorgeous plants of *Meconopsis integrifolia* with flowers six inches across were prominent on the moorland as well as lower down in the rhododendron forest, whilst *M. simplicifolia* was also in some abundance on the north side of the pass as well as the Ivory Poppy. A rough track brought the travellers to the waters of the Rong Chu which they followed to Tumbatse (11,600 feet) through meadows fragrant with the yellow *luna* form of *Primula alpicola*. At Tumbatse, in appalling weather and with Ludlow dosing himself with 15 grains of quinine to keep a slight fever at bay, they reorganised their resources for their return to the main collecting ground on the Himalayan range.

Their route to the Tsangpo valley led over the Nyima La (15,200 feet), so easy a pass that Ludlow rode the whole way from Tumbatse to the summit where they were greeted with a violent hail and rainstorm. Through the rain screen the snow-capped main range was clearly visible — and in sunshine — and after the rain and hail there were fine views of the snowy range beyond Tongkyuk as well as a fleeting glimpse of the cone of Namcha Barwa. It was on the Nyima La that Ward had discovered the hybrid Ivory poppy, but only six plants of it scattered over a wide area. Ludlow and Taylor were more fortunate, finding dozens of it on both sides of the pass wherever the two parents intermingled. Apart from *Meconopsis integrifolia* (5146) and *M. simplicifolia*, *M. horridula*, *M. speciosa* and *M. impedita* were also in perfect flower. Of primulas there were two which Ward had discovered in 1924. There was the pretty geranium-leaved, rose-pink-flowered *Primula latisecta* (5144) which, almost past its flowering, was carpeting the shady moss-covered forest floor, much of it under fallen timber. And there was what Ludlow called, immediately he saw it, 'the curious and somewhat unattractive *P. maximowiczii* (5138)'. The flowers are curious in that the plum-purple lobes are completely reflexed along the tube. Now *P. maximowiczii* originally was found 400 miles to the east, in the mountains in the neighbourhood of Peiping in the province of Chihli in Northern China before 1874, and since then has been recorded from Shansi and Shensi. As Ludlow and Taylor's plant (and Kingdon Ward's before them) differed from the N China plant in details of the winter resting bud, the leaves and the sepals, it was regarded as a geographical form of *P. maximowiczii*, was given the name *euprepes* — and might well have been described as a new species.

That evening, 5 July, they reached Timpa (9,700 feet) on the north bank of the Tsangpo and enjoyed superb views of the Himalaya. 'A temporary break in the monsoon had dispelled the

> clouds from the main peaks and the magnificent cone of Namcha Barwa, with its sierra-like northern spur falling away to the Tsangpo gorge, was an unforgettable sight. We lingered over our evening meal in the open, enthralled by the changing colour effects on the snow as the setting sun crept up the flanks of the mountain. Later, the peak was bathed in soft moonlight and remained clearly etched against the starry heavens when we retired reluctantly to rest.' [31]

The next day they ferried across the Tsangpo in one of the tru

dug-outs and camped at Pe (10,000 feet) at the mouth of the valley
leading to the Doshong La. Before making the exploration of the
Doshong La, whose floral richness to some extent was known to
them through the writings of Kingdon Ward, they decided to break
new ground and ascend to a neighbouring pass a little to the east,
the Pero La — only to be grievously disappointed. Although the
two passes practically adjoin, few of Ward's wonderful Doshong
La plants were in evidence on the Pero La. Possibly the travellers,
and the season, were too early for this particular valley. Certainly
much snow still remained in the gullies and a bitterly biting wind
and stinging rain swept down from the pass. And the pass itself was
choked with snow. 'A huge cornice overhung on the north side and
we climbed up the ridge above the pass to get a glimpse to the south
of the main range. Below was a seething cauldron of mist, swirling
round the great cliffs which were weathered in places into ghostly
pinnacles. A penetrating wind drove us back to the lee side and we
skirted the rocks and scree-slopes round the head of the valley.'[32]
With nearly empty presses they returned to Pe on 12 July and the
next day made camp on the Doshong La in the upper fringe of the
abies zone, there to stay for five days in the most deplorable of
weather. 'The Doshong La seems to be a place of perpetual rainfall

> at this time of the year (snow at other seasons). The clouds,
> driven by a violent wind, sweep over the pass and drop their
> moisture in a constant spray. We descended to a marsh at
> the foot of the tree zone and just below a lengthy waterfall
> but the wind was piercingly cold and the camping ground a
> swamp so we camped in the forest just below. And a very in-
> different camp it was — all on an incline . . . . The Doshong
> La seems to be all that Kingdon Ward described it to be [in
> *The Riddle of the Tsangpo Gorges*]. It is very rich in flowers,
> especially rhododendrons, most of which are in flower now,
> even though the season is late. We wandered up towards
> the pass whilst a camp was being prepared in the forest below
> and soon came across *Primula falcifolia*, so strikingly like *P.
> elizabethae* in appearance and habitat. The leaves however are
> very different and the flower not quite so imposing, nor does
> it grow so gregariously as *P. elizabethae*.'[33]

Kingdon Ward had called the Doshong La 'a rhododendron
fairyland' and even he, so fluent of pen, had confessed to the difficulty of describing the rhododendrons as he saw them in June. For
all that, he succeeded pretty well. '. . . . the valley, flanked by grey
> cliffs, roofed by grey skies, with the white snowfields above,
> spouting water which splashed and gurgled in a dozen bab-

bling becks; and everywhere the rocks swamped under a tidal wave of tense colours which gleam and glow in leagues of breaking light. The colours leap at you as you climb the moraine: Scarlet Runner [*Rhododendron forrestii* var. *repens*] dripping in blood-red rivers from the ledges, Scarlet Pimpernel [*R. forrestii* var. *repens*] whose fiery curtains hang from every rock; Carmelita [*R. chamae-thomsonii* var. *chamaethauma*] forming pools of incandescent lava; Yellow Peril [*R. campylocarpum*] heaving up against the foot of the cliff in choppy sulphur seas breaking from a low surf of the pink "Lacteum" [*R. doshongense*] whose bronzed leaves glimmer faintly like sea-tarnished metal.' [34]

From this 'rhododendron fairyland', Kingdon Ward had gathered at least a dozen fine species. But he had by no means exhausted the treasures of the pass for, from the almost inextricable dwarf thicket-tangle of the rhododendrons which in places dominated the boggy hillsides, Ludlow and Taylor collected two which Ward had not taken; the small, greenish-yellow tight-trussed *Rhododendron cephalanthum* var. *nmaiense* (5240) and the deep magenta-pink *R. charitopes* (5237) whose 2–3-flowered trusses were now almost past.

Primula falcifolia (5226), which had reminded Ludlow of *P. elizabethae*, was an extraordinary bog species which Ward had discovered on the pass in 1924 and had called the Daffodil primula. He had found it growing abundantly in open bogs on the steep alpine turf slopes facing south, a great winter resting bud sometimes nearly two inches long producing efarinose linear leaves up to six inches long and a scape only an inch or two longer and carrying sometimes a single, or sometimes up to four, daffodil-yellow, fragrant flowers over an inch in diameter. Ludlow, Sherriff and Taylor made every effort to introduce the Daffodil primula into British gardens. Later in the year they gathered seeds but unfortunately these did not germinate. Resting buds were also gathered and flown to Britain and although one or two of these allowed gardeners a glimpse of the yellow flowers, all had been lost to cultivation by 1943.

The southern slopes of the Doshong La lead to the Pemakö district and it was on these slopes that Taylor refound another of Ward's discoveries, *Primula chionota* (5284, part of 5284a). It inhabits grassy alpine slopes either in bogs or amongst boulders near melting snow. At the end of June, when Ward had discovered it, the rosettes of small leaves and the solitary primrose-yellow or white, greenish-yellow-eyed flowers were bursting from the winter resting bud. Now, half way through July, the leaves and the flowers were fully developed, the flowers on some plants being violet and carried on a conspicuous sturdy four-inch scape — var. *violacea* (part of 5284a). The variation in colour of *P. chionota* was as nothing compared to that in *P. dickieana* (5262, a, b, c, d) of which Taylor gathered specimens with white, yellow, mauve, violet or purple corollas, the variation occurring sometimes in the same mass of plants and at other times being more localised. Among the countless thousands of *P. valentiniana* (5229) there was no deviation from the crimson norm — 'Cherry Bell' as Kingdon Ward aptly named it.

The foul and cruel weather could not deter Taylor from scouring the slopes of the pass in search of his beloved meconopsis. But strangely enough this genus seemed to be represented on the Doshong La only by its most inconspicuous member, *Meconopsis lyrata* (5228), usually no more than four inches tall and with pale watery-blue, 4-petalled flowers. Even Taylor was much more interested in the sight of two British plants inhabiting the peaty bogs, the bogbean, *Menyanthes trifoliata* (5265) and a white form of the Marsh marigold, *Caltha palustris* (5241), both of which are much more worthy garden plants than the meconopsis. And even more worthy still, could it but have been introduced into cultivation, was another caltha, a most beautiful one with deep magenta-pink petals and red stamens which proved to be a new species and was appositely named *Caltha rubriflora* (5263). By way of compensation for the dearth of the poppies Taylor discovered a new deep blue gentian, which Sherriff was to gather a week later on the Kucha La, and which was appropriately named *Gentiana taylori* (5268).

On 17 July the little expedition returned to Pe where for three days it rained hard, and it still was raining hard and showed no signs of abating as the party set out for Gyala (9,300 feet) at the head of the Tsangpo gorge. However by the time they reached Tripé (10,000 feet) at the foot of a beautifully wooded glen dominated by the ice-bound peak of Namcha Barwa (25,445 feet) the clouds had lifted and the awesome splendour of the snowy peak stood before them. This one sublime sight was recompense enough for all the foul cruel weather of the Doshong La. But not for this one day but for seven whole days the skies were clear. 'We had struck a break in the monsoon at the very time we ourselves would have chosen had Providence vouchsafed us the choice.'[35]

The journey to Tripé was memorable from another point of view. 'At Pe the Tsangpo is still a placid river half a mile wide. There is, as yet, no hint of the astounding change that is to follow. But three miles below Pe the path leaves the river and

rises to a terrace 800 feet above it. Immediately there comes a muffled sound of rushing waters, and in the distance the Tsangpo is seen to enter a rapidly contracting valley where its waters first become ruffled and finally break into rapids. This is the entrance to the gorge. A few more miles and the path leaves the terrace and descends to the level of the river at the village of Kyikar, where two glacial torrents come hurtling down from Namcha Barwa. The scene here held us spellbound. The mile-wide Tsangpo we had seen at Tsela Dzong was here confined to a narrow gorge, 100 yards in width, down which it leapt in one appalling cataract. Where giant boulders choked the bed, great waves were flung high into the air, to fall hissing and seething into the cauldron below, and over the river hung a permanent cloud of spray in which rainbows danced in the sunshine.' [36]

The next day's fifteen-mile march from Tripé to Gyala (9,300 feet) was a tiring albeit rewarding one. For most of the time they were threading their way through the thick riverain forest with the roaring Tsangpo almost doubling on its tracks and flowing now westwards and now north-eastwards. The most interesting find of the day was somewhat fortuitous. Near Gyala, in the dense shade of the oak forest, they had halted for a wayside lunch and soon realised that they were sitting on a fruiting paeony (5350). The fruits were green and immature and although flowers were not to be seen the natives affirmed that they were white. Two months later, when Ludlow and Taylor returned to collect mature fruits, they found that all the seeds had been shed. However, nine years later, on 24 April 1947, Ludlow gathered beautiful white-flowering specimens at the same spot, and in August 1949, at Tamnyen, near Sang, the paeony was in mature fruit and Ludlow collected indigo-blue seeds from out the bright red capsules. The species was a new one and was named *Paeonia sterniana* (5350) in honour of Sir Frederick C. Stern, an authority on both the taxonomy and cultivation of the genus.

Gyala is the last Tibetan village on the Tsangpo and on the clear days such as Ludlow and Taylor now enjoyed, and especially from the sulphur springs three or four miles beyond, they had the most perfect views one could imagine of the northern spur of Namcha Barwa — and even of Gyala Peri — and could hear the avalanches thundering down the steep slopes. Ludlow had been anxious to

65. The Doshong La from the Nyima La
66. Namcha Barwa from Tripé

investigate the avifauna at this point where the Indian Himalaya terminates rather expecting that certain birds from the south would ascend through the gorge. To his surprise, however, he found that the avifauna didn't differ materially from that of the Tsangpo near Tsela Dzong and that any infiltration of birds up the gorge from the south was not in evidence — at any rate during this short visit and at this time of year.

In a week's time Ludlow and Taylor were due to be reunited with Sherriff at Tse, and thus on 24 July they returned to Tripé, there spending a couple of days to enable Taylor to investigate the valley leading to Namcha Barwa, 'a thrilling and very profitable expedition' as Taylor described it. 'We climbed steeply through bamboo, evergreen oak and juniper into the *Picea* and *Abies* zone, and then emerging from the forest we were confronted with the full grandeur of Namcha Barwa. Countless frozen cascades seamed the face of the mountain and there was a continuous tinkle as ice particles sprayed down from the topmost rocks. Two large crevassed and gravel-strewn glaciers, one from the base of the main peak and the other from a lateral valley, merged below the spot where we camped. The dirty tongue of ice thrust down the forest-clad valley towards the Tsangpo between large lateral moraines. This glacier terminates barely a mile from the Tsangpo and on its surface, some distance above the snout, it supports a thin conifer forest. The small emergent stream was glassy green.

'The icy north-facing cliffs of Namcha Barwa looked entirely barren and it did not seem possible for plants to become established on its inhospitable rocks or gain a foothold on its huge scree chutes. What a contrast was afforded by the south-facing slopes which teemed with glorious alpines and where the rocks were draped with choice plants.' [37]

Of the nine different primulas Taylor gathered one was new to science. With tufts of scabrid leaves barely two inches long and with trusses of up to six pale lavender or deep blue-violet white-eyed flowers on slender stems up to six inches tall, it was an arresting sight on the grassy rock ledges where it grew amongst bushes of the yellow-flowered form of *Rhododendron lepidotum*. Taylor named it *Primula aliciae* (5423) after his wife and it has never been found in any other locality save in this one valley above Tripé. Moreover, with *P. xanthopa*, which Ludlow and Sherriff had collected in Bhutan in 1933, it is the only non-Chinese representative of a group of closely allied primulas classified in that section of the genus known as Souliei. Towards the end of October Taylor was able to gather seeds and from these Mr R. B. Cooke, and others, raised flowering plants in 1940. Only Mr Cooke's plants set seed which he distributed but by 1953 the charming species had been lost to cultivation.

The party rejoins

From Tripé, four marches along the Tsangpo brought Ludlow and Taylor to Tse on 31 July, and within a few minutes of pitching their tents, Sherriff joined them. At Tse, six busy days were spent reading and answering their mail, developing films, drying their collections, comparing notes on them and preparing them for the homeward journey. A wealth of botanical and horticultural material of some 3000 gatherings, representing 15 coolie loads, had been preserved and of most there was sufficient to allow a generous distribution of duplicates to various herbaria. On all their collecting grounds plants had been marked for future seed collecting and in some four—six weeks' time these would have to be revisited. In the interval the three friends would once again split their resources. By this time Ludlow had a fairly comprehensive conception of the avifauna of the Tsangpo valley and was anxious to explore other areas. There were two particular biotypes he was now eager to investigate, the high plateau region on the Kham border and the low semi-tropical region in the gorges. He therefore decided to leave his two companions and visit the Pasum Kye La at the head of the Shoga Chu, the largest tributary of the Gyamda river, and from thence turn east into Pome and work the humid forests at Trulung, at the junction of the Tongkyuk and Po Tsangpo rivers. In the meantime Sherriff and Taylor would work the lower side valleys on both banks of the Gyamda or Nyang Chu and rendezvous with Ludlow early in September.

With Ludlow into Pome (Map 15)

Thus on 7 August they all crossed to the north side of the Tsangpo in coracles and camped below Tsela Dzong on the west bank of the Nyang Chu shortly above its delta-like confluence with the Tsangpo. And on the 10th, Ludlow's 53rd birthday, whilst Sherriff and Taylor journeyed north-west towards the rolling plateau in the neighbourhood of the Mira La, he marched north for twelve miles to Mape (9,800 feet) up the well cultivated valley of the Gyamda Chu. Thence along an excellent track for eleven miles, with *Codonopsis vinciflora* trailing over almost every shrub, to Chomo Dzong (9,800 feet) where Ludlow received from the dzongpen a present 'of a couple of dozen inferior eggs and I returned the compliment with an inferior Homburg hat'.[38] A long march of twenty miles, mostly through mixed evergreen oak and pine forest on the floor of

which ground orchids were common, brought him to Nyarlu (9,900 feet), but not without a good deal of trouble from the bullock transport. 'I hate bullock transport. Apart from the fact that it is terribly slow, it is also terribly hard on the loads. Bullocks have no sense. They just quietly and determinedly force their way past rocks and through bushes, wrenching off locks and the iron bands on the yakdans, and tearing the covering of tents and bedding.'[38] In fact this bullock transport was 'the very devil' throughout his journey up the Gyamda Chu to Drukla Gompa (11,000 feet) where he arrived on 18 August. Throughout his journey there were always alluring side valleys inviting exploration 'and I could not help wondering how long it would take to exhaust the botanical possibilities of this huge valley. A decade probably! So the young botanist can take heart for if it is going to take a decade to explore the Gyamda Chu properly, a generation at least must elapse before we know all there is to be known about the flora of SE Tibet.'[39]

The Drukla monastery is a large one, situated at the mouth of a wide valley coming in from the Yigrong range, and for several miles above Drukla almost perpendicular granite cliffs descend into the valley, the teeth of their immensely serrated summits inclining in all directions. Beyond the cliffs Ludlow's little expedition entered a long marshy valley, teeming with migrating teal and snipe, and reached Pangkar (11,800 feet), the last village in the valley, on 2 August. Thus far the birds had been few and uninteresting, but the Pasum La (17,250 feet), whence he was bound, was only two marches away so Ludlow ascended the Shoga Chu to its source on the pass, optimistic of there meeting with plateau birds. But he was to be disappointed for none of the birds he had hoped for was on the pass. From its summit he could see the dry plateau stretching away to the north and another couple of marches probably would have brought him to the birds he sought. But he could not afford the time; if he was to keep his appointment with Sherriff and Taylor at Tongkyuk, not a day was to be lost.

However if the birds on the Pasum La were a disappointment the flowers were a great joy and in a single day Ludlow added thirty species new to his collections. There were several saxifrages: yellow cushions of *Saxifraga saginoides* (6893); *S. punctulata* (6910) in two colour forms, sometimes pale yellow but more often white, and always red-spotted within; the yellow *S. diapensia* (6916); the white, red-streaked *S. melanocentra* (6917); and the purple *S. bergenioides* (6896); they were all in fairly rocky habita-

67. Yakherds, west of the Nyang Chu

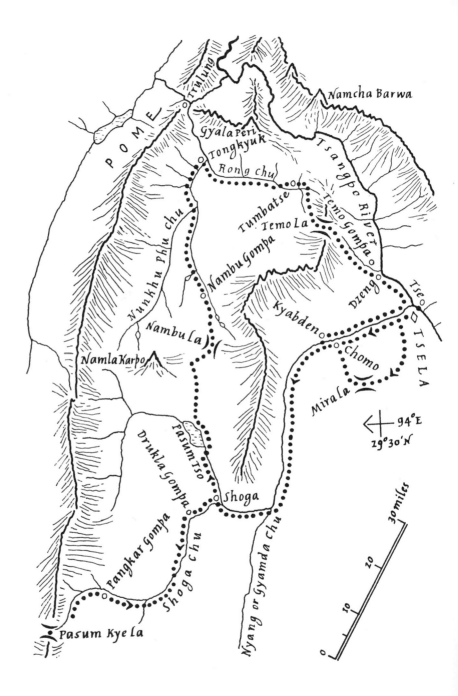

Map 15. With Ludlow into Pome

Namcha Barwa

POME

Trulung

Gyala Peri

Tongkyuk

Rong chu

Tsangpo River

Tumbatse

Temo La

Temo Gompa

Nunkhu Phu chu

Nambu Gompa

Nambu La

Dzeng

Tse

Kyabden

Namla Karpo

TSELA

Chomo

Mira La

94°E

29°30'N

Pasum Tso

Drukla Gompa

Shoga

Pangkar Gompa

Shoga chu

Nyang or Gyamda chu

30 miles

20

10

Pasum Kye la

214

tions between 16,000 and 17,000 feet. There was *Cyananthus spathulifolius* (6919) dyeing the grassy meads with sulphur-yellow, and the elegant yellow *Cremanthodium plantagineum* (6915) in the crevices and shade of boulders. Louseworts added colour to the wettish grassy hillslopes: *Pedicularis latiloba* (6895) in pink and crimson; *P. elwesii* (6894) in purplish-red; *P. trichoglossa* (6905) in claret. *Meconopsis horridula* (6897) clung to the crevices of the rocks or sought the shade of the rocks, sometimes with flowers of blue, sometimes of pink and sometimes of a mixture of the two; and *M. integrifolia* was there in masses, at this time of the year in fruit of course. Delphiniums for the most part sought the open meadow land or the partial shade of the under-shrub growth in the thin juniper forest at 13,000 feet; the purple *Delphinium potaninii* (6920) and the white, mauve-streaked *D. pylzowii* (6921). On the rocky grassy hillslopes from 14,000–16,000 feet gentians were abundant and a splendid spectacle: bright blue *Gentiana calliops* (6908); *G. aglaia* (6909), the corolla deep rich blue ribbed with dark purple on the outside and sometimes streaked with greenish-yellow; the white *G. leucomelaena* (6913); and the cream, greenish-purple-striped *G. algida* (6914).

Ludlow returned to Pangkar on 24 August and was about to leave the next morning when a jangle of bells announced the arrival of one of Sherriff's Tibetan servants who bore a disquieting letter to the effect that Taylor was seriously ill with what possibly was appendicitis. This alarming news prompted Ludlow to decide to do double marches to Kyabden, in the Gyamda Chu, where Taylor and Sherriff were halted. However, at Shoga Dzong the following day, he was relieved to receive a further note from Sherriff telling him that Taylor was much better and that they should all reunite at Tongkyuk as previously arranged. Thus Ludlow now adhered to his original plan and marched eastwards into Pome via the Pasum Tso and the Nambu La.

The Pasum Tso is a fine and pleasant sheet of water with a little monastery on an island about a couple of miles from the western end of the lake. A large river enters it from the Nambu La direction and away to the north-east are fine rugged glacial peaks, the valleys leading up to which looked promising for flowers. Certainly there were flowers in plenty in the woodlands along the southern shore of the lake; *Primula latisecta* (6924) and *P. jaffreyana* (6928), their flowers now past, were in great abundance, the former in the shade of the forest, the latter on dry banks by the roadside; the dainty lily-like *Streptopus simplex* (6932) had also lost its small white fragile flowers and was now in fruit; on the other hand the tubular

flowers of *Cyananthus lobatus* (6930) in bright purple-blue, and of *Codonopsis tubulosa* (6931), yellow-green and purple-veined, were much in evidence. An abundance of seeds of the streptopus was collected and living plants were sent to Britain by aeroplane in an attempt, successful in the garden of Mr R. B. Cooke of Corbridge and of Major and Mrs Knox Finlay of Perthshire, to introduce the lovely plant into cultivation.

On the last day of August Ludlow crossed the Nambu La (14,970 feet), where junipers and rhododendrons grew to within a few hundred feet of the summit, and descended to Nambu Gompa (13,800 feet) through a wide valley which reminded him of Tsari. Nambu Gompa consisted of one house and one monk, but there were a number of herdsmen and cattle in the vicinity. Ludlow was now in Pome, heading for Tongkyuk Dzong, and on 2 September he camped near the junction of the Nambu Chu and Tongkyuk Chu where five or six years before a large lake, which had been impounded some distance up the latter valley, had broken its bonds and a terrific flood had ensued. Uprooted trees lay everywhere in the scoured river bed and the high-flood mark was clearly visible 40 feet above the normal bed of the river.

Ludlow at last was about to fulfil his long cherished desire to march down the Tongkyuk river to its junction with the Po Tsangpo. He anticipated a lot of interesting birds on the march as, in three stages, he would descend from 8,500 to 6,500 feet and strike the semi-tropical avifauna at Trulung. But it was not to be, on this expedition anyway. On 3 September, at Tongkyuk Dzong, he received a further letter from Sherriff telling him that Taylor once again was seriously ill, so instead of descending the Tongkyuk river to Trulung, with close on 200 specimens he posted up the Rong Chu to Tumbatse (11,600 feet), digging up 100 bulbs of *Lilium taliense* on the way, crossed the Temo La (14,000 feet) and rejoined Sherriff and the ailing Taylor at Dzeng (9,500 feet), near Temo Gompa, on the 7th.

With Sherriff and Taylor to the Mira La (Map 15)

Taylor had been periodically unwell since he and Sherriff had left Ludlow, for the Mira La, on 10 August. On 13 August at their camp south of the Mira La Sherriff rather ominously entered into his diary: 'Taylor felt rather rotten yesterday evening and was not looking well this morning. But he went off up the hill to the south, not getting much and having no energy at all. He looked bad when

68. *Saxifraga punctulata*
69. The Nyang Chu from Kyabden

he came in at 3.3o and an hour or two later went to bed without any food. I think he has a cold in the tummy.' [40] And several days later, at Kyabden, Sherriff wrote again: 'Taylor has been very ill here with some unknown complaint. So much so that I sent off to Lhasa for help if possible, and also to call Ludlow. It appeared to be appendicitis, with awful pains, nausea, vomiting and a high rate of pulse. This went on till yesterday evening [21 August] when, after vomiting, he said he was alright and that the pains had gone. Shortly after this the pain seemed to have localised to the left of the navel. What it has been I don't know, though I should think that a gastric ulcer is possible. He is ever so much better today and I have sent off [a letter] to Ludlow and hope he will send on my letter to Lhasa countering anything that may have been done there. That has all kept me on the move here and we have done no collecting since arriving. It will still take Taylor some days to recover and I would not be surprised at a further week's halt here.' [41]

Actually, at Kyabden, Taylor was unable to leave camp for a fortnight. Fortunately, on 14 August, though feeling well below par, he had summoned up the energy to explore the Mira La (15,800 feet) with Sherriff and to find some of his beloved meconopsis. He had found the little yellow poppy (6064) which Sherriff had gathered at Go Nyi Re on 21 July and which he was to regard as a previously unrecorded yellow form of the white *Meconopsis argemonantha*, giving it the varietal name of *lutea*. He had found another of Sherriff's discoveries, the yellow-flowered form of *M. horridula* which Sherriff had seen in very small quantity on the Shagam La in Tsari, about 80 miles to the south. But here, on the Mira La, *M. horridula* var. *lutea* (6062) was in abundance between 15,000 and 16,000 feet in block boulder-scree on a very steep grassy hillside. Plants were up to $3\frac{1}{2}$ feet high, with pale sulphur-yellow petals, and were growing in association with the short-styled form of *M. integrifolia* (6082) which was in immature fruit and commonly had had but one flower. Not far away, amongst dwarf rhododendrons, there was *M. simplicifolia* (6083) whilst on grassy cliff ledges and on the loose granitic scree *M. impedita* (6052, 6084) still carried a few blue-violet flowers.

It was always a red letter day for Sherriff when he found a primula new to him and especially one which he thought might prove new to science, and Taylor must have shared his friend's pleasure when they found one such on the Mira La on 14 August. Though most of the primulas were past their flowering period two quite distinct species, seeking the shelter of huge boulders and growing in dry moss, were still carrying magnificent flowering stems. One they recognised straightway; it was the very farinose, pale mauve-pink, white-eyed, fragrant *Primula littledalei* (6045). The other (6061) neither Sherriff nor Taylor had seen before. Like its companion, the lower side of the rather flaccid, long and narrow and finely toothed leaves, the 8–12 inch scape, bracts, sepals, all were thickly dusted with white farina, whilst the usually deep blue-violet flowers had a prominent large white eye. It was a new species whose nearest of kin is *P. obtusifolia* of the N W Himalaya, was named *P. youngeriana*, and has never since been recollected.

The floral spectacle which earlier would have been provided by the primulas was now given by the gentians. Let Taylor describe them: 'The turfy hill-slopes were bright with hosts of gentians which seemed to vie with each other in their lavish display. The variety of species on the Mira La was astonishing. Here, plants which were hailed with delight when introduced from s w China grew with those usually associated with the eastern Himalaya. In no other area visited was this mingling of the eastern and western floras so forcibly illustrated, but it was not, of course, confined to the gentians. Blue was the predominant colour amongst the autumn flowering alpines and the most vivid splashes of colour were provided by species of *Gentiana* . . . . At 12,500 feet on a damp grassy flat near the stream, *G. sikkimensis* [6126] spread over mossy hummocks. The corolla-tube was green and mottled with greenish-blue on the outside towards the top; the segments were slaty-blue with intervening white plicae. Close at hand, amongst *Salix* on a steep bank and on the open hillside, was *G.* [*stictantha* — 6043]. The corolla-tube and lobes were white and speckled with greenish-blue or flushed with blue-purple on the outside; the lobes were evenly spotted with greenish-blue on the inside and the plicae were white. In some plants the inflorescences were congested and only a few inches in height, but usually they were more open and up to a foot. *G.* [*prolata* — 6027, 6072] with bright blue flowers formed rosettes on open meadows from 13,500–15,000 feet. A splendid form of *G. trichotoma* [6081], over a foot high in places, grew amongst dwarf rhododendrons and on open damp grassy meadows. Its corolla-tube and segments were blue with a tinge of green and the plicae were mauve. *G. tsarongensis* [6047], a neat and attractive little plant with miniature leaves along short prostrate branches, was growing on open grassy hill-slopes. The corolla-tube is short, concealed by the green calyx, but the flowers expand as beautiful blue-violet stars. On the higher hillsides from 14,500 to 16,000 feet, in damp hollows amongst boulders, were mats of *G.* [*namlaensis* — 6049]. The cobalt-blue flowers, for all the world like bubbles, are sessile and so restricted at the top that the mouth of the corolla is closed by the small white-margined lobes which conceal the white plicae. Even in full sunshine the flowers did not appear to open to any extent. *G.* [*aglaia* — 6053] occurred very sparingly on the grassy slopes. Dense tufts of *G. infelix* [6038] grew in mossy situations and in damp grassy meadows. This is a small prostrate, tufted plant with slaty-blue lobes at the apex of the blue-veined white tube. One of the neatest and attractive gentians in this valley was *G. filistyla* [6040]. From the miniature rosettes arise dark-blue trumpets an inch and a half in height and about half an inch in diameter. Another charming member of the *Gentianaceae* on the Mira La was *Lomatogonium* [*stapfii* — 6039], three to four inches in height, with blue-violet darkly veined petals forming shallow cup-shaped flowers about an inch across. Sheets of this plant gave a gay touch to the steep rocky hillsides.' [42]

Rivalling the gentians in their diversity, though not perhaps in their number and general effect on the landscape, were the saxifrages of which Sherriff and Taylor gathered a dozen different kinds in two days, including three new species. One of these latter was the yellow *Saxifraga montanella* (6068) which Ludlow had first found at Lukuthang in 1934; another was *S. miralana* (6078) whose lemon-yellow, orange-spotted flowers brightened the granitic screes; and the third the velvet crimson *S. haematochroa* (6070) growing in clumps in moss on the large block boulder screes. They were all at between 15,000 and 16,000 feet, along with the buttercup-yellow, orange-speckled *S. diapensia* (6071).

Having spent so profitable a time on the Mira La, Sherriff and Taylor moved on to Kyabden, where the latter became confined to camp, and then on to Dzeng, where they were joined by Ludlow. Taylor was still a pretty sick man and neither of his friends could diagnose the trouble. It might have been dysentery; it might have been liver; fortunately, it certainly wasn't appendicitis. His friends decided to starve him and for a week he survived on milk and egg flips and cholera pills. And not only survived but improved to the extent that he began to clamour for something more solid. He was given a Tibetan chicken with no adverse effect. At the end of a fortnight his ration was two chickens a day, one of them in the form of soup; and at the end of the third week the ration was increased to three per day. Fortunately for the chicken population of the Kongbo region, at the end of the fifth week Taylor returned to a normal diet.

Seed collecting (Map 16)

By 16 September Taylor had made such splendid progress that the expedition was able to put into operation its proposed plans for the seed harvest. Sherriff returned to Molo via Lilung to collect on the Lo La and the passes on the main range in the Langong valley, whilst Taylor and Ludlow drifted down the Tsangpo in kowas to Lusha (9,500 feet) which they made their base for a week of hurried expeditions to the passes visited in June and July to collect the autumn flowers and seeds from the plants that had been marked. Whilst the frustrated Taylor convalesced in the immediate environment of the camp, Ludlow visited the Lusha La, and their collectors the Tamnyen La and the Doshong La. The weather on all these passes was deplorable; even so a rich harvest of seeds was reaped and several living plants were gathered for sending to Britain, by air, when the travellers reached Calcutta.

They left Lusha on 23 September and, following the Tsangpo, on 1 October reached Lilung to find that the bridge over the river half way to Molo had been cut the previous day. But Molo had to be reached for large collections and the main expedition's winter clothing had been left there and this had to be rescued. So, whilst Ludlow and Taylor marched up the Tsangpo to Kyimdong Dzong, gathering fruits of the beautiful silvery-haired, pale blue-mauve *Cyananthus sherriffii* on the way, an intelligent Bhutanese servant and a party of porters travelled up the Lilung Chu on a difficult hunter's track to Molo, claimed their loads and delivered them at Kyimdong Dzong on 12 October. On that same day Ludlow and Taylor left for Tsari by the pass which leads into this district, the Bimbi La, where they saw the most glorious display of autumn colouring either of them had ever seen. 'On the south-facing slopes

> above Sumbatse there grew a *Berberis* — I am proud to think it bears my name [*Berberis ludlowii* — 6337] — not in hundreds or thousands but in tens and hundreds of thousands, whose fiery red leaves glowed so vividly that the whole hillside for mile upon mile seemed to be ablaze. Here and there, in the midst of this burning fiery furnace shone the rich orange-bronze of a rose; but the dominant colour was red — the red of glowing embers seen at night.' [43]

This was Taylor's first visit to the Bimbi La and Ludlow was happy to show him some of their findings on the 1936 expedition.

Taylor was especially excited to be shown two meconopsis new to him in their natural surroundings, *Meconopsis argemonantha* var. *genuina* (6343) and *M. bella* (6344). Both species were fruiting on the grassy cliff ledges and both collectors were thrilled to gather seeds to send to Britain. Though there was a degree of success with those of *M. bella*, there is no record of any success with those of *M. argemonantha*. But failure in cultivation never deterred Ludlow and Sherriff — nor their companions — from harvesting as much seed from as many desirable plants as possible, often under the most frustrating of circumstances. Ludlow describes their attitude to the matter. 'To label, press and preserve plants for the herbarium was

> never the sole object of our botanical expeditions. Few, save the expert, ever spend much time in herbaria, but the living plant, that grows in our gardens and parks, is a joy to all who behold it. And so we experienced just as big a thrill when we garnered the seeds of some dry and shrivelled plant in autumn as when we plucked it in all its floral loveliness in spring — for others, perhaps, would now be able to enjoy its beauty.

> 'The trials of a seed hunter are numerous and we often used to wonder how many of the people who received our seeds ever paused to think of the labour entailed in their collection. Only a few, I fear.

> 'Yet the labour at times was great, not a mere snatching of capsules and berries by the roadside as we passed by. To climb 3,000 or 4,000 feet and find the seeds all shed, or green and unripe, eaten by grubs or birds or cattle, or buried 'neath a blanket of snow, were some of the difficulties with which we had to contend.

> 'And even when we *had* collected good ripe seed we could never be quite certain that it would germinate.

> 'Perhaps the three loveliest primulas we obtained on this journey were Kingdon Ward's *Primula falcifolia*, the Abbé Delavay's *P. calliantha* and our own beautiful *P. elizabethae*. We collected ripe and abundant seeds of each and we sent these seeds to at least fifty expert gardeners in Great Britain. All failed; not a seed germinated. [A few seeds of *Primula calliantha* did germinate.]

> 'In addition to seed we sent home living seedlings of these three primulas by air, and when I was home in the summer I went to see them in their new abode. Some were alive, a few had even flowered, but all looked unhappy, and none, I fear, will survive.' [44]

Having crossed the Bimbi La they had left the Tsangpo drainage

Map 16. Seed Collecting

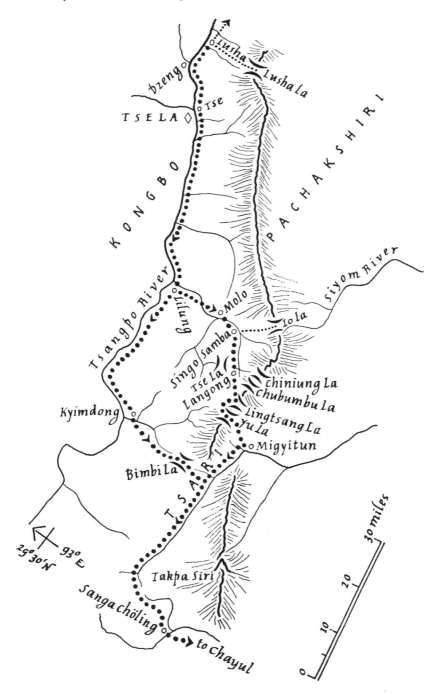

Lusha
Lusha La
Dzeng
Tse
TSE LA
KONGBO
PACHAKSHIRI
Siyom River
Tsangpo River
Lilung
Molo
Io la
Singo Samba
Tse La
Langong
Chiniung La
Chubumbu La
Lingtsang La
Yu La
Kyimdong
Migyitun
Bimbi La
TSARI
93° E
29° 30' N
Takpa Siri
Sanga chöling
to Chayul
30 miles
10
20
0

system and were now in Tsari where most of the natives had already left on their annual begging pilgrimage. However transport *was* procurable and Ludlow and Taylor reached Sanga Chöling on 18 October only to find that 'Rosy Cheeks'' lovely 1936 complexion was hidden under a horrible smear of caoutchouc — applied to ward off an attack of neuralgia.

Ludlow and Taylor now had to travel quickly for the latter had a boat to catch and the Diwangiri railhead was still a good three weeks' journey distant. Thus leaving Sanga Chöling on 21 October, they reached Chayul Dzong on the 24th — having been unable to resist stopping on the Drichung La, the only known locality for *Meconopsis sherriffii*, to take fruiting material of Sherriff's lovely pink-flowered poppy — and arrived at Tsona five days later, by the route Ludlow had taken with Sherriff, in 1936. They travelled in the most frightful weather conditions which Ludlow graphically described. 'Every day we rode in the teeth of a violent wind which

> raised clouds of grit that seared our faces like sandpaper. Our lips cracked and bled, the skin peeled from our faces, and day after day we rode with heads bowed on our breasts in stony silence. The climax came on the summit of the Nyala La where we met an icy wind of such tempestuous violence that we had to lean forward at an angle during the descent and had the greatest difficulty in breathing. That evening I was afflicted with a sneezing fit which lasted more than an hour and left me quite exhausted.' [45]

Soon after sunrise on 30 October they left Tsona and well before noon were on the summit of the Po La. Down they sped into the fir and rhododendron forests above Trimo, thankful that the plateau lay behind them and that the last pass on their long journey had been crossed. Two weeks later, on 14 November, they reached Diwangiri and looked down on the plains of Assam half hidden in a smoky haze.

Whilst Ludlow awaited the arrival of Sherriff, Taylor left for home, tired and still not really fit but content that he had seen, growing in the wild, sixteen different Himalayan poppies. Nothing else really much mattered.

Sherriff rejoined Ludlow at Diwangiri on 25 November bringing with him a harvest of over 300 different lots of seeds as well as a dozen and a half plants, mostly primulas, with their attached soil.

> 'All these are at the moment timed to arrive in London on 12 December, and in the RBG [Royal Botanic Garden] Edinburgh on early morning of 13th! Ludlow and Taylor also have a bundle of roots too, and altogether we have far more roots,

bulbs, and tubers than ever brought back before. So I hope we get them successfully home this time. I intend to send all by KLM freight and shudder at the thought of how much they will cost.'[46]

Though Sherriff seemed fit enough he had had his moments of anxiety regarding his health. When he left his colleagues on 16 September he had returned to Molo prior to harvesting seeds on the Lo La and the passes on the Langong valley. And at his camp on the Lo La on 5 October he confessed; 'I have been feeling my heart the last two days and must go quietly for a bit. It is a pity now, just when there is such a lot to do, which means a lot of climbing. We still have Sari Sama (2 passes), Tse La, Chiniung La, Chubumbu La and the two to Migyitun to do, and one at Migyitun and Chikchar.'[47] Fortunately his heart stood the strain of these high passes, but he wasn't to be so fortunate some years later.

The two passes to Migyitun were the Lingtsang La and the Yu La which Sherriff crossed whilst exploring a new direct route from Langong to Migyitun — and he had poor weather for his crossing. At Langong he wrote; 'It seems too much to hope for fine weather, as I have seldom seen the weather so set wet-looking. It snowed all day where we were, but wet snow which did not lie much during the day. I have always thought the monsoon ended quite suddenly up here on the 18th October. But that cannot be so in this longitude. This is the 24th of the 8th month [Tibetan month] and locals say quite definitely that during the 8th and 9th months they expect a lot of rain here, and snow up the mountains. In the 10th month, they say, it does not snow much, but they have their really heavy falls in the 11th and 12th months, when everything is shut up. They retire to their huts then, and remain in them for four months, seldom ever going out to visit their next door neighbour, 50−100 yards away. All the animals too are shut up in the huts and feed on the dried grass the locals are now busy storing. It is a great problem how to dry our seeds in this weather. I have most laid out in blotting paper, which is changed and dried as often as possible, and the seeds moved about on it. If out of the capsule they dry thus pretty quickly, but rhododendrons take a long time.'[48] Rain and mist and sleet and heavy snow accompanied him all the way to Sanga Chöling where he had his first fine spell for nineteen days and where 'Rosy Cheeks' looked 'as much a peach as ever' having removed the 'kutch' from her face. Ludlow and Taylor ungallantly must have taken her unawares — shame on them!

And a great shame too that the outbreak of World War II largely nullified the expedition's efforts to introduce to gardens in Britain, by way of hundreds of packets of seeds and dozens of living plants, so many of the floral treasures of the high passes of South-Eastern Tibet. Gardeners now had more to think about than the cossetting of rare alpine plants. But if so few of such Ludlow, Sherriff and Taylor plants remain with us, the four thousand gatherings of dried plants, now deposited in various herbaria throughout the world, are a lasting memorial to the tremendous scientific value of this remarkable ten-month expedition.

Naturally, with the outbreak of the war Ludlow and Sherriff had to abandon, temporarily at any rate, their ideas for further expeditions in SE Tibet. They were both very active men, Ludlow aged 53 and Sherriff 41, and both were desperately anxious to help Britain's war effort as best they could. They were both accepted for military service in India where Sherriff married Betty, the youngest daughter of the Rev. Dr Graham of Kalimpong. In the following Interlude Mrs Sherriff writes of life in Lhasa where she and her husband, and before them, Ludlow, were stationed.

Meconopsis sherriffii

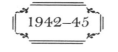

LHASA,
The War Years

o

The outbreak of the Second World War found Ludlow and Sherriff in Scotland, staying at a Shooting Lodge in Perthshire with Geordie's brother, Chris Sherriff, and his family. Obviously all plans for further plant collecting expeditions in the foreseeable future had to be abandoned, the shooting party broke up, and Ludlow and Geordie hurried south to London. Ludlow set off for Kashmir as soon as he could make his travel plans and Geordie started to apply for a Medical Board to get back into the Army.

Unfortunately Geordie had strained his heart rather badly on the 1938 expedition to South East Tibet when helping one of the coolies to carry his load over a high pass. As a result, and on his doctor's advice, he had reluctantly agreed to go slow for some months and before joining his brother in Scotland had been staying with his sister Mary and her husband in London. After failing several Medical Boards in London he decided that he might be more successful in India and so made plans to return there as soon as he could get a passage. To his great relief and joy he passed the Indian Medical Board and was posted to Digboi in Assam to command an AA Battery of Indian Troops.

Ludlow, returning to Kashmir, was appointed Joint Commissioner in Ladakh in 1940 and continued in this post until his appointment as Additional Assistant Political Officer in Sikkim. Early in 1942 he was sent to Lhasa with the special task of persuading the Tibetan Government to allow supplies to pass through their country to help China's war effort and to organise the transport for this undertaking — by no means an easy assignment.

About this time Sir Basil Gould, Political Officer in Sikkim, Tibet and Bhutan, applied to the Army for Geordie's release so that he might organise the Tibet wool trade in Gangtok and Kalimpong and also help in the administration of the quota of such things as cloth, kerosine oil, sugar etc. which HMG had agreed could be exported to Tibet — another far from easy job.

I was living with my father in Kalimpong at this time and helping him in the work at the Dr Graham Homes for needy Anglo-Indian children which he had founded in 1900. Geordie and I had very good mutual friends in the Dorji family who had their headquarters in Kalimpong, and so it was very appropriate when we married on 3 November, 1942, that Raja Dorji, familiar in this story as Tobgye, the Prime Minister of Bhutan and a friend since my childhood's days, should be our best man. We were married very quietly in the Homes' Chapel built in memory of my mother.

For a time we moved between the Residency in Gangtok and our house, Tashiding, in Kalimpong, until, to our great joy, early in 1943 Geordie was appointed to take over from Ludlow in Lhasa. We had a great many preparations to make before leaving including a visit to Delhi for Geordie to be vetted and to be given instructions by the Viceroy, Lord Linlithgow. This turned out to be a very informal interview for a catapult lying on the Viceroy's table made a bond between them and I believe some sort of competition took place out of the Viceregal window!

The Journey to Lhasa

We set out on 10 March and reached Lhasa on 9 April, a date which had been agreed on by the Tibetan Government as being a propitious one for our arrival. Our caravan was a mixed and fairly large one consisting of 42 pack mules, 5 porters, 4 riding mules, 4 servants, 2 ducks called *Tunis* and *Bizerta,* 2 guineafowl, *Jill* our labrador, our two ponies *King Rabden* and *Gallingka*, and our two selves. After final farewells in Gangtok we started on the two-and-a-half day climb to the Nathu La on the Sikkim-Tibet frontier, with all our adventure before us.

As it was early spring we expected a good deal of snow and the conditions to be difficult and experienced these in full measure on the last march before crossing into Tibet. Here we had to have a gang of about 20 coolies clearing the road before us right up to the top of the pass and down the other side. Our ponies sank into the snow and we had to dismount and struggle through the drifts urging and pulling them along. The pack mules had to be unloaded and the loads manhandled over the pass and down the other side. How thankful we were to reach Chumpitang Bungalow and the warmth of a glorious fire of rhododendron wood. Our expected scratch meal and the bungalow curtains as bedding did not materialise as our transport arrived about 6 p.m. Thanks to the coolies' stout efforts we enjoyed a splendid meal and a comfortable bed.

From Chumpitang we marched to Yatung where we halted for

four days and where we stayed in the Tibetan Trade Agent's bungalow and were very glad of the chance to relax and organise before setting off again. As yet there were not many flowers to be seen but *Primula gracilipes, P. denticulata* and an attractive little gentian gave us great pleasure and encouragement.

Leaving Yatung we were very fortunate with weather and had two perfect days as we rode up the Chumbi Valley and on to the high plateau and to Pharijong. The sun beat hot on our backs and the sky was an indescribable Tibetan blue. There was always something new to look at — Laughing thrushes, grossbeaks, Himalayan blackbirds and, encouraged by the warm spring sunshine, even some Camberwell Beauties.

Phari has often been described as the dirtiest town in the world and it is certainly true that the entrances to some of the houses, with the rubbish of years piled up around them, are reminiscent of those of subterranean caves. But at least the windswept plain on which Phari stands is ringed by a magnificent panorama of snow mountains which were arrestingly beautiful in that still sunny weather. Chomolhari, 'Goddess of the Mountains', took pride of place. There were no flowers as we travelled on but we saw hares, gazelle and a fox and one day a herd of kyang — Tibetan wild asses. It was a relief to reach the Rest Houses each afternoon for the wind and very unpleasant dust storms became very troublesome and it was bitterly cold. Happily a blazing fire always awaited us; dried yak dung pats were now used as fuel and very effective they were.

We rode on to Dochen where the lake was every shade of turquoise and green and thousands of Bar-headed geese, pintail, mallard and Brahminy duck haunted the water's edge. We tried to get near enough for Geordie to take ciné shots but all rose in a cloud at our approach and the whole sky was filled with the beat and swish of their wings and with the honking of the geese. Later a few braver birds returned and were photographed.

In continuing good weather we rode along the side of Dochen lake and down the valley to Kala. We passed one village, a haphazard sort of affair, in which it was difficult to distinguish between the crazy-looking stone walls put up to provide shelter for the sheep and yaks and the equally crazy-looking walls where, surprisingly, someone's front door appeared! There were a few fields round the village and up in the hills flocks of sheep and yaks were grazing — I could not think what on!

Gradually, as we approached Gyantse, villages became more frequent and at some the villagers were busy carrying manure to the fields on their donkeys. We passed several caravans of yaks and

bullocks and one mule caravan carrying wool down to India. One day's march from Gyantse we rode into 'The Red Idol Gorge', a fierce looking bit of country where a battle was fought in the 1904 Younghusband Expedition. The surrounding country was dry and rocky and we saw very little in the way of flowers but there were Rose finches, magpies, dippers, hoopoes and Tibetan partridge along the way.

As we were about to enter Gyantse, which is the third largest town in Tibet and the headquarters of a detachment of the Indian Army as well as the end of the Trade Route over which the British had control, a mounted infantry escort led by Colonel Gloyne and Captain Davis with Captain Saker of the Indian Political Service rode out to meet us. The Union Jack floated against the blue sky and the mounted infantry made a brave show. We felt very travel-stained and dingy in comparison. After Geordie had inspected the Guard of Honour we remounted and rode on to the Agency where we spent five very happy days.

Apart from a nightly hot bath in a decent sized tub, I think that during these days the event I enjoyed most was a Variety Show produced by the troops as part of a series of farewells to Colonel Gloyne who was about to return to India. This was staged one evening in the courtyard of the Agency. Unfortunately a tearing sandstorm blew unmercifully during the whole proceedings but the actors carried on undeterred. The whole show was fairly impromptu anyhow and there were long pauses while the blowing curtain was held down firmly against the whirlwind and the next act of the drama was rehearsed. Meanwhile rum was handed round and quickly became a turgid fluid with clouds of sand blowing into it! The actors all came from the heat of South India so their fortitude in carrying on under these arctic conditions was very remarkable. It was a really good show.

Another vivid recollection was the visit we made to the very fine monastery with an impressive eleven-storied chorten nearby and an ancient fort behind.

On 1 April we started out on the remaining nine marches to Lhasa with a considerably increased caravan of stores for the Lhasa Mission. Our transport animals were a mixture of donkeys, bullocks and yaks with an odd mule or two all in the charge of a few porters. Except for the first night at Gobshi, where we slept in our

70. The Potala, Lhasa
71. Golden roof of tomb of the 13th
    Dalai Lama, the Potala

tent, we now put up in Tibetan houses. In these the ground floor is usually given up to animals, mules, ponies, pigs and chickens and is very dark. A precipitous ladder leads up to the next storey where the kitchen, sleeping quarters, and very often a small chapel, open off the central area open to the sky. Here weaving, butter making, mending etc. take place. A strong smell of rancid butter mixed with the smell of yak dung fires hangs over everything. We soon discovered that it was a good idea to pull our camp beds into the middle of the room and well away from the cracks in the walls. The 'loo' could either be a slit in the floor of a small cupboard-like room or the same sort of accommodation up on the roof. The different houses we stayed in varied considerably in the matter of cleanliness but never in the hospitable welcome we were given.

Two marches took us up a narrow gorge-like valley where nothing grew except a scrubby juniper. The little river was frozen and as we came to the Karo-la Pass (16,600 feet) there was a magnificent glacier to the north from which came a biting cold wind which made us hurry down the long and stony track to the Nang-kartse plain and the vivid beauty of the great Yamdrok Tso Lake. We rode along by the shores of the lake for one-and-a-half days, watching great numbers of geese, mallard, pintail, Brahminy, teal and gulls swimming in the brilliant blue water and feeding on the water weeds near the lake's edge.

6 April found us over another high pass and down to the welcome height of 12,000 feet. Here we found spring had arrived and the peach trees were in full bloom. It was a relief to pull off layers of woollies and to have a good wash and clean up.

Next day we crossed the Tsangpo river in the flat-bottomed wooden boat called a 'tru'. It was surprising how many laden mules and passengers were ferried over at a single crossing. That night we slept in an exceedingly nice house at Chushul lent to us by Tsarong Shape, one of the most important Lhasa officials and a good friend of ours. Peaches, walnuts, poplars and willows flourished in the garden and made a wonderful contrast to the hard windswept country we had left behind. Villages were more frequent now and gaily dressed yaks were being used to plough the little fields.

Lhasa

One more day's march and then our final ride up the Kyichu river until we turned the last corner and found the wide Lhasa valley, ringed by high hills, opened up before us. Dominating everything was the Potala, the Dalai Lama's Winter Palace, dramatically built on top of a small hill and crowned with the golden roofs of previous Dalai Lamas' tombs gleaming in the spring sunshine. It was a thrilling sight. Quite soon we were met by Ludlow, who, accompanied by Reggie Fox (Wireless Operator) and other members of the Mission Staff, by Tibetan friends, and by Daud, Geordie's Kazak cook from old Kashgar days, had ridden out to welcome us. We all rode on to Dekyi Lingka which was to be our home for the next two years.

I was relieved to find that Dekyi Lingka, a typical Tibetan building, was situated on a backwater of the Kyichu and had ample open ground around it. It was about a mile from Lhasa City. The Norbu Lingka, the Dalai Lama's summer residence, attractively set in a walled garden, lay down the valley. Behind us was the magnificent Potala, and the Chok Po Ri or the Medical College, at the top of a precipitous rocky hill, a little downstream. But there was barely time to take in our surroundings for almost at once officials, gloriously attired in every shade of Chinese satin brocade, started to call. They brought farewell presents for Ludlow of Tibetan rugs, skins, etc., and welcoming presents of grain for our ponies, and sundry sheep and eggs (90 per cent bad) for us. Every caller had to be regaled with cups of tea and cream cracker biscuits —no other brand of biscuit was considered correct!

In between callers, Ludlow tried to hand over to Geordie, and I tried to unpack, but before long we were all back at tea and cream cracker biscuits. In addition to all this there were farewell parties for Ludlow and welcoming parties for us and the eight days we had together passed all too quickly in a daze of hospitality.

Ludlow was not sorry to leave Lhasa. He had become very tired of all the parties and social engagements without a wife to assist him and being very busy with Government affairs had not been able to escape from Lhasa for even a few continuous days of respite. Of course he had made many good friends among the officials some of whom had been his pupils in 1923–26, when, in Gyantse, he had taught the sons of Tibetan officials the rudiments of western education. We were to benefit greatly from the good relations he left behind and all that he had done to improve the house and garden. Feeling was growing distinctly more tense between the Tibetans and the Chinese and Ludlow's experience, wisdom and tact had been given full play. For instance I recall how during the autumn of 1942, Ludlow had entertained two Americans, Tolstoy (a grandson of the famous Russian author) and Brook Dolan, who were emissaries from President Roosevelt from whom they brought a letter and presentation to the Dalai Lama. At first the Tibetan Government had been unwilling to give permission for the Americans to come to Lhasa, but finally having done so, due to Ludlow's persuasion, they

treated them very hospitably and later gave them permission to travel to Jyekundo and on to China.

Throughout his year in Lhasa Ludlow continued to collect plants. Sometimes he himself would make a short day expedition to the surrounding hills. At other times he would send his faithful Kashmiri servant, Ramzana, and a companion, to areas further afield. Their most dramatic find was *Meconopsis torquata*, a magnificent blue-flowered meconopsis growing at a height of 15,500 feet in boulder scree. This meconopsis had originally been found by Captain H. Walton on the 1904 expedition and the flower had been described as being red. Geordie and I found it again in July 1943 north of Lhasa growing in similar boulder scree. It was much more common here and grew between 14,000 and 16,000 feet — sometimes in open situations and sometimes in complete shade. Unfortunately it has not survived in cultivation.

It was a sad moment to watch Ludlow drifting down the Kyichu in a yakskin kowa on the start of his long journey back to India and Kashmir. Later he resumed his work as Joint Commissioner in Ladakh.

However there was little time for regrets for there was so much to be done. Geordie's meetings with the Foreign Office, sometimes held in Lhasa and sometimes in Dekyi Lingka garden, kept him very busy as did the decoding of the long wireless telegrams from the Government of India. Reggie Fox, or 'Foxy' as we called him, operated the wireless from a little house at the side of Dekyi Lingka where he lived. He had been in Lhasa for several years and was to stay several more. He dined with us each evening and sometimes we played three-handed bridge.

One of the aims of the Mission was to maintain good relations with the Tibetans and help them in every possible way. To further these aims the Mission ran a school and small hospital both of which were well patronised. Dr Terry, ex-Medical Superintendent of Burmah Shell and now a refugee from Burma, was in charge of the hospital for some time and did a fine job.

One of our first official engagements was to visit the Potala to pay our respects to His Holiness The Dalai Lama, a boy of eight at this time. It was a memorable experience. Before we entered the Audience Chamber we had a short wait in an ante-chamber where rows and rows of red-robed monks sat on either side of the long central aisle leading up to the massive red door of the Audience Chamber. Two of the Dalai Lama's Personal Bodyguard sat on

72. *Meconopsis torquata*

either side of the entrance. They were enormous men with well padded shoulders, vast boots with well stuffed toes and suitably deep giant-like voices. Every now and then they left their seats and stalked through the crowd to keep order. Suddenly the great door was flung open and we entered the Audience Chamber. The boy Dalai Lama was seated on a throne at the far end. He wore a yellow pointed monk's cap and his red robes were partially covered by a fur-trimmed golden brocade cape. The atmosphere was hazy with incense, and the painted walls and ceiling, with the dull gleam of images and the vague outlines of beautiful tunkas (religious paintings) set in brocade banners in the background, were only faintly visible.

We processed slowly towards the Dalai Lama. Several monks were ahead of us and immediately behind me as I followed Geordie came one of the Potala servants carrying our presents of various silver dishes, a Hornby train set and a clockwork speed boat. When Geordie was within some fifteen paces of the throne he opened his white silk scarf, placed it across his arms, and was given a silver dish containing images made of butter. This he carried to the Dalai Lama who just touched it with his fingers and waved it on to a monk standing on the right of the throne. Geordie was then given in turn, an image, a book, and a chorten, all of which he presented in the same way. Finally he presented his scarf and saluted. The little Dalai Lama's eyes sparkled with fun and pleasure at the sight of Geordie's uniform and scarlet Gunner cap and was obviously much diverted at seeing us. And his eyes opened wide with surprise and delight at the sight of the Hornby train and the speed boat. I then approached the Dalai Lama and presented my scarf, bowed, and followed on to do the same to the rather tired-looking old Regent, after Geordie had finished his presentation of tokens. We then walked past the Dalai Lama's father to sit on two cushions opposite the Regent. Our Tibetan guide now stepped forward before the Dalai Lama and tasted a cup of tea, which we had provided, to show that it was not poisoned. He then prostrated himself three times before joining us. The First Ceremonial Tea was now handed round and we took a sip of it. The whole ceremony usually is carried on in complete silence but on this occasion the Dalai Lama said, in Tibetan, that he hoped 'the Representative of the British Government, Major Sherriff, was in good health'. After this short speech had been translated for us we rose from our cushions, saluted and bowed, and received a very nice smile. A great basin of rice was carried round and we all took a few grains and threw them in the air as an offering. This was followed by the

Second Ceremonial Tea and then one of the 'Giants' announced the end of the audience by saying 'Arise, Arise' in a very deep, gruff voice. The simple dignity with which this boy of eight carried out his duties of blessing the long procession of monks and pilgrims was very remarkable and it has left an unforgetable impression on my memory.

Before we were properly settled in we had to begin a long round of return calls to the various officials. Some of them lived in large houses with gardens on the outskirts of the city but others lived in Lhasa itself. We never lingered longer than was necessary when we rode through the streets for there was very little attempt at a proper sewage system and the stench was nauseating — only the cold, dry climate saved appalling epidemics. The plight of the beggars and the pie-dogs was pathetic to see. The officials' houses were clean and well kept and we received a kind and friendly welcome always. We gave them the presents we had brought from India — lengths of serge, tea sets, rifles, watches, make-up for the ladies, etc. I was disappointed never to know if the officials were pleased with our gifts for it was considered bad form to look at the presents or show any enthusiasm.

Our Tibetan was slowly improving and we were glad that we did not have to rely entirely on our interpreter.

Before long the season of parties started. A Tibetan party can be rather a daunting business for it may last for as long as a month. We only accepted luncheon engagements though even this would usually mean a four to six hour visit, starting with the usual tea and cream cracker biscuits.

One of our first parties was at the Foreign Secretary's house, Surkhang Dzasa. The menu, written in both Tibetan and English, was as follows: (1) Sweet peaches, (2) Fish stomach mixed with milk, (3) Chinese bacon; bread, (4) Sea Fish, (5) Sea snake, (6) Fried Chicken, (7) Dumpling with meat inside, (8) Sea slugs, (9) Black fish, (10) Yellow flowers soup, (11) Rice pudding. Bowls of Gyatu noodles were also part of the meal. Chang, the local barley beer, flowed freely and on this occasion Enos Fruit Salt was passed round to add to the Chang because they liked the fizz!

After the meal the guests played mahjong or cards or just chatted.

At one party where we had been invited not only for lunch but also for an evening meal for which we could not stay, we were given both meals at the same time! A terrific challenge.

When we started to give our own parties we found that not only had we to feed our guests, say 30 officials and their wives, but also

their servants and attendants who would bring the numbers up to about 70. On one occasion we had four consecutive days of parties and we estimated that the number entertained was 250 including servants. To feed the multitude we used 2 sheep, 60 lbs yak meat, 4 yak tongues, 30 dozen eggs, 240 lbs rice, 25 lbs butter, 20 lbs sugar, 40 lbs flour and various other minor items. Daud, our cook, was in his element on these occasions. He had built a splendid oven in which he could roast a sheep whole and this was brought round on a sort of stretcher. Daud, complete with a chef's hat and a voluminous white apron stretched across his ample frame, with carving knife in hand, would accompany the sheep and would slap his leg or shoulder or point to his ribs to find out which cut the guest would like. One of Daud's assistants followed with a wash-hand basin of gravy and a large soup ladle!

We had brought a large quantity of seeds and bulbs for the garden and these proved most successful. An old man called Agoo (his right hand had been damaged by a bullet in the 1904 Expedition, which made me feel very guilty) and his assistant, Pema, spent days journeying to Kundeling Monastery and carrying back basket loads of black soil and sweepings. The Abbot of Kundeling, who was our landlord, gave us permission to extend the garden and to include a small stream. Using this water, Geordie constructed a most ingenious water wheel made from two old pony trap wheels and old cigarette tins and this was a great help in watering the garden. The stream was very pleasant too, especially for our two ducks *Tunis* and *Bizerta*. The latter, however, rather defeated our hope of nice fresh eggs by changing her sex and becoming a drake! A small greenhouse we constructed was a great asset for bringing on seeds in the winter and was copied by several of the keen gardeners among the Officials. We carpeted the banks of a small water channel we made with turves of *Primula tibetica* and a small starry gentian and made a corner for different plants dug from the surrounding hills. These included *Primula jaffreyana*, *P. waltonii*, *P. sikkimensis*, *Gentiana waltonii* and a variety of androsaces and irises. Our Tibetan guests did not recognise these as being components of their own flora for they did not wander about the hills searching for plants. The exceptions were some of the monks who collected seeds and herbs for use as medicine. Vividly do I remember a time when visiting the Chok Po Ri we saw a row of monks sitting cross-legged

73. Serpang ceremony, Lhasa
74. Serpang ceremony: the silken Tunka of
    the Buddha on the wall of the Potala

in the sunshine making up rather sinister-looking pills, each adding some item as the pills were handed along the line.

Our garden became a great asset. Though the Tibetans did not collect plants from the countryside they love flowers and many of the poorest little houses in Lhasa had odd pots and tins planted with such flowers as geraniums, marigolds and antirrhinums. A monk once remarked to me that the temperature of Lhasa had risen considerably since brilliant scarlet geraniums became so popular!

When we visited the Norbu Lingka we were shown the peacock whose life had been saved by Ludlow. He had been called in to see the poor bird when it was very low and dispirited and was expected to die. Ludlow found that it was being fed on a very wet mash of tsampa (parched barley meal) and advised that a good handful of small gravel be left in the cage. The result was spectacular and Ludlow's fame was established.

Each year a number of Processions and Ceremonies take place and all Lhasa turns out to see them. In the autumn and spring the Officials, monk and lay, gorgeously garbed and mounted and each with an attendant leading his horse, accompany the Dalai Lama from the Norbu Lingka to the Potala and vice versa. We watched this mile-long procession called the Chipgyur, a glorious riot of colour and pageantry, from the slopes of the Chok Po Ri hill that first spring. Fortunately for us no Tibetan is allowed to look down on the Dalai Lama and so the crowd which surged up towards us was quickly dealt with by a fierce little man who bounded up the hill wielding a whip and a long bamboo pole.

The scented smoke of incense burning along the roadside as well as the dust which rose in clouds as the procession passed gave a dream-like softness to the whole pageant. The crowds lining the route were completely silent. Relays of palanquin bearers gaily dressed in bright green coats, scarlet hats, white trousers and black velvet boots, were stationed along the route. Early in the procession came an escort of mounted monks headed by one of the 'giants' of the Dalai Lama's Personal Bodyguard. They were in charge of the Dalai Lama's luggage and his pony covered with a cloth of gold brocade. Lower-ranking lay officials, beautifully dressed and wearing an amazing variety of headgear, ambled past and their monk counterparts followed with yellow satin showing below their sober red robes. They wore papier-mâché hats decorated with gold. And now came the sound of drums and trumpets as the Dalai Lama's procession drew nearer. Outriders, brilliantly dressed in yellow satin and in scarlet hats rather like fringed lampshades, were followed by more monk officials with saddle cloths of multicoloured brocades. Hard on their heels rode a long line of horsemen each bearing a brilliant banner and each with a large hat of a different colour. Mounted drums preceded a line of red robed lamas with their sacred books, wrapped in saffron yellow, strapped on their backs.

The Golden Palanquin, with the figure of the young Dalai Lama just discernible, passed in a riot of colour. Servants in yellow and saffron walked and ran alongside and members of the Cabinet and high-ranking officials followed in strict order of rank.

Words fail to describe the rich beauty of their satin robes and splendid hats as well as those of the lower-ranking officials who followed. We were left speechless and quite transported by this pageant of such amazing colour and beauty.

We were brought back to earth, however, by companies from the three Tibetan Regiments who passed below us as they played themselves back to their barracks. Each had a band and there was a strange medley of bagpipes, fife and drums and Tibetan music; one of the tunes they played was 'God Save The King'.

Later in the year, in the 10th Tibetan month, is held the 'Festival of Lights' which celebrates the death of Tsongkapa, the founder of the Yellow Sect of monks. On this night Lhasa takes on a fairyland appearance with hundreds of little lights, in earthenware containers, flickering along window ledges and edging the flat roofs of the houses. Special lamps are also lit on the family altars. How we wished we could have floodlit the Potala!

At the end of the Tibetan New Year celebrations, in our February, we attended the Serpang Ceremony which takes place in a courtyard below the Potala. A magnificent 300 feet silken tunka or banner is hauled up the wall of the Potala and a splendid day-long spectacle of monk processions and dances takes place. The highlight of the day is the arrival of the Nechung Ta Lama, the Government Oracle. He is helped up the steps to the arena by his attendant monks and is hurried across to a position in front of the marquee where the Cabinet Ministers are seated. They rise to greet him and watch as he works himself into a trance, tossing his head with its 30 lb. feather headdress and executing a funny little dance. One of the monks wipes the sweat from his face, as the officials, led by the Cabinet Ministers, queue up to present their scarves. Then he is hurried out of the arena and the ceremony continues.

In between parties and processions we were kept very busy at Dekyi Lingka. A sun room was added and this was a great asset in every way. It seemed only sense to make full use of the warmth of the sun when it wasn't easy to procure sufficient wood for our one stove. We made great use of this addition.

As long as the weather allowed us we had most of our meals in the garden and a good many of our smaller parties as well — Tibetans love picnics and meals out of doors so long as they do not sit in full sun. *Tunis* and *Bizerta*, *Jill* and the guinea fowl also loved these parties. They wandered round the tables with a riveted expectant air and were not disappointed.

In the summer we entertained our guests by introducing croquet and table tennis. The former went down especially well with the monks who found their long robes wonderful cover for manipulating the ball into a better position !

Occasionally we were fortunate in being able to take short trips out of Lhasa. One of the most interesting of these was a five day visit to Ganden Monastery which lies about 24 miles due east of Lhasa and is the third largest monastery in Tibet. We crossed the Kyichu in kowas at a ferry four miles upstream, but our ponies had to swim. It was late September, most of the flowers in the valley were past, the villagers were busy harvesting their barley, mustard and peas, and several caravans of mules were bringing tea from China.

The 2,000 feet climb to the monastery was on a rather dry stony hillside but once we got over the brow of the hill we came into much more promising country. We pitched camp on a grassy terrace beyond the monastery where there were some trees and a magnificent view of the Kyichu valley winding away eastwards in the direction of Kham and China. During the evening we had visits from three groups of monks who brought us presents of eggs and grain and had tea with us. We had been told in Lhasa that Ganden water was bad and so Geordie arranged for a daily supply to be brought up to us from the main river nearly 2,000 feet below !

Next day we visited the monastery and were met on the outskirts by two Simgas, monastery policemen, with well-padded shoulders and yellow hats; they reminded us of Roman centurions. Two attendant monks carried their maces made of polished iron decorated with copper and brass and with a bunch of different coloured strips of silk tied to one end. The Simgas led the way into the monastery where three rather wild looking lay-servants, with sticks, walked beside us shouting something which sounded like 'Ho Chow Chow' and which we were told was a warning to all the monks (5,000 of

75. Monks making pills at the Chok Po Ri Medical College
76. Sherriff's water carriers *en route* to
the Ganden monastery

them) to keep out of the way. Two abbots, the heads of the monastery, ushered us into a little chapel where tea and the inevitable biscuits and some rather partworn-looking boiled sweets were laid out for our refreshment. Bowls of boiled rice topped with a sprinkling of sugar were produced and before we started to eat this we duly threw a few grains into the air as an offering.

Conversation was a little difficult but there was a nice friendly atmosphere and the monks told us that they were glad to hear the war was going well for us and that they prayed daily for a speedy and final victory. Geordie then handed them the present of cash which he had brought for the monastery and a succession of monks came in bearing trays of the usual aged presentation eggs. Before we took our leave and returned to camp we were shown round some of the buildings but unfortunately did not see the famous Tsong-kapa's tomb.

We spent the next day in the best possible way — out on the hills with Tsongpen and Khanden. After riding the first couple of miles we sent our ponies back. The going was perfect and the grassy slopes most beautiful with a blue haze of small gentians and cyananthus. Higher up we found two more gentians, *Gentiana tubiflora* and *G. przewalskii*, and collected seeds of *Incarvillea younghusbandii*, *Paraquilegia, Meconopsis integrifolia* and *Primula sinopurpurea*. The autumn colours were very fine, glorious splashes of yellow, orange, rust and red, vivid gainst the grey rocks. We were now at about 16,000 feet and the panorama of blue hills and far reaching valleys was magnificent. To the west, down the shining ribbon of the Kyichu, were the hills beyond Lhasa. To the north, across endless folds of blue hills, we could see the snow covered peaks of the Nyenchengtangla Range. To the east, Geordie picked out the great snow masses of Gyala Peri, Namcha Barwa and the Himalayas in Kongbo. All Tibet seemed mapped out at our feet and over all the land was stretched a canopy of brilliant blue sky.

Our shadows grew long as we walked down the grassy slopes to our camp, weary but exhilarated with all the sun and beauty of the day.

Next morning we walked down to the river where two kowas were waiting to take us back to Lhasa. It was amazing how we all fitted in! ; eleven people, seven mule loads of kit, *Jill* and the two boatmen and their two sheep. Each kowa carried a sheep as crew, so that when the return journey upstream, along the bank of the river, was made, the boatman carried the yakskin coracle on his back and the sheep carried the boatman's blanket and food! I was amazed how tirelessly the boatmen rowed and only occasionally took a breather for a pinch of snuff or a cup of river water or chang. They either sang or kept up a sort of wheezing noise all the time. The sheep lay peacefully at the bottom of the boat and shared his master's tsampa when he had his meal. We reached Lhasa in 6½ hours.

Though the summer months were the busy ones we were by no means idle in the winter. We spent much time curling on the frozen Kyichu using flat stones we picked up on the river bank. The game became very popular and 'Foxy' and our Tibetan friends became very keen and we enjoyed some splendid sport. While we curled *Jill* dug for mousehares, unsuccessful but undaunted!

Cinema shows also were very popular although our supply of films was rather limited owing to wartime restrictions. Our guests never tired of Charlie Chaplin; and never tired of seeing themselves in the films which Geordie took of them. The resuscitated tennis court and the introduction of basket ball for members of the Mission Staff and schoolboys were other activities enjoyed.

Nothing much was enjoyed in January and February when we had horrible dust storms and the sand penetrated everywhere.

At all times of the year we found endless interest and entertainment in watching the many different kinds of duck and the geese which visited our backwater just over the garden wall. These included teal, Brahminy, mallard and goosander as well as gulls and cormorants. One October evening a flock of about 100 Black-necked crane flew over the house and landed by the river. They flew off next morning. In December thousands and thousands of Bar-headed geese swept over us on their migration to India, and the sky was filled with their haunting cries.

Very occasionally more than birds swept over us and at the end of October 1943, the sound of an aeroplane flying over Lhasa at night was the source of great surprise and excitement for the Tibetans who had been told by the monks that no plane could fly over Lhasa and look down on the Dalai Lama and escape disaster. So, of course, no one was surprised at the news that the 'object' had dropped out of the skies near the little town of Tsetang and had crashed on the banks of the Tsangpo. Later the 'object' was identified as an American plane. The Americans had been on a routine flight from China to Assam, flying the 'hump', and had lost their way in very bad weather. Finding they were running short of petrol they had baled out just before the plane crashed and fortunately only one airman was injured — and that slightly. Reports came to the Tibetan Government that the airmen had Chinese writing on their backs and this led the Government to believe that they were Chinese and not American. The writing, it was said, was to ask anyone who found the airman to look after him as he was an ally of China. When this matter was cleared up the five airmen were allowed to come to Lhasa and to stay with us. We were very amused at their surprise and dismay when they found they could not just 'whistle up a jeep' but had to ride and walk for 24 days to reach the nearest railhead. We much enjoyed having them with us.

In fact we much enjoyed the company of anyone who came to stay with us. I remember particularly Sir Basil Gould paying us a ten-day visit during which we had a tremendous surge of parties in his honour. Other guests included Colonel Hislop of the Indian Medical Service and Mrs Hislop, as well as Hugh Richardson who came to relieve Geordie during a short trip we made to India.

Our two years in Lhasa passed all too quickly and we were very sorry when the time came for us to return to India. We had found the Tibetans exceedingly friendly and kind, their sense of humour very like our own, and we had been happy working and living among them. In the years to come we were also to be happy showing the ciné films of the various Ceremonies and Processions which Geordie had made.

Fortunately our departure was made easier by the fact that the longed-for end of the war was at hand and that Geordie and Ludlow were already planning their next plant hunting expedition. They were hoping it would be to south-east Tibet.

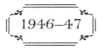

Chapter Six

The Gorges of the Tsangpo,
S.E. TIBET

°

The war at an end, Ludlow met the Sherriffs in Kashmir, in 1945, to discuss yet another expedition to those rich plant-collecting grounds in SE Tibet. In 1938, Ludlow and Sherriff, with Taylor, had reached Gyala at the entrance to the Tsangpo gorge in the province of Kongbo. How splendid if the three friends could now carry their explorations still further eastwards into the gorge country in the almost unknown province of Pome.

Those who have read so far will have realised that the main Tsangpo river, which has its origin near Mount Kailas in Western Tibet, flows eastwards behind, and parallel to, the main Himalayan range for a thousand miles, seeking to pierce the huge mountain barrier and to escape to the sea. In the vicinity of 95° E the mighty river, a mile wide near Lhasa, becomes confined to a narrow gorge down which it thunders in a stupendous cataract. Some fifty miles below Gyala there comes in from the north the Po Tsangpo, a tributary worthy of the Tsangpo itself, draining the whole of Pome and Potü. It is born of two rivers, the Nagong Chu and the Yigrong Chu which, flowing respectively from the south-east and north-west, meet almost head on at Tangme. Such was the area Ludlow and Sherriff now proposed to investigate, planning to reach their base at Tongkyuk Dzong by Christmas so as to be in time for the earliest of the flowers in the low-level gorge areas.

Unfortunately Taylor was not free to join them; he could not be released from work connected with the reconstruction of the Botanical Department of the British Museum. In place of Taylor, Colonel Henry Elliot of the Indian Medical Service most willingly agreed to join the expedition.

But first, permission had to be sought from the Tibetan Government to enter the country the following year and the precious passport, or *lamyig*, was issued on the '21st day of the 7th Tibetan month of the Fire-Dog Year' — 17 September, 1946. Translated it read thus. 'All the dzongpens and the headmen in and along the route from Gyantse to Gongkar, Tsetang, Tsegang, Tong-Jug, Po-Me and Yirong should know that British Officers Mr Ludlow, Major and Mrs Sherriff, Doctor Sahib and nine servants have been permitted to visit the above mentioned places for sight-seeing and to collect flowers. Their transport requirements of eight riding ponies with saddles, and sixty two pack animals (donkeys or yaks) should be supplied immediately from dzong to dzong as a special case, by all concerned, at the rate of twelve sangs per pony and six sangs per pack animal for a day's journey and receipt of the hire paid should be noted at the foot of this passport. Accommodation, kitchen and stable coolies, fuel, eggs, milk, and grass and grain for private ponies, which are actually required, should be supplied on payment of reasonable price at the local prevailing rates without any restriction for once only during the journeys to and from these places.

'The visitors and their servants must not visit any other places excepting those specified above and must not shoot or trap animals and birds existing on sacred and other hills, valleys and in rivers. They are also not allowed to beat or ill-treat the Tibetan subjects during their visit. All the dzongpens and headmen should see this passport and carry out the orders faithfully.

'Given this day the 21st day of the Tibetan month of the Fire-Dog Year (corresponding to 17th September 1946).'

Unfortunately the Tibetan Government was unwilling to issue the passport in duplicate and this was to have serious consequences.

When Ludlow and Elliot joined the Sherriffs in Kalimpong in the autumn of 1946 they found that practically everything was in readiness for the expedition, the Sherriffs having spent a busy six months organising stores and equipment and in other ways preparing for the long journey to Pome. By 14 October all plans had been finalised and with their staff of a Tibetan headman, two Nepalese cooks, four Lepcha plant collectors, the two Labrador dogs Joker and Jill, and travelling by way of Sikkim's capital, Gangtok, and the Nathu La, they reached Gyantse in Tibet on 29 October.

'All told we had a hundred loads, which sounds rather luxurious travelling, but it must be remembered that we were a large party going away for at least a year, and that a quarter of our baggage was composed of botanical and ornithological equipment. However, although we did not carry many of the luxuries of camp life, we certainly did not omit many of its necessities. There is a happy medium in most things and Sherriff and I have always tried to strike this balance in our travels. If necessary, we were always prepared to wash our faces in the frying pan and sleep on the cold cold ground, but we never rejoiced in such austerities, and preferred to forego them whenever it was possible to do so. Transport and funds are, of course, the crux of the whole problem.[1] If transport is abundant, as fortunately it is in SE Tibet, it is a good policy to be as comfortable as funds permit, provided always mobility is maintained. No Tibetan thinks the better of you for travelling like a tramp. "Only beggars walk" is a Tibetan saying. So, on this particular journey, we had our own riding ponies, camp-beds, wash-basins, and so on, and travelled in tolerable comfort.'[2]

From Gyantse the expedition followed the 1938 route along the Lhasa road and down the right bank of the Tsangpo to the 1938 base at Tse, near Tsela Dzong, which they reached on 12 December, having been on the road for two months, travelling eastwards through six degrees of longitude and having left the arid plateau region of Tibet far behind. 'Our transport varied greatly. Some-times we had ponies or mules, sometimes donkeys, oxen, or even yaks, but for the most part our loads were carried by villagers. Of all the forms of transport, I think we appreciated mules the most. With these animals we moved gaily along to the joyous tintinnabulation of bells at a good three miles an hour. Ponies were slower and less sure-footed. With oxen we crawled along at half this pace, while the damage these brain-less creatures did to our baggage filled us with dismay. Yaks, too, were most provoking at times, but, unlike oxen, they had their good points; on a really dangerous track or snow-bound pass, when conditions are at their worst, the yak is supreme.'[3]

In Tibet there are two systems of transport. One is known as 'sadzi sadzi'; transport is provided only from one village or group of villages to another and, as has been seen, when villages are close together changes of transport have to be made at frequent intervals, sometimes four or six times in the day, much to the exasparation of travellers. The other system is called 'dzongya dzongya'; demands are levied on the inhabitants of a whole administrative area and the

overleaf
77. Red Idol near Gyantse

78. Camp in poplar park at Tsetang
79. The Tsangpo near Tsetang. Elliot is in the foreground

traveller proceeds from Dzong to Dzong with the same animals. Although this system is better, it too, has its drawbacks, for although progress is smooth once the animals have been procured it may take several days for them to appear, as demands may be made on villages as much as a hundred miles apart. This was the system of transport to which the expedition's passport gave entitlement and it was not surprising therefore that there was a hold-up at Tse for several days pending the arrival of the necessary ponies.

As soon as the Sherriffs, Ludlow and Elliot had settled down at Tse, they crossed the Tsangpo in coracles and called on the Tsela dzongpen. He was a monk official, and a very charming one, who was to give the visitors much help and hospitality throughout their stay in Kongbo. He asked to see their passport and straightway agreed that the payment of twelve *sangs* per day for a riding pony and six *sangs* for any other transport animal was quite exorbitant. Instead, he insisted that no more than four *sho* should be paid for a riding pony and two *sho* for a pack animal — which was exactly thirty times less than the rate stipulated on the passport! Fully did he merit the expedition's gifts of a pair of binoculars, saffron, a photograph of the Dalai Lama, and some cloth.

Although the delay at Tse was not without profit in that it enabled Ludlow to study the birds from the point of view of assessing which were the winter visitors to this area — he was surprised to find the Three-banded rose finch (*Carpodacus trifasciatus*), the Crested tit-warbler (*Lophobasileus elegans elegans*), the Himalayan crossbill (*Loxia curvirostra himalayensis*), a nuthatch (*Sitta leucopsis przewalskii*), the paroquet (*Psittacula derbyana*) and the Eastern Spot-billed duck (*Anas poecilorhyncha zonorhyncha*) — the zero temperatures they were experiencing made them all anxious to descend to lower altitudes. But when the transport did arrive it was sufficient only for half the party. Thus on 18 December the Sherriffs, with half the expedition's baggage, left for Tongkyuk in Pome where they arrived on 22 December, the day Ludlow and Elliot were able to leave Tse with *their* transport. Both parties, Ludlow and Elliot especially, were given a superb view from the summit of the snowless Temo La (14,000 feet) of the Namcha Barwa-Gyala Peri range with the gap of the Tsangpo clearly visible, and all were able to marvel at the sight of *Primula whitei* (L S & E 12021)[4] about to flower on a shady mossy bank beside a frozen stream where the minimum night temperature was −5°F. In fact it was only with difficulty that plants were hacked from the frozen earth, Sherriff damaging his kukri, and causing sparks to fly, in the process. '*Primula gracilipes*, *P. atrodentata* and *P. pumilio* all flower in winter

Map 17. The Journey to Pome

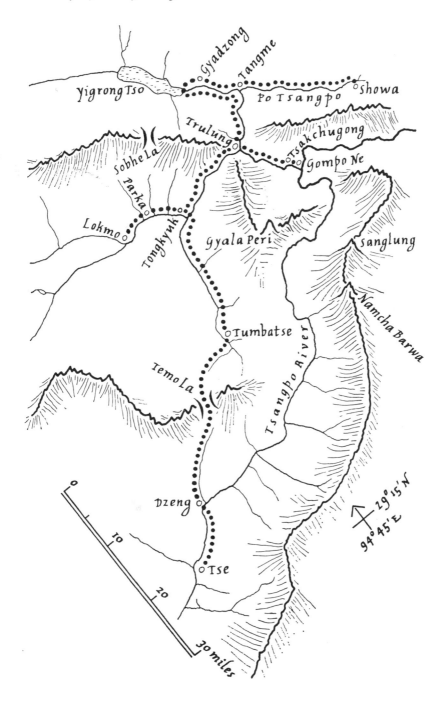

or very early spring, but I cannot recall any primula that blooms so happily amidst such intense cold as *P. whitei.'* [5]

On Christmas Day the whole expedition was reunited at Tongkyuk (8,600 feet) and dined on clear soup, roast stuffed goose, plum pudding, and a bottle of champagne, 'Pome's first Pomeroy' as Betty Sherriff described it. Had the local monk official dined and wined as elegantly he might have been more agreeable and less avaricious, for when he was told that the expedition wished to descend the Tongkyuk river to Trulung and thence to visit the Po Yigrong and Po Tsangpo valleys, he agreed to the proposals but insisted that payment for transport should be made directly to him, in advance. Suspecting that those who carried the expedition's loads would never receive their dues Ludlow and Sherriff demurred. But the dzongpen insisted and there was no alternative but to accede to his demands.

Before journeying to Truling Sherriff and Ludlow left the main party to pay a short visit to the Sobhe La (16,500 feet), a pass leading into the Yigrong valley which Kingdon Ward had discovered in 1935. They were deeply interested in the temperatures to which plants were subjected and they intended to instal a thermograph on the pass. On the main Himalayan range, at this time of year, there was a great deal of snow; indeed, above 13,000 feet, or even less, there had been snow for a month or two so that plants were protected under a warm blanket. But north of the main range, on the Temo La for instance, there was no snow at all and at nearly 14,000 feet plants must have to survive temperatures of 15–20° below zero.

This was all contrary to Ludlow and Sherriff's expectations. 'When I visited SE Tibet in the summer of 1938 I tried to picture the scene in winter and I had visions of snow lying everywhere . . . . and the whole world dumb and lifeless. The winter I spent [in Kongbo and Pome] completely disillusioned me. When we reached Kongbo all the passes on the main range, it is true, were deep under snow and would not be open until the end of June at the earliest. But this snow did not extend much below 11,000 feet and the Tsangpo valley itself was quite bare. Villagers informed us that snow rarely fell in the valley, and the little that did fall disappeared almost immediately. The range of mountains on which the Temo La, Sang La and Nyima La are situated was practically devoid of

80. Gyala Peri from the Temo La
81. Sunset on the Himalaya, near the Tsangpo gorge

snow at Christmas and on the Temo La there was not a particle to be seen anywhere. During February the Sang La and Nyima La became closed but the Temo La remained open throughout the year. After crossing the Temo La I did not see more than an inch of snow anywhere in Pome below 11,000 feet. On the last day of 1946 I climbed to within a thousand feet of the summit of the Sobhe La before I got into snow. During January, February and March in Pome the sky was often overcast and snow fell high up on the ranges, but never, except in powder form, in the valley bottoms. This absence of snow in the Tsangpo, Po Yigrong and Po Tsangpo valleys undoubtedly accounts for the large numbers of birds which are found wintering in these areas.' [6]

From a little below the entrance to the Sobhe La, a pass which Ludlow and Sherriff immediately saw would be good for flowers, there were entrancing views of the snow peaks around Gyala Peri which alone made their visit worthwhile. But their main concern was with the thermograph which was placed beside a huge boulder and amidst plants of primula, gentian, swertia, aconite and other things. It was set to work at mid-day on Tuesday 31 December, one thermometer being six inches under the ground and marking green, the other on the surface and marking red. Sherriff had arranged for a lama to change the graph every twenty-five days and to pay him thirty *sangs* for each visit. Unfortunately, for various reasons, the experiment was something of a disappointment; most at fault was the lama who managed to disarrange the red pen almost at once.

The Journey to Trulung

Back again on New Year's Day to Tongkyuk where snow-white fragrant flowers were covering the bare branches of *Lonicera setifera*, the travellers prepared for their journey down the Tongkyuk river to its juction with the Po Tsangpo at Trulung. It is but a score of miles from Tongkyuk to Trulung and in this short distance the expedition descended 2,000 feet along a track just possible for pack animals. As winter is the season of travel in Tibet and all roads lead to Lhasa, every day they met pilgrims and traders all of whom were ready to talk and happy to barter their belongings, including local wild honey, for tobacco, empty tobacco tins and whisky bottles, and matches. As honey is an excellent substitute for sugar which one longs for at high altitudes they consumed a good deal of it during the journey.

So far the vegetation had been rather dull but approaching Trulung it became more interesting, dense temperate rain forest being on all sides with open woods of *Pinus armandii* and bracken on the south-facing slopes. *Ceratostigma willmottianum* (12097) was carrying both seeds and the last of its blue flowers; the bright yellow panicles of *Senecio scandens* (12093) were draping themselves over, and seeking the support of, many shrubs in the forest; and most interesting of all was a mahonia, with a good deal of red tinting in the yellow flowers, which proved to be a new species and was named *Mahonia pomensis* (12092).

But also approaching Trulung Ludlow began to feel very unwell. 'The last day's march into Trulung [10 January] was a tiring one and when I reached camp I attributed my exhaustion to the difficulties of the road. As time passed, however, my malaise increased. I felt cold and giddy, and had a curious numbing pain at the back of my neck, and could neither see nor write properly. Eventually I lay down on my bed as I was unable to stand and sent my servant for Elliot, but he came back a few minutes later to say that "Doctor Sahib" was also unwell. After a few minutes, however, Elliot gallantly struggled across to my tent, accompanied by Sherriff who quickly summed up the situation and sent both of us supper-less to bed.

'Next morning we had completely recovered and the topic of conversation at the breakfast table naturally turned to our indisposition the previous evening. Elliot now began to suspect that we had been poisoned by the honey we had eaten, although Sherriff found difficulty in accepting this theory as he also had eaten the honey and had been unaffected. Orders, however, were issued for the honey to be thrown away. A few hours later Elliot's suspicions were confirmed in a most convincing manner, for one of our Nepali cooks, being reluctant to waste good food, disobeyed instructions, ate a large quantity of the honey, and was in due course assailed by symptoms similar to, but more severe than, those which had afflicted Elliot and myself. There can be no doubt that the honey was the root cause of all the trouble, although it is difficult to account for Sherriff's immunity, except on the assumption that, like the curate's egg, the honey was good in parts.

'The local Pobas, whom we consulted on this matter, asserted that wild honey was occasionally poisonous and added that honey deposited on rocks in the spring of the year is particularly dangerous. It is impossible to say what plant, or plants, are responsible for this honey, but a species of rhododendron is likely to be the cause. Xenophon's account of the poisoning of troops during the retreat of the "Ten Thousand" is generally attributed to the consumption of honey derived from *Rhododendron ponticum*, but this species, of course, occurs nowhere in the Himalayas.

'Bees are particularly abundant in Pome and they must be a hardy race for swarms were encountered at 9,000 feet with the thermometer registering 20 degrees of frost. They also appear to be a vicious race for on a warm afternoon in March, on our return journey, we were suddenly attacked by a swarm and had to flee to our tents for shelter. On this occasion my Labrador [Joker] was stung so severely that he became paralysed and had to be carried to my tent, where he lay in a critical condition for some time.'[7]

Trulung was an insignificant village of three or four houses, at an altitude of 6,500 feet, at the junction of the Tongkyuk river with the Po Tsangpo whose waters Sherriff described as being the colour of copper sulphate and Betty Sherriff, more poetically, 'of a wonderful deep jade green'. During the past few days Sherriff had remarked on the curious lack of rhododendrons on the hills and not until the expedition reached Trulung did they find their first rhododendron of distinction. It was not in flower of course, but its leaves, eight inches or more long with the nerves impressed on the upper surface, as well as its great, slightly curved capsule over two inches long, proclaimed it to be what for a time was called *Rhododendron sino-nuttallii* (12117) but which is now regarded as being a magnificent form of the Bhutanese *R. nuttallii*, itself a most glorious plant; in four or six weeks' time the fat flower buds of 12117 would have opened to six-inch, white, yellow-flushed, lily-like trumpets which would perfume the air with their fine fragrance.

The trees in the neighbourhood of Trulung were quite splendid and what Ludlow called 'the pride of the Tsangpo' was a cherry up to 100 feet in height — Kingdon Ward's 'Carmine Cherry', *Prunus cerasoides* var. *rubea* (12222); it was beginning to flower and with its sepals of brilliant crimson and petals of pink in a week or two's time its rosy glow would be seen on the hillsides a mile distant. Ward named it the 'Carmine Cherry' when he collected material in North Burma in 1931 and sent home seeds from which it was introduced into cultivation in Britain, receiving the Award of Merit from the Royal Horticultural Society in 1946. But the marvellous cherry had been known long before 1931; it had, in fact, been collected by J.S. Gamble in the Darjeeling hills as early as 1876 and had been noted by J.F. Cathcart in the Darjeeling district twenty years before that. Now, it is known to occur in N Bengal, Bhutan, Assam, Upper Burma, SE Tibet and W Yunnan.

82. The Tongkyuk river and Gyala Peri

In stature, the Carmine Cherry was dwarfed by other trees at Trulung and especially at the expedition's encampment one march beyond Trulung and opposite the village of Tangme (7,000 feet) where the Po Yigrong and Po Tsangpo meet. Here, on a spur overlooking the junction of the rivers, were some tremendous monarchs of the Bhutan Cypress, *Cupressus torulosa* (12141), a Himalayan tree which also occurs in Szechwan. They were at least 200 feet in height and as straight as a rod, and whilst several were at least 20 feet in circumference one giant had a trunk which measured 36 feet in girth at 5 feet from the ground. Though they grew but sparingly they were of such splendour that, towering over everything else, they were discernible from afar.

Around Gyadzong

The expedition was now bound north-west, for Gyadzong. Apart from takin which the local hunters killed whenever they wanted meat there were few or no supplies at Trulung and thus this rather large party had to settle in a more prosperous region where supplies would be more readily available, pending the beginning of the flowering season. Though a rope bridge spanned the Po Yigrong just above the Tangme junction they did not cross it but for two days marched up the right bank of the river in the direction of the Yigrong lake through dense rain forest, and experienced strange vicissitudes in the weather. 'Ramzana woke Elliot and me at 5 a.m.

with tea and the information that there were four or five inches of snow on the ground. These kaleidoscopic changes in the weather are very disconcerting. Last night when we went to bed the meadows were dry and brown and the stars shining; this morning all was white and with powder snow still falling. However, after a hearty breakfast we set out and the snow soon ceased. The march through the forests was a very easy and pleasant one. After passing the village of Sangyü we proceeded 5 or 6 miles and then ascended gently the great earthen dam created by the Tralung river [a tributary of the Yigrong] about 45 years, or more, ago. This earthen dam, which held up the waters of the Yigrong river for a long time and created a lake 10 miles long by $\frac{3}{4}$ to a mile wide, eventually burst and the resulting flood was disastrous, wiping out whole villages and destroying thousands of acres of cultivation. The high flood level, though now covered with quite respectable pine trees, is still to be discerned 200 feet above the present bed of the Yigrong. We camped on the shore of the lake near a ferry of hollowed out pine trees [tru]. This lake is now silting up. I should think it is now only 2−3 miles long. Our camp is a pleasant one on the grassy shores of the lake with a lovely view (reminding one of a Norwegian fiord scene) looking up the valley to some magnificent snow peaks.'[8]

In fact the effects of the flood caused by the rupture of the earthen dam were felt as far away as the plains of Assam, where trees, hitherto unrecorded from the Assam Himalaya, were deposited on the banks of the Dihang.

On 16 January the expedition crossed the lake in a dug-out and descended the left bank of the Yigrong, past the Tralung Chu, to the large scattered village of Gyadzong (7,250 feet). Here they decided to halt for a month or so, by which time, no doubt, the early flowers would be in bloom. They found the locals friendly enough, but rather guarded and, at first, not very forthcoming with supplies.

'They are afraid of officials, knowing only the "Phokpon" here, who is an official changed every three years . . . . For the last 3–4 years the Phokpon's treatment of locals in collecting taxes has been such that many families have left for Kongbo. He supplies tea and salt and takes in place 6 bos for every 1 he gives. The result of all this has been that whereas there were 104 "Kangs" in Tombe Dzong there are now 34; in Be there were 9, now 2½ [Tombe and Be were nearby villages]; in Gyadzong a similar diminution from 11 to 4½. A "Kang" is a measure of what is available in barley and equals 40 bos.'[9]

The locals appeared to be not only poor but in bad health as well, goitre and syphilis being very prevalent. But when they learned that the expedition paid well for its stores, supplies began to come in — a few hens, but no eggs, a little milk and butter, some beans, turnips, potatoes and radishes, and meat and flour.

As it was still too early for intensive plant collecting at Gyadzong, Ludlow, with a small party, decided to march south to Showa Dzong, the capital of Pome, with the object of ensuring whether or not the Po Tsangpo valley would be a profitable area for botanical work. Ludlow had the idea that the Sü La range of mountains which separates Pome from Pemako would prove to be very rich floristically, much more so, in fact, than the lower Yigrong which Sherriff was proposing to visit. Even if this snow-capped range was not an extension eastwards of the main Himalayan range it seemed to Ludlow to act as a rain screen, in exactly the same way as the main range did to the west of Namcha Barwa; and experience had shown that the main Himalayan range holds a far richer flora than any of the drier fold ranges to the north. Moreover the Sü La range was virgin country to the botanist — another excellent reason for visiting it.

Leaving Gyadzong on 28 January Ludlow's little party descended the Po Yigrong to Tangme, where *Rhododendron nuttallii* grew in vast profusion, and thence ascended the Po Tsangpo. Though there were a few deep silent blue-green reaches on the great river, for the most part it was a boisterous foaming cataract most of the way to Showa. The difference in the vegetation of the two banks was very marked; on the right bank thin open forest of oak and pine with no snow; on the left dense conifer forests with snow descending to within a few hundred feet of the river. Ludlow was not only impressed by the botanical possibilities of the district but by other things as well, notably by the nomenclature of the passes, and by a wonderful piece of engineering. 'The range near Showa looks promising for flowers. The snow high up is heavy. Two passes cross the Sü La range from near Showa (called colloquially Shoaka). One just south of Showa is the Showa or Dokar La. About 5 miles east of Showa the Sü La takes off. Both passes are only practicable for coolies. The Sü La seems the one most used, especially by Lobas. These Lobas are apparently Mishmis and give no trouble like the Abors and Daphlas further east in Pachakshiri and Chayul. Passes often have two names accorded them by the people on either side of the Sü La range. Thus the Showa La is so-called only by the Pobas of the Po Tsangpo. The people on the other side of the range at Dokar call the pass the Dokar La. Similarly the Dashing La, further up the valley, becomes the Chindro La when spoken of by the inhabitants of Pemako. The cantilever bridge over the Po Tsangpo at Showa is the largest of its kind I have ever seen. It is roughly 160 feet in length and 80 feet above the river. Nine wooden cantilevers jut out on each side one above the other and at each end of the bridge are block houses substantially built of stone. ["There is not a nail, screw, or piece of metal in the whole structure." [10]] On the whole a fine bit of engineering work when it is considered that the local Pobas have had no technical training nor are they familiar with plans and estimates. But what are we to say of the Tibetan people who constructed the 8th wonder of the world — the Potala, the Palace of the Dalai Lama at Lhasa ?' [11]

Ludlow camped on a bracken-covered terrace near the bridge from which there was a beautiful vista to the mountains overlooking Dashing monastery and to two striking peaks in the Yigrong range; one, overlooking the capital, a sharp spire called Changlung Chago, the other a rounded mass called Tsetan — or Satang — Peri. He crossed the bridge to take photographs of these fine peaks, passed through a lot of rhododendrons and bamboos in the process, and became quite convinced that there were a number of novelties to be taken from the Sü La range in the flowering season. On 10 February he returned to the rest of the expedition at Gyadzong well pleased with his trip and with the three specimens of a most interesting bird, the White-eared pheasant (*Crossoptilon crossoptilon* subsp.) which he had shot near Showa Dzong. Though shot in the same place, and though members of the same flock, they showed marked variation. Two were male; one was pure white except for the crown of the head which was black and white and the tips of the tail and wing feathers which were purplish-black and grey-brown respectively; the other was darker in the tail and wings. The third, a female, was much darker than the others and had a pale ashy back. 'One bird that passed over me at a good pace in bright sunlight was a vision of beauty with its white body outlined against the blue of the sky.'[12]

It was now mid-February and the days were appreciably warmer, so the whole expedition began the return journey to Tongkyuk, Ludlow and Elliot making a prolonged halt at Trulung whilst the Sherriffs and a small party descended the Po Tsangpo to its junction with the Kongbo Tsangpo at Gompo Ne (4,950 feet). As the track to Gompo Ne lay along the left bank it was necessary for the Sherriffs, with their fifteen coolies and their loads, to cross the river by the rope bridge or 'dring'. This consists of a strand, or sometimes two, or in exceptional cases, as here, four, of twisted bamboo rope slung across the river and suspended from two stone piers built on eminences on either bank. The single strand sags pretty horribly in the middle, four strands rather less so. This is the bridge. For the rest there is a wooden runner, about the size and shape of a boomerang, which travels over the dring ropes. Pobas merely tie themselves by the waist to this runner and slither across, upside down, like monkeys, using the runner for support and their hands and legs for propulsion. But other human beings, and some animals of course, are given different treatment. They are bound hand and foot so that they are unable to move, attached to the runner, and then pushed off into space. Down they go with a rush to the bottom of the sag; then there is a sickening pause whilst they dangle above the swirling waters in mid-stream. In the end, with a series of violent jerks, they are hauled to safety on the far bank by a life-line attached to the runner. '[When] my turn came I felt like a well-trussed fowl with

> bamboo ropes lashed round my thighs and waist and then attached firmly to the triangular piece of wood which slides along the dring ropes. I held on to the bottom of the triangle

with my hands, shouted "Cheeroh" to Ludlow, Henry, Jill and Joker, and then launched off across the foaming waters of the Tsangpo. In actual fact I felt amazingly secure and the coolies at the far end pulled with a will so that the ordeal was quickly over.' [13]

On the left bank of the Po Tsangpo there was much more bamboo than on the other. But there were also large areas of open pine forest showing a strange mixture of wet and dry zones close together. Whenever a valley came in the north face would be covered with bamboo, hydrangea and many broad-leaved trees, and the south face with nothing but grass and pines and a few rhododendrons. There still wasn't a great deal to collect but occasionally plants were put into the presses, and fruits and seeds were gathered including, from the pine forests at Lubong, the bluish-purple globose fruits, with their glaucous bloom, of *Gaultheria wardii* (12238) which Kingdon Ward had discovered at Tongkyuk in August 1924 amongst the bracken on the pine clad slopes. Subsequently he collected it again on two other occasions, in July 1926 in the Lohit valley of NW Upper Burma and in October 1928 in the Delei valley in Assam. Seeds of both these later collections introduced this fine plant into cultivation and in November 1933 fruiting specimens received the Award of Merit from the Royal Horticultural Society when exhibited by Lt-Col. L.C.R. Messel. Apart from the gaultheria Sherriff was impressed by the 15-foot tree-like specimens of *Luculia gratissima* (12240) which was fairly common in the wet forest zone at 7,000 feet, as was an aralia-like tree with digitate leaves and creamy flowers which proved to be a new species of brasseiopsis and appropriately named *Brasseiopsis karmalaica* (12244) as it was collected on the Karma La, a pass only a little over 8,000 feet on the way to Dzama. And on this same Karma La was a purple and white flowered deliciously scented daphne (12245) which Sherriff was quite sure was also a new species. His optimism was understandable for the plants he found were of remarkable proportions with stems between 10 and 15 feet in height and at least 6 inches in diameter. However, except in size, the plant is indistinguishable from the Yunnan *Daphne papyracea* var. *crassiuscula*. Unfortunately fruit was not available and, more unfortunate still, the expedition was never able to return to these lower reaches of the Po Tsangpo.

Apart from Kingdon Ward and Cawdor no Europeans had previously visited Gompo Ne and the former gives a typically vivid account of his visit in *The Riddle of the Tsangpo Gorges*. 'Gompo Ne is little more than a name. Amidst a wilderness of gneissic monoliths, rasped and scoured by the shock of the river, then idly cast aside, their raw wounds abandoned to the sly healing jungle, stands a tor whose shape suggests a natural stupa or pagoda; while hard by, crowning another gigantic rock, on whose face are cup-marks, is a real *chorten* [shrine]. Leaning against this rock are a number of long poles, notched into steps, so finely cut that no human being could possibly climb them, and one might be puzzled to account for their presence, did not one recollect that we are now in the land of *nats*, those elfish spirits which live in the trees and in the lakes, and rivers, and mountains of the twilight forest land. Strong spirits mount quickly to the head, by these ladders. There is also an open shed, where pilgrims such as ourselves sleep the night; nothing more. Once there was a monastery here, we were told, but it fell into the river; and now the great grey rocks, quarried by the river which storms by 50 feet below, lie around in confusion, while tangled mats of orchids help to conceal their bald heads, and the crawling jungle slowly buries them.' [14]

As with Kingdon Ward in 1924, so now, Sherriff found no habitation at Gompo Ne which is a place of pilgrimage. But 1,500 feet above the confluence of the Po Tsangpo and Kongbo Tsangpo was Tsakchugong, the finest situated village the Sherriffs had ever seen and one which offered awe-inspiring views of the great Namcha Barwa-Gyala Peri range to the south-west. 'I studied the mass of ice to see if I could follow the route of the Kongbo Tsangpo, but it appeared quite impossible that any river could cut through what appeared to be an endless wall of ice and rock. But this evening, with the sun behind the range, there appears a route, though it seems almost impossible that any river should cut so low. I think this is the finest sight I have ever seen.' [15]

Having left Trulung on 20 February, the Sherriffs began the return journey on the 27th and even this one week had seen a great change in the vegetation. Trees were breaking into leaf and here and there catkins were visible. *Rhododendron virgatum* in several colour forms was in flower and another rhododendron from which they had already gathered seeds (12231) was now unfolding its fragrant apple-blossom-pink trusses. This was *R. scopulorum* (12264) which Ward had discovered in the Tsangpo valley in 1924 growing, as the specific name implies, on boulder screes and on steep rocky slopes either in full sun or in thickets. Later Ward collected seeds and introduced the lovely species into cultivation as a plant for the cool greenhouse; as such it received the Royal Horticultural Society's Award of Merit in 1936 when exhibited by

Mr Lionel de Rothschild. A third rhododendron, with flower trusses of deep crimson, was *R. tanastylum* (12280), another Kingdon Ward discovery this time in Eastern Upper Burma and the adjoining parts of Western Yunnan. But not only the plants, even the sky was showing a marked change; there were more cumulus clouds and this Sherriff took to be a sign of spring. The Sherriffs rejoined their colleagues at Trulung on 2 March and the entire expedition returned to its base at Tongkyuk to reorganise and formulate its plans for collecting in the flowering season — and, incidentally, to discover the two forms of *Primula whitei* (12299) growing intermixed, the typical plant with crenulate petals, and the form with the petals usually tri-dentate which had been incorrectly described as a new species with the name of *P. bhutanica*.

The fact that so-called *Primula bhutanica* is merely a form of *P. whitei* wasn't the only discovery made at Tongkyuk, as well as at Layoting which was passed on the way from Trulung. At both places Przewalsky's thrush, *Turdus kessleri*, was common in the fields and Ludlow and Sherriff procured a fine series of specimens, two females at Layoting, three females and seven males at Tongkyuk. Previous to these being shot for the National collections this thrush was represented in the British Museum only by two others, both from Sikkim, and consequently had been considered a rare bird.

'What strange pranks Nature plays! Here is a bird, a *rara avis*, if there is such a thing, and yet at this particular time and in this particular place it occurred in such numbers that I could have filled a museum drawer with its skins had I so desired. Often has this happened to me in SE Tibet. Every now and then Nature has lifted the veil and shown me some of her "rarities" in all their abundance — birds such as Temminck's tragopan, Sclater's monal, Prince Henri's Laughing thrush, the Himalayan crossbill, Beavan's bullfinch, the Tibetan siskin, and a dozen others which I had only seen in my wildest dreams until I descended the Tsangpo.' [16]

'Pome is a district which is much favoured by the genus Turdus in winter, probably because of the abundance of food and the comparative mildness of the climate. Sherriff, in his descent of the Po Tsangpo to Gompo Ne, encountered considerable numbers of *Turdus rubrocanus gouldi* [the Abbé David's ouzel], whilst on my trip to Showa I came across *Turdus ruficollis* [the Red-throated thrush] and *Turdus albocinctus* [blackbird] literally in their tens of thousands. The two latter

83. Gyala Peri from Gyala

were gorging themselves on the crab apples and cotoneaster berries that grow so plentifully in Pome.' [17]

But to return to the expedition's future plans. As on past expeditions it was vital to cover as much ground as possible in the limited time available. Consequently it was necessary to split up into various parties, in this instance conveniently into three. The Sherriffs would work the Sü La range, Elliot the main range between Tsela Dzong and Gyala, whilst Ludlow would proceed to the Yigrong. But first Ludlow and Elliot were resolved to visit the great gorge of the Kongbo Tsangpo below Gyala and accordingly left the Sherriffs on 20 March. 'We said goodbye to the Sherriffs whom we shall not see again until October unless something very unforeseen happens . . . This Rong Chu valley is full of rhododendrons which should be a glorious sight when we return six weeks hence.' [18]

With Ludlow and Elliot

Ludlow and Elliot had an easy and pleasant march to Tumbatse, passing myriads of *Primula atrodentata* about to flower in the grassy meads, before crossing the Temo La on another lovely day but without being blessed with the marvellous view from the summit which had been theirs on 23 December last. They halted at Dzeng, near Tsela Dzong and a couple of miles beyond Temo, in practically the same spot as in 1938. Ludlow was anxious to arrange for transport and supplies with the friendly dzongpen and discuss with him the possibility of returning to India, in the late autumn, by a new and quicker route via the Doshong La and Pemako. As the dzongpen was away from home and wasn't expected back for another week at least there was no alternative for Ludlow and Elliot but to wait for his return. There was no great hardship in this for life was very pleasant; the days were now much warmer, there was very little frost at night, the peach trees were in blossom and the willows rapidly coming into leaf.

It was while they were thus waiting that a most disquieting letter arrived from Sherriff on 4 April. During the 1938 expedition, Sherriff, who never spared himself on the hillside, had badly strained his heart and this had troubled him from time to time since then. However, by riding whenever possible and taking life as easily as possible, he was quite certain that he would be able to stand up to a year of living in Kongbo and Pome where the altitude was not excessive. But it was not to be. Sherriff's letter unemotionally said that his heart had begun to give him so much trouble that he and Betty had decided to return to Kalimpong and were intending to leave Tongkyuk on 11 April, arriving at Dzeng three days later.

The only mention of this sad state of affairs in Sherriff's diary was a brief entry on 30 March: 'I have felt my heart rather a lot and have had to decide to go back to India, a blow which I haven't realised fully yet.'

Ludlow hastily wrote to the Sherriffs to say that he and Elliot would await their arrival at Dzeng. The time would not be wasted for there was little to collect and Ludlow still had to discuss plans with the dzongpen. This was satisfactorily done during a four-hour interview on 11 April and the Sherriffs arrived on the 14th.

After Ludlow and Elliot had left Tongkyuk on 20 March, the Sherriffs had paid a brief visit up the Nunkhu Phu Chu valley and over the Parka Phu Chu to Lokmo, a village of some thirty houses and considerable cultivation. Though plant life was not so far advanced as at Tongkyuk — even *Primula whitei* (12311) was still not in flower — they did collect a lovely bright yellow cushion-saxifrage covering the boulders on the steep hillsides which proved to be a new species and was named *Saxifraga elliotii* (12309). They also arranged in their plant presses the compact trusses of a fine form of *Rhododendron anthosphaerum* (12313), with pink, darker spotted, bells, which was very common in the area; this species was one of Forrest's discoveries in NW Yunnan in 1906 and is a well-known plant in cultivation. From Lokmo, Sherriff sent Tsongpen up to the Sobhe La to try to correct the thermograph which had been installed there on 31 December last but which had been disarranged by the lama who was being paid to change the graph on his first visit. Tsongpen found eighteen inches of snow at the instrument and reported that when he corrected the red pen it read at the bottom of the scale, $-20°$F or below. He further reported that precious little was in flower, even *Primula whitei* being in much the same state as on 31 December. On the other hand, returning to Tongkyuk, after an absence of eight or nine days, the Sherriffs found the landscape greatly changed. 'Parka village was trans-

> formed. From being a quite attractive but really rather ordinary little village it had become a place of beauty with its rough little wooden houses embowered in clouds of pale pink [peach] blossom. And Tongkyuk is lovely. The whole place is a peach orchard with carpets of tender green barley stretched beneath the trees. We had no idea of the number of peach trees there are when we were here in the winter. We look out on to drifts of blossom all around us here in our camp varying from deep to pale pink.' [19]

Elliot's examination of Sherriff when he arrived at Dzeng showed how wise Sherriff was to decide to play no further part in the expedition. So on 20 April the Sherriffs sadly set out on their long journey back to India via Tsetang, Tsona and East Bhutan.

With Ludlow and Elliot to Pemakochung (Map 18)

Obviously it was now necessary to reconsider the plan of operations. Ludlow was anxious to disturb these as little as possible so decided to adhere to the original programme as far as he and Elliot were concerned, and to send the experienced Lepcha collector, Tsongpen, to the Sü La range in place of the Sherriffs. But first Ludlow and Elliot would make their journey to the Tsangpo gorge and on 18 April set out eastwards for Gyala.

They crossed to the right bank of the river in the conifer dug-out, the tru, and halted at the village of Tamnyen, pitching their tents in the same spot at which Ludlow, with Taylor, had camped in 1938. Growing in great profusion on a shingly alluvial fan, they found the new white-flowered paeony, *Paeonia sterniana*, which Ludlow and Taylor had accidentally discovered in 1938 when sitting down to a wayside lunch at Gyala. Later on in the year Ludlow collected ripe seeds (14231) from which the species was introduced into cultivation in Britain. Several rhododendrons were now in flower, the most conspicuous being one which Ludlow knew well for he had collected it on many occasions in 1936 and 1938 and had noted its variation in colour from white to rose, *Rhododendron vellereum* (13524); at Tamnyen it formed prominent belts on the hillside along the 11,000 feet contour line. It was one of Ward's 1924 discoveries, above Nang Dzong in the Tsangpo valley. Another of his discoveries now much in evidence was 'Ward's Mahogany Triflorum', *R. triflorum* var. *mahoganii* (13546), the zygomorphic normally yellow flowers, in trusses of 2–4, having a mahogany-coloured blotch and spots or being in varying degree suffused with mahogany. Still another distinguished Tamnyen rhododendron was a white-flowered form of the variable and common Yunnan and Szechwan *R. uvarifolium* (13521). Primulas, too, were coming into flower, the very early *Primula vernicosa* (13529) of course, and the lovely, though stinking, *P. calderiana* (13530).

At Gyala they found the village headman most obliging; he willingly agreed to their proposed visit to the gorge and promised every assistance. They were lucky, for quite unwittingly they had reached Gyala at the most propitious time in the whole year, the time when the villagers were about to start their annual pilgrimage to the little monastery at Pemakochung, whither Ludlow and Elliot

84. *Paeonia sterniana*
85. *Rhododendron leucaspis*

were anxious to go; naturally the locals were quite willing to com-
bine religion with business and carry the visitors' loads. Except for
the tiny monastery of Pemakochung there were no habitations in
the gorge so Ludlow and Elliot drastically had to reduce their loads
and leave most of the servants behind. Even so, there were twenty
coolie loads when they left Gyala on 27 April with twenty very
cheerful men and women porters.

The first day's march of about six miles to Kumang (10,300 feet)
was full of interest — of various kinds. 'We got away early and
rode our ponies for the first mile until we reached a chorten on
a spur where the path degenerated into a mere goat track.
Almost immediately we entered [bamboo] forest we were
assailed by ticks in their thousands. We soon reached a
stream — the Kenta Chu — where there was a strong smell of
sulphur and here we sat down to pick off as many ticks as we
could. From this stream the path led steeply up the hillside into
burnt forest. Soon we came to a cold sulphur spring and a
little further on to a nullah where sulphur is mined. Above the
shaft of the mine, on a perpendicular rock face, were three
species of rhododendron — one pure white [13549], another
pink with five radiating lines of dark pinkish-red [13551], and
a third which seemed to be a *virgatum* [13550] was also pink.
There now followed a stiff, almost vertical, climb of
1,000 feet or more up the hillside. After this we came to a
small glen with a little water trickling down over mossy stones
and here among moist moss was a primula [13552] which I have
never seen before but which, from the deeply cleft calyx, I take
to be *Primula sonchifolia*. The flowers are vinca-blue with a
white eye and they grow in umbels. The scape is 4–6″ long.
The flowers are rather past their best and many fell off at the
touch. A ¼ mile further and we came to the Kumang hut in
Picea [*lichiangensis*] forest where we camped [and where the
fiery-red pheasant, *Tragopan temminckii*, was particularly
plentiful].' [20]

Rhododendron 13549 was *Rhododendron leucaspis* which Kingdon
Ward had discovered in this very locality in 1924; 13551 was *R.
hirtipes*; and, as Ludlow suspected, 13550 was a blush-pink form of
the variable *R. virgatum*. The primula, however, was not as Ludlow
suspected, *Primula sonchifolia* which frequently has been collected
in Yunnan, Szechwan and NE Upper Burma and twice, by Rock, in
Tsarong in SE Tibet. How easy it is to be misled! Ludlow had seen
many many thousands of *P. whitei* in the field, both in SE Tibet and
in Bhutan, and yet he failed to recognise it now that it had developed

Map 18. With Ludlow and Elliot to Pemakochung

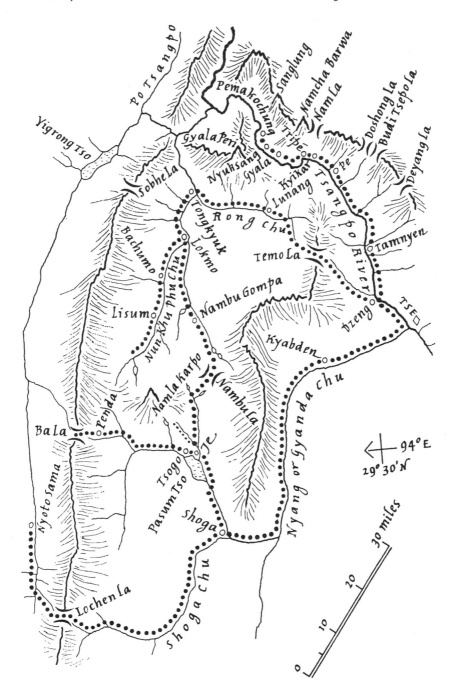

a considerable sized scape, as it sometimes does in cultivation at late flowering and fruiting time, and now that the calyx was a little more deeply cut than perhaps is usual.

The next morning, 28 April, they continued their ascent, through mixed bamboo and rhododendron, to a cliff face which had to be scaled with ladders of notched logs. Soon they reached the pass on the summit of the spur, appropriately called the Musi La or the Sulphur Pass, and here they found a primula (13557), quite new to Ludlow, growing on grass-covered rock faces in the ravine bed. It was a pretty thing, rather reminiscent of *Primula jaffreyana*, with leaves 2 inches long and 1 inch broad covered below with cream-coloured farina, and with a cluster of eight violet, pale yellow-eyed flowers on a scape up to 4 inches long. In the same locality more material (13628) was gathered a few days later, and seeds and seedlings (13250) in September. The seedlings, along with those of other primulas, were flown to the Royal Botanic Garden, Edinburgh, where one flowered, obviously out of season, in August 1948. Other seedlings under the same number flowered more profusely and more in character in March 1949, as indeed did plants which had been raised from seeds of 13250. Quite clearly here was a species new to science and it was named *P. tayloriana*. Plants received the Award of Merit from the Royal Horticultural Society in March 1949 when shown by Lord Aberconway and Mr David Livingstone. Unfortunately the species was not as amenable in cultivation as it promised to be and was soon lost.

Apart from *Primula tayloriana*, *P. whitei* was in great masses in the damp ravines, not only in moss on the ground but in moss on fallen tree trunks. And among the bamboos was a most distinctive paris with a dark brown perianth, long pinkish-yellow stamens, and with leaves dark green and conspicuously white-veined above and purplish-brown below; it, too, proved to be a new species and was named *Paris marmorata* (13564).

Descending from the Musi La, towards Nyuksang, they dropped 2,000 feet in a couple of miles passing first through pure rhododendron forest. Two of Forrest's Yunnan discoveries, *Rhododendron uvarifolium* (13567) and *R. anthosphaerum* (13559) were in great abundance and showing marked variation in flower colour. The former, which Ludlow regarded as perhaps the commonest rhododendron in the Tsangpo gorge, sometimes carried trusses of white flowers, sometimes of white flushed with rose, and sometimes of pale rose; the flowers might be spotted and blotched with crimson or quite free from these markings. Similarly the flowers of *R. anthosphaerum* varied from rose to rose-magenta to mauve, and a

dark crimson basal blotch was usually present. The other outstanding and very common species was one which Ward had
discovered in the Tsangpo gorge in November 1924 and which *he*
had described as the most abundant rhododendron throughout the
gorge, in the region of Pemakochung forming practically pure
forest, *R. ramsdenianum.* Ward never collected flowering material
and until Ludlow and Elliot now gathered its blood-red trusses
(13561) the only flowering material known to botanists was that
taken from plants grown from seeds of KW 6284 in the garden of
Sir John F. Ramsden, at Munchester Castle, in 1934.

Having descended through the rhododendron forest, and then
through an almost inpenetrable growth of bamboos, they reached
Nyuksang (8800) very tired and weary for though they had
covered but eight miles at the most they had been on the march for
eight hours. Nyuksang is the name given to an encampment at the
base of a cliff where the Tsangpo descends in a mad, roaring, leaping
cataract, and here the excitements of the plant finds of the day were
almost forgotten at the sight of thousands and thousands of *Cardiocrinum giganteum.* 'Never have I seen this giant among lilies growing
in such riotous profusion before. The plants were about a foot
high and for a mile we literally walked on a carpet of lilies,
crushing them under our feet at each step. I tried to picture the
scene at a later date when all would be in bloom and the air
heavy with their fragrance, but I never saw them again in all
their splendour and so missed the lily picture of a lifetime.

'The next day's march to Senge Dzong, the Lion's Fort
(8,500 feet), also a mere encampment, was gruelling, and the
rain persistent. The track was non-existent and even our
porters lost their way. I have no idea how many miles we
reached—not many—and when we halted at dusk I had
barely the energy to assist in the pitching of camp. But again
we had a memorable day and if our feet were heavy so were
our presses, and our hearts were light. We were never far
from the river on this march and the thunder of its waters rang
in our ears the livelong day. Down, down it came in a
demoniacal cataract so that the earth trembled beneath our
feet and we stood at times appalled by the fury and violence of
the scene.' [21]

As with *Rhododendron ramsdenianum,* the only flowering material
in herbaria of Kingdon Ward's Tsangpo gorge *Rhododendron auritum*
was that taken from plants in cultivation grown from seeds of KW
6278, gathered in November 1924. These plants had flowered both
in Sir John Ramsden's garden as well as in the Royal Botanic

275

Garden, Edinburgh, in 1930. Ludlow and Elliot now found the species sprawling over the cliffs (13570) and took its smooth coppery-red stems and creamy-yellow flowers for their plant presses. And into their presses they proudly placed the great trusses of white or pale mauve, purple-spotted flowers, from the splendid 15-foot trees of *R. mollyanum* (13568). Ward had collected fruiting material of this fine plant at Pemako in October 1924 (KW 6261) under the tentative name of *R. sinogrande*, a species which Forrest had discovered on the western flank of the Shweli-Salwin divide, in SW Yunnan, in 1912. Not until plants flowered in gardens in Britain, notably in the garden of the Duchess of Montrose at Brodick, Isle of Arran, under the number KW 6261, was it realised that Ward's plant was different from Forrest's and was named *R. mollyanum*. It is of interest that the herbarium specimen Ward collected under 6261 is not of *R. mollyanum* but of *R. exasperatum*, another of Ward's new species which Ludlow and Elliot found a few days later.

The next day, 30 April, they reached their destination, Pemako-chung (8,807 feet) after, for the most part, a very trying, even dangerous, march during which they were forced to descend to the river bed and step carefully from one boulder to another. Approaching Pemakochung the track improved and finally ascended a wide alluvial fan to the monastery, a little wooden erection perched on a mound and surrounded by one or two miserable fields choked with weeds. Dark conifer forest hemmed it in to the south, beyond which rose the snowy splendour of the great Sanglung (23,018 feet) peaks and ridge descending from Namcha Barwa (25,445 feet). The only inhabitants of the monastery were an old lama and a boy, and they gave their visitors the most kindly of welcomes. A room was set aside for their use and a little present of butter and flour was proffered with a white scarf of greeting. In return Ludlow gave the monk a photograph of the Dalai Lama which Sherriff had taken in Tibet. The old man obviously was deeply moved; for a time he gazed at the photograph in mute admiration, then held it reverently to his forehead, and finally gave it the place of honour on his altar.

On taking his customary evening stroll Ludlow was given a fore-taste of the rhododendron riches of Pemakochung. Forming loose straggling tangled thickets on rock faces and in rather swampy ground in the mixed forest there was a species (13584) with compact trusses of up to ten fleshy tubular-cup-shaped deep crimson flowers, each with five conspicuous black nectaries at the base. It was the species of which Kingdon Ward had gathered seeds in this same locality in November 1924, under the number KW 6285.

When plants from these seeds had flowered in various gardens in Britain, and had gained the Royal Horticultural Society's Award of Merit in 1933 when exhibited by the Hon. H. H. McLaren, they had been named *Rhododendron venator*.

The following day, and in a fairly short time, they found no fewer than nine different species on a cliff-face which was aglow with pink and crimson. 'At every few steps I uttered some rapturous outburst as my eyes lighted on the splendour of some glorious plant. It seemed as though Pemakochung was the birthplace of the genus, the very epicentre from which it had sprung.'[22]

There was the well-known *Rhododendron thomsonii* (13589) not only with a port-wine-red corolla but with a blackish-red calyx cup as well. There was *R. ramsdenianum* (13596), this time not with blood-red trusses but with flowers of dark reddish-pink, the posterior petal with darker spots. There were at least two distinct forms of Forrest's *R. anthosphaerum* — one (13593) with the pink flowers striated with reddish-pink on the posterior petal, the other (13594) with all the petals of the pink flowers, especially the posterior one, densely streaked with purplish red. There was another of Forrest's Yunnan discoveries — he found it on the mountains of the watersheds of the Yangtse, Mekong and Salwin in 1914 — the very variable *R. glischrum* (13590); the specimens Ludlow and Elliot collected carried fine trusses of pink, dark purplish-blotched flowers. There were three of Kindon Ward's discoveries. The rich pink compact-trussed *R. lanigerum* (13591) formed a splendid tree up to 20 feet high, its leaves carrying a brown woolly indumentum on the lower surface; Ward had found a great deal of it on rhododendron-clad slopes in the Delei valley of the Assam Himalaya on his 1927–28 expedition. Also with deep rich pink flower-trusses — as well as with bristly shoots and petioles — was Ward's *R. exasperatum* (13595); he had found it on three different expeditions — here, at Pemakochung, in November 1924, at Seinghku Wang in Northern Burma in June 1926, and in the Delei valley in May 1928; like *R. glischrum* it is a member of the Barbatum series of rhododendrons and Ward's discovery of it in Assam served to link the Barbatums of the Himalaya with the Burmese and Yunnanese members of the series. The other Kingdon Ward plant was *R. uniflorum* (13592) a shrub no more than 2 feet tall with solitary, or sometimes paired, purple or mauve flowers; Ward had discovered it on the Doshong La in 1924. The other rhododendron Ludlow and Elliot found on this memorable day and on this remarkable cliff face was a five-foot shrub with blood-red flower-trusses which no one, save Ludlow and Sherriff, had ever seen; it was the new species, *R. populare*, which they had discovered at Chayul Chu during their 1936 Tibetan expedition; Ludlow and Elliot now collected more of it under 13598.

On each of the next four days they filled their plant presses with more rhododendrons; more forms of *R. glischrum*; more forms of *R. anthosphaerum*; more forms of *R. ramsdenianum*; several forms of *R. oreotrephes* (13613, 13614, 13622), which Forrest had discovered on the Lichiang range in Yunnan in 1910, from the palest of pink, to darker pink with a purplish tinge to apricot-yellow; two of Ward's new species from the Doshong La in 1924, the small twiggy *R. kongboense* (13633) with small, tight, bright rose flower-trusses, and a form of *R. parmulatum* (13612) with rather loose trusses of flowers the ground colour of which was pale yellow with apple-blossom-red towards the tips of the petals, 'a very strikingly coloured and handsome shrub.'

The rhododendrons, of course, were the most prominent plants, but there were many other fine things Ludlow and Elliot could not resist collecting including the new *Primula tayloriana* (13600) as well as another primula (13607) which was obviously akin to the now well-known *P. atrodentata* and to the universally known *P. denticulata*, but which had a more lax flower head than these, each blue-violet flower sitting on a short stalk. This interesting plant also proved to be a new species and was named *P. laxiuscula*. Also of great interest was the finding of the magnificent *Omphalogramma souliei* (13599) whose rich purple velvety flowers carried yellow streaks half way up each petal; discovered in Yunnan by the Abbé Soulié in 1890 or 1891, Pemakochung represents the western limit of its distribution.

One memorable morning was spent not in hunting for flowers but in sight-seeing at Kinthup's Falls, so named after the famous lama Kinthup who visited Pemakochung in 1881. The falls are also known as the Rainbow Falls and are about a mile from the monastery and about 800 feet below it. 'They are a wonderful sight.

Immediately above the actual falls the river comes tearing down in a cataract for 200–300 yards. The waters are then confined between a cliff face and a series of gigantic boulders into a hissing seething cauldron of foam, not more than 20 yards in width, which falls 30 feet or more into a veritable maelstrom below. The noise is deafening and the spray forms a mist which hangs above the falls to a great height. It is in this

86. The Tsangpo nearing the gorge
87. The Tsangpo at Pemakochung

spray, when the sun shines, that rainbows can be seen. We were fortunate in our day for the sun shone brightly. All around the falls . . . . are enormous boulders and rock faces where in many places sulphur springs bubble out of the rock. On a great cliff overlooking the falls are half a dozen shrubs of a white Maddenii rhododendron (13603) with blooms as large as lilies and of intoxicating fragrance. White orchids cling to the inaccessible cliffs.' [23] The rhododendron with the lily-like flowers was still another of Kingdon Ward's discoveries, and one of the finest he ever made, *Rhododendron megacalyx*; he originally collected it in 1914 on the southern part of the Nmai-Salwin divide in Eastern Upper Burma, and from here it ranges north-west through the divide to the Tsangpo gorges and to Tsarong on the Tibet-Yunnan frontier. The white orchid was the Himalayan *Coelogyne corymbosa* (13616).

Ludlow and Elliot could only afford five days at Pemakochung and in this short time were able to gather a mere tithe of the vast plant treasures of the area. They had to leave the great Sanglung valley which descends from Namcha Barwa quite untouched. To have reached it would have meant hacking a track through the forest for there were no paths save those made by herds of takin. They could only speculate on the richness of the flora of this unknown and uninhabited valley. 'I wonder what this virgin country holds as regards flowers. This gorge country must be a very rich one botanically. During the past four days alone I have obtained 30 different species of rhododendron [some were forms of species]. I imagine if I came here, say in mid-June, I should be able to add another 30 rhododendrons to my collections by working the higher altitudes and all would be different from those we have been collecting during the past four days. The trouble about this place is that there are only a few tracks here and there, and that it would be impossible to camp higher up unless one had a number of permanent coolies to cut a road. These would have to come from Gyala and they would have to feed themselves as no supplies are available locally. This means that other coolies would have to be engaged to carry the permanent coolies' food. There is a hot spring near Kinthup Falls. Tsongpen . . . . went down yesterday to have a bath. What do all these sulphur springs and deposits and hot springs betoken ? Volcanic activity of some sort. I cannot help feeling that the Tsangpo is hereabouts cutting right through the main Himalayan range and not merely outflanking it as geologists contend.' [24]

The question of how best to organise botanical exploration in this almost virgin country continued to exercise Ludlow's mind for the rest of the expedition, to the extent that 'in 1948 I made plans to spend an entire flowering season in this gorge country with Pemako-chung as my base, but the Tibetan Government, fearful of communist China, withheld their consent — for the first time — to this last request I shall ever make of them.' [25] No doubt some day the area will yield up its botanical secrets.

Reluctantly they left Pemakochung on 5 May and reached Gyala four days later, there to repack and reorganise their kit and supplies for the departure next day to Pe whence Elliot was to work the passes on the main range which Ludlow, Sherriff and Taylor had been unable to explore in 1938 and whence Ludlow was to hurry off to Tongkyuk on his way to the Yigrong range. Leaving Tripé on 11 May they witnessed a strange phenomenon. The whole of the north face of the Namcha Barwa massif was hidden in a cloud of dazzling brightness and time and time again, as they ambled along, their attention was drawn to it. Suddenly the shadow of Namcha Barwa and the satellite peaks was thrown on to the cloud as on a screen. The configuration was perfect and clear but before Ludlow could set up his camera the vision had vanished.

They reached Pe (10,000 feet) by mid-day on 11 May, arranged their dried plants and made plans for their separation on the 13th.

> 'We seem to be continually planning and replanning, packing and repacking. Boxes have to be reopened and repacked, stores divided up amongst various parties. It is all very difficult and people who imagine that an expedition of this nature is just "Roses, Roses" all the way are sadly mistaken. And when I get back to Kashmir people will say "Oh, Mr Ludlow, here you are again. And did you have a nice time and get lots of flowers? Oh how I would have loved to have been there." And I suppose if these good people had been with me they would have returned to Kashmir before they ever got as far as Gyantse.' [26]

With Ludlow to Tongkyuk and the Yigrong

Ludlow and his little party left Pe on 13 May, with nine pack ponies and one riding pony, en route for Tongkyuk, via Tamnyen, Dzeng and Lunang. As the Nyima La (15,200 feet) was still closed with snow they had to cross into the Rong Chu valley by way of the Temo La (14,000 feet) which is never closed. This was the third time during this expedition that Ludlow had crossed the Temo La in fine weather and, as on the first occasion, the view was 'sublime'. In the Rong Chu valley some of the rhododendrons were at their splendid best. Along the banks of the river, growing quite in the open and no more than 3 feet tall, the Yunnan *Rhododendron primuliflorum* var. *cephalanthoides* (13698, 13699), in several forms, was a common sight, its small compact flower-trusses, sometimes yellow, sometimes pale rose, and occasionally white. Common too, in the same situation, was the small solitary purple-flowered *R. nivale* (13701), one of Hooker's Sikkim discoveries of 1849–50. On the other hand *R. dignabile* (13703), one of the new species from the 1936 expedition, grew not in the open but sought the shelter of the abies forest; it was a most handsome plant of ten feet and the white flowers frequently had a prominent pinkish-red suffusion and a dark purple splash at the base of the posterior petal. All the meadows were spangled with myriads of the dark blue compact heads of *Primula atrodentata* whilst the woodland paths in the picea and abies forest were gay with the maroon *P. calderiana*, a great profusion of the orange-red and orange-yellow *P. chungensis* (13706) one of Kingdon Ward's Yunnan discoveries of 1913, and of the rose, geranium-leaved *P. latisecta* (13710). *Meconopsis betonicifolia* and *M. integrifolia* were some six inches high, the latter carrying fat flower-buds.

Ludlow arrived at Tongkyuk on 18 May and there had to halt for five days, mostly to pack and repack and to discuss with the mean and greedy dzongpen the question of transport and supplies for his visit to the Nambu La and the Yigrong range. But during his enforced halt he sent Tsongpen to the Sobhe La to retrieve Sherriff's thermograph since the lama who was still being paid to remove the monthly recordings had decided no longer to do so. Tsongpen spent a profitable day on the pass and returned, not only with the thermograph, but with some lovely plants as well. There was a form of *Rhododendron vellereum* (13747) with handsome pink flower-trusses, each flower darkly spotted on the posterior petal; a very distinct and distinguished form of *R. anthosphaerum* (13755) with full trusses of sulphur-yellow bells — the first yellow form of this species to have been collected; a form of *R. dignabile* (13760) with salmon-pink flower buds and with open flowers of the palest yellow with salmon-pink lines along the petals; a form of *R. oreotrephes* (13761) with pinkish-purple trusses; a form of *R. wardii* (13754) with a pinkish tinge suffusing the lemon-yellow flowers, each with a dark purple basal blotch. These were all familiar species to Ludlow.

But not so familiar were forms of three other species which

88. Kinthup's Falls

Forrest had discovered in Yunnan; a form of *R. roxieanum* (13746) with thick narrow leaves covered with a dense dark fawn indumentum below and with compact trusses of pure white flowers except for a few pink spots; a form of *R. agglutinatum* (13753) with pale pink, purple-spotted flowers and with a rusty-brown leaf indumentum; and a most attractive form of *R. stewartianum* (13756), each flower in the loose truss being suffused and spotted with pink.

Of primulas there was a very vigorous large-flowered *Primula calderiana* (13757); there was *P. jaffreyana* (13758, 13762) always inhabiting open dry situations, sometimes with more violet, sometimes with more red in the pinkish flowers; and growing in great abundance on rock faces at 12,500 feet, a species (13749) which neither Ludlow nor anyone else had seen before. It was an attractive dwarf with leaves thickly plastered below with cream or yellow meal, and with one-inch flowers, a rich violet-mauve in colour, either solitary or 2–4 together on a short scape. It was a new species which was named *P. candicans* and has never been found in any other locality but the Sobhe La. Such a splendid day's haul of plants augured well for Tsongpen's imminent visit to Showa and the Sü La range in Pome, on the first stage of which, to Layoting, he proceeded on 24 May whilst Ludlow marched up the Tongkyuk Chu with the Yigrong range as his objective.

Ludlow and his men at the end of their first day's march had reached the small village of Lokmo where the river divides, one branch coming in from the Nambu La and the other from a pass on the Yigrong range called the Nunkhu La. Ludlow was anxious to explore this latter valley for two reasons. In the first place it seemed to offer the quickest route to the alps of the Yigrong, and in the second place he was anxious to ascertain, if possible, the cause of a gigantic flood which had swept down the valley some twenty years before. Thus, the following day he set out north-westwards up the scoured-out valley. The level, smooth and straight road led through a forest of buckthorn which had colonised the gravel which the flood had brought down and deposited in flat terraces on either bank of the river. The river itself seemed to be glacier-fed and ran in a prolonged rapid the whole way with no sign of a single quiet reach. After a march of twelve or fourteen miles they pitched camp at a yak bothy called Bachumo. 'I got *Primula jaffreyana* and saw *P. latisecta*. All around Bachumo *P. chungensis* grows in masses. *Rhododendron triflorum* in all its various colour phases, greenish-yellow, yellow, apricot, orange, pink and mahogany-red grows to perfection in this valley.' [27]

Continuing in the same direction they gathered a couple of

handsome species of pedicularis, one with yellow flowers, *Pedicularis cryptantha* (13775), the other with lilac-pink, *P. diffusa* (13773), as well as marvellous specimens of *Cassiope wardii* (13768) and *C. selaginoides* (13767) of which Ludlow had never seen finer clumps; clearly they appreciated the alluvial gravel, brought down by the flood, on which they were growing. Arriving at yet another encampment called Lisum they were told that the 'fons et origo' of the flood was but a few miles distant and therefore they optimistically set out for the scene early next morning, 27 May. They must have trudged some ten miles up the Nunkhu La valley before reaching the shores of a lake which Ludlow at first thought must have been the cause of the disaster. However, a quick investigation convinced him that such could not have been the case and that the flood waters must have come hurtling down a side valley where a colossal gash fully 2,000 feet in height had been cut in the mountainside.

Therefore up *this* valley Ludlow toiled, stepping from boulder to boulder for there was no semblance of a track. Finally he reached an unclimbable rock face and reluctantly had to conclude that he could go no further and that he would never definitely solve the secret of the flood. However he was now pretty certain that the lake which must have caused the flood lay somewhere near the summit of the Nunkhu La at an altitude of at least 14,000 feet.

> 'There can be no doubt I think that this cataclysm, like many another in the Himalayas, was due to the sudden liberation of the pent-up waters of a lake caused by a glacial dam, or landslide, on the heights above. The almost perpendicular gash, half a mile wide, which had been cut in the mountain side by the descending flood, was terrific, and the sight and sound of the mountainous wall of waters falling 2,000 feet must have been truly appalling, transcending Niagara a hundredfold in its awful sublimity.' [28]

Thus, foiled in his attempt to reach the source of the flood, Ludlow returned to Lokmo somewhat disappointed with his first excursion to the Yigrong. Even the flora had hardly met his expectations although he had found many plants of great beauty, including several forms, all of them handsome, of *Rhododendron dignabile*, some (13792) with the cup-shaped corollas pure white at the base and a lovely rose-pink in the upper half, others (13793) pure white except for a dark purple blotch at the base of the posterior petal, and still others (13794) of a very fine pink; a form of *R. calostrotum* (13780) with almost violet flowers; *R. hirtipes* (13782) with an abundance of reddish-purple spotting on the posterior petal of the otherwise white flowers; and *R. puralbum* (13781) of Yunnan which he had last gathered in 1938. And for the first time, in fact the first time it had been seen in SE Tibet, he had gathered the lovely violet-flowered *Omphalogramma vincaeflorum* (13785), a common species in Yunnan and Szechwan.

He had expected, and still expected in the days which followed, to find more plants completely new to him. 'There was such a vast area of the Yigrong to explore, and such a short time in which to carry out the work, that perhaps I was tempted to do things too hurriedly. Had I been more patient, more methodical, the true floral wealth of the Yigrong might have been revealed to me. But who can be patient and methodical with the *ignis fatuus* of 100 miles of an unknown mountain range tempting one's every footstep, luring one on to investigate each peak, glacier and river ? And so from week to week I hastened from valley to valley, snatching the treasures that grew by the wayside, thrilled as a child with visions of the unknown and the great joy of exploration.' [29]

Ludlow, now having returned to Lokmo, on 3 June crossed the river, by cantilever bridge, into the valley leading to the Nambu La, and marched westwards to Nambu Gompa which he had visited at the end of August 1938 and where, now, yaks innumerable, from Tongkyuk and other villages lower down the valley, were grazing the rich uplands. Here he halted for a week and botanised in the valleys around the Nambu La (14,970 feet), the pass on a spur projecting southwards from the Yigrong. From the hills to the south of the monastery there were fine views of the Yigrong range whenever it was free from cloud. Two peaks stood out conspicuously; one was somewhat truncated like Gyala Peri and the other more pointed, like Namcha Barwa. One of these was Namla Karpo (24,000 feet) but *which* one Ludlow couldn't quite decide for they both looked much the same height.

He was rather too early in the year to find the flowers at their best but consoled himself with the thought that Elliot, when he passed this way in a few weeks' time, would gather a richer haul. Rather early he may have been but not too early to glean several handsome plants from the Nambu La. Pride of place, and certainly one of the most striking, Ludlow gave to the Chinese *Incarvillea mairei* (13879) which at 12,000 feet plastered the rock faces and the stony hillslopes with its flat rosettes of dark green pinnate leaves and with its great gloxinia-like reddish-pink, white-striped flowers

89. The flood-eroded valley of the Nunkhu Chu

sometimes 3 inches in diameter. In the middle of August Elliot found this splendid plant in fruit, at Nyota Sama in the Kongbo district, and collected seeds under number 15614. There was some variation in the resultant seedlings, and one, opening its light pink flowers in late May or early June, has been given the cultivar name of 'Nyoto Sama' and received the Award of Merit when exhibited before the Royal Horticultural Society by Major and Mrs Knox Finlay in May 1963.

Primulas, not unexpectedly, were very strongly represented. 'Exquisite in its loveliness was *Primula waltonii* [13895] glowing like burgundy in the sunshine. For many years I was under the impression that this primula was always wine-red in colour, but during the war I found yellow and white forms in the mountains near Lhasa, and here on the Nambu La all these colour forms were present. Other species were *P. amabilis, P. alpicola* (also in a variety of colours), *P. bellidifolia, P. tsariensis, P. baileyana, P. dryadifolia, P. chungensis, P. jaffreyana* and a delightful new species belonging to the Minutissimae Section which has been named *P. rubicunda.*' [30] *Primula amabilis* (13888) was inhabiting sodden peaty soil soaked by the melting snow at 14,900 feet and this was the first time Ludlow had seen its beautiful mauve, white-eyed flowers. Sherriff *probably* had collected the species at Go-Nyi-Re in July 1938 but all the plants he saw were in fruit. *P. rubicunda* (13852) was another high elevation plant, growing on rock-faces at 14,500 feet, with the efarinose leaves, not more than half an inch long, forming a compact rosette, and with a one-inch scape carrying a solitary, erect, pinkish-red bloom over half an inch in diameter. The species seems never to have been found since and thus is known only from this one locality, the Nambu La.

P. rubicunda was not the only new species which Ludlow snatched on his hurried visit to the Nambu La. There were two beautiful yellow-flowered Kabschia saxifrages growing in clumps on the rocks, *Saxifraga buceras* (13840) at 12,000 feet, and *S. nambulana* (13850) at 14,000 feet. And the grassy banks of the conifer forest were gay with the lilac flowers of a new lousewort, *Pedicularis stenotheca* (13835). Both Ludlow and Elliot later found the genus *Pedicularis* very well represented in the Yigrong and between them they gathered twenty-two different species, six of which were new to science. Of poppies, *Meconopsis betonicifolia* (13821) decorated the woodland glades at 11,500 feet; a marvellous dwarf form of the Lampshade poppy, *M. integrifolia* (13853), with flowers four inches or more in diameter, half hid itself amongst the willow-scrub at 13,500 feet; and at slightly below this altitude a starved looking, reddish-purple form of *M. impedita* (13878) grew here and there in the boulder scree.

In places, for half a mile or more and from 12,500 to 13,500 feet, the hillsides were stained with the colours of the rhododendrons. Not a great diversity of species was present but such species as were showed a marked degree of variation. Even the usually pure white *Rhododendron puralbum* (13842) showed flowers which were tinged with yellow. Some forms of *R. dignabile* (13857) carried creamy flowers which were faintly suffused with purple, whilst still others had a purplish tinge only on the posterior petal of the otherwise lemon-yellow flowers (13843). Most variable of all was *R. agglutinatum* (13855, 13858); there were forms with pure white flowers, white with small reddish streaks on the posterior petal, through all shades of pink and all degrees of red spotting. These were all growing intermixed and wove a beautiful pale tapestry of colour.

Ludlow and his companions left Nambu Gompa on 13 June and in three marches reached Tsogo at the head of the Pasum Tso which Ludlow remembered vividly from his brief visit in 1938. During the journey they hastily gathered a few items including the striking greenish-yellow and reddish-purple-slippered *Cypripedium guttatum* (13912), several potentillas the most impressive of which was the russet-red-sepalled form of the white-petalled *Potentilla glabra* known as *rhodocalyx* (13886), *Lilium nanum* (13898), the dark damson *Thermopsis barbata* (13904), but otherwise nothing 'belonging to the aristocratic orders'. The Himalaya abound with lakes of glacial origin which as a rule are small and not very significant. The Pasum Tso (11,800 feet), however, is of considerable size, some twelve miles long and two miles wide, and is, moreover, a holy lake, those who perform its two-day circuit acquiring great merit. On Ludlow's arrival it was greatly swollen by the monsoon rains and the village of Tsogo was surrounded by water. Draining the Yigrong, two large rivers enter the lake at Tsogo and Je and at the head of the former river a pass known as the Ba La was reported to lead into the Yigrong valley. This pass Ludlow determined to explore and after buying three sheep for the present equivalent of 40 pence each, he turned north again towards the snowy Yigrong on 19 June.

The first day's march was a very wet one leading up a flat flooded valley. They were continually fording streams and marshes, where in some places the water was three feet deep, and often had to off-load the ponies and manhandle their boxes. A horrid place for

a march, it was, at the same time, a splendid place for potamogetons and Ludlow collected half a dozen for his friend Taylor who had lost all his 1938 collections when the Germans had dropped incendiary bombs on the botanical wing of the British Museum. Not himself interested in 'pots', Ludlow was more pleased with his finding of one of the most handsome barberries he had ever seen, *Berberis agricola* (13937, 14071) with pretty pendent tassels of yellow. Although it was quite common in the hedges in this particular valley he found it nowhere else and as he was unable to return in the autumn to gather fruits he was never able to try to introduce it into cultivation. The following day they turned up a subsidiary valley to a yak encampment called Penda, a short march of three or four miles mostly on the flat with *Primula alpicola* in all its colour forms growing luxuriantly with golden trollius.

All this area in the region of the Po Yigrong range was quite unexplored and it was almost impossible to obtain any reasonably accurate information about routes and passes from the local inhabitants, and only with the greatest difficulty was transport to be had. However, on 21 June Ludlow had secured enough transport to enable him to march up the Ba La valley and to climb to what appeared to be the summit of the Ba La, although he was by no means convinced that he had reached the true watershed since his carriers told him that it was still three marches to the Po Yigrong.

The pass, as far as they went, presented no great obstacles even though the track was very undefined, indicating that the pass was seldom used. But further north it was said to be very difficult necessitating the use of ropes and pitons driven into the rocks. Every small valley in the vicinity held a hanging glacier and the Ba La itself was a-glacier pass. They camped on a grass flat at 13,000 feet where there was a lovely cool spring and here they stayed for the next six days, adding to their already large collections, Ludlow being kept pretty busy constantly changing and drying papers. 'We haven't got enough presses and straps. I have now collected 500 specimens in triplicate since I started in March last and I suppose I shall have over 1000 before the end of this expedition. I collect everything and do not specialise in the pretty pretties.' [31]

For all that it was mostly the 'pretty pretties' he mentioned in his diaries. On 22 June it was 'all sorts of nice things including Rhododendron [*forrestii* var.] *repens* [13969] in all its scarlet

90. *Incarvillea mairei*
91. *Cirsium eriophoroides*

splendour,' *R. puralbum* (13981), *R. hirtipes* (13982), *R. calostrotum* (13974); *Diapensia himalaica*, the pale yellow form (13970) and the pink (13978); the beautiful pink *Diplarche multiflora* (13973); *Primula dryadifolia* (13975); *Omphalogramma vincaeflorum* (13976); a new species of saxifrage with pink flowers, like a much enlarged *Saxifraga oppositifolia*, which was fittingly named after him, *Saxifraga ludlowii* (13968) and which has never been collected since; and 'a strange looking red Sorbus' which was *Sorbus microphylla* (13979).

The next day it was 'a dwarf pale yellow rhododendron [*Rhododendron lepidotum* (13985)], *Primula baileyana* [13989], *P. waltonii* [13990] and a little nivalid primula (purple) I can't quite make out [13988].' This last was identified as *P. sinoplantaginea*, resembling *P. sinopurpurea* but with very narrow leaves, a fairly common plant in w Szechwan to northwards of Tatsienlu, and in Yunnan restricted to the NW corner and to the adjoining Tibetan frontier. On 25 June Ramzana 'got a few nice things including a lovely cushion potentilla [the yellow *Potentilla microphylla* (14001), a shrub 2 feet in diameter growing on a huge boulder], a dwarf lonicera [*Lonicera hispida* (14000) with very pale yellow flowers] and a very fragrant primula belonging to the Rotundifolia section [13998] which may, or may not, be *Primula littledalei* [it was].' And on 27 June 'Ramzana went out after flowers and got one or two rhododendrons but nothing else of importance'. Ludlow did not realise that Ramzana had also collected a new species of pedicularis with handsome reddish-pink flowers which ultimately was very cumbersomely named *Pedicularis tantallorhynchoides* (14025).

From Penda Ludlow returned to Tsogo on 1 July. '*Primula alpicola* in all its colour forms of yellow, white, purple, and various shades of purplish-red, was a lovely sight, its scent and that of a lilac lonicera filled the air for mile after mile. The day was dull and rainy but the sight of these lovely primulas at every step made the march memorable.' But he was worried about his servant's footgear. 'The boots bought in Kalimpong are in tatters. The rubber-soled plimsolls only last about ten days and then have to be thrown away. Laboo's boots are in a hopeless state and he is now wearing his last pair of plimsolls. I cannot purchase local Tibetan boots for love or money and even if I could the untanned leather soles would only last a few days on the rough hillsides. I must see what I can get at Gyamda when I send in for our mail. I expect Elliot's men are just as badly off.' [32] And about a mile from Tsogo there was another worry when they had to cross flood water a little higher

Primula tsongpenii

than the ponies' bellies. 'But we hitched up our loads and got through without soaking our boxes. The only nightmare I have is the ruination of weeks and weeks of work because one's precious museum plants and birds have been immersed in water. The fording of these rivers and lagoons is nothing and it matters not if one's bedding *does* get wet and one's food sodden. One can always dry bedding and buy more food. But you can't replace specimens, collected at various altitudes and in various localities, when ruined by water.'[33] Fortunately Ludlow's nightmare was never realised.

'Tsogo is a muck-heap in summer and knee deep in mire water. Why the village was ever built here, when a much drier and better site is available in the hillslopes close by, I really cannot understand.'[34] For all that, it had its redeeming features. In the hedges the lilac-belled *Notholirion bulbuliferum* (14079) grew in vast profusion along with a white gloriously fragrant briar, the Chinese *Rosa sweginzowii* var. *inermis* (14056) and the Himalayan rose-red *R. macrophylla* (14055). And in the fallow fields along the shores of the Pasum Tso were two of the most attractive of all louseworts: *Pedicularis garnieri* grew like a weed, its pale yellow, golden-throated flowers at a casual glance giving the impression of a vast carpet of primroses; and the white, purple-spurred *Pedicularis fletcheri* (14058), the new species which Ludlow and Sherriff had first found at Singo Samba, near Molo, in 1936, was also an uncommonly abundant and attractive plant in the Pasum Tso meadows.

From Tsogo, on 7 July, Ludlow marched along the eastern shore of the lake to Je, a pleasanter place by far, its wooden chalets and surrounding fir trees reminding him of Switzerland. 'A young man of means lives here who took care to impress on me that Je was independent of Showa and paid no taxes to the Dzong. Also that he was the owner of the big Penam Chu valley up which I want to look for plants. So I took pains to propitiate him and was promised every help in the matter of transport and supplies. We leave tomorrow.[35]

Thus, on 8 July, they left for the headwaters of the Penam Chu, a glacier-fed stream which rises on the slopes of Namla Karpo, and spent three days at the head of the valley, greatly hampered in their movements by the torrential rain and a river so swollen that it had destroyed all the bridges. Though the flora seemed not to be a rich one Ludlow added another Chinese rose to his collections, the pale sulphur-yellow *Rosa omeiensis* (14117) which grew very sparingly near the tree-line at 13,500 feet. Between this altitude and 14,500 feet he gathered several pretty saxifrages, including the white *Saxifraga pallida* (14104) and *S. granulifera* (14109), and a completely new yellow-flowered species which formed lovely rounded cushions on the moss-covered boulders, *S. anadena* (14092). And at 15,000 feet vast sheets of *Primula baileyana* (14114) covered the scree and the grass-covered rocks in a nullah which seemed to descend from the vicinity of Namla Karpo.

Whilst encamped in this valley Ludlow received disquieting news. Thamchen, the Tibetan servant Ludlow had engaged to accompany his Lepcha collector, Tsongpen, to the Sü La range in Pome, arrived one evening and announced that they had been forced to leave the district because the dzongpen of Showa, after accepting the presents Tsongpen had offered him, had refused to give them any assistance in the matter of food and transport until he had seen the expedition's passport, which was in Ludlow's keeping. Ludlow understandably was greatly dismayed. 'There is only one passport

> and this can't be split up amongst four parties (Sherriff, Elliot Tsongpen and self). We did ask . . . . for a duplicate passport but [were told] it was not customary to grant duplicates. What the dzongpen should have done if he had doubted [Tsongpen's] assurance that I held a valid passport was to have written asking me to send it to him and not to have peremptorily refused all help just because the passport did not happen to be in Tsongpen's keeping. The dzongpen has now ruined the whole of our plans for collecting on the Sü La range. It is of no use sending Tsongpen back with the passport. I must think of some other area for Tsongpen to work. Perhaps it would be best if I returned with him to the Tsangpo valley and worked the passes on the main range [for the autumn flowers and seeds], and leave Elliot to cross the Lochan La and work the upper Yigrong near Nyoto Sama . . . . I am sending Thamchen, tomorrow, with a letter to Elliot asking him to come along to Shoga Dzong as soon as he possibly can, for a consultation.' [36]

Ludlow was desperately disappointed for he had seen enough of the Sü La range in February to convince him that it would be a most profitable hunting ground. Moreover, Tsongpen's short sojourn there had more than justified Ludlow's optimism for he had gathered upwards of twenty rhododendrons, including Forrest's Yunnan discovery, the deciduous precocious greenish-yellow-flowered *Rhododendron oulotrichum* (13118), close ally of the well-known *R. trichocladum*, as well as a new species, low-growing and pink-flowered, *R. pomense* (13177); a new lily, up to 18 inches tall, with a solitary purplish-red unspotted flattish flower up to 3 inches across, *Lilium paradoxum* (13114); and a new species of omphalogramma akin to *Omphalogramma elegans, O. tibeticum* (13141). There can be no doubt that Tsongpen's collection would have been a notable one had he been granted reasonable facilities for working the area.

Thus, rather low in spirit, Ludlow returned to Je and thence proceeded to Shoga Dzong where, on 22 July, Elliot joined him. There was a great deal to talk about including the important item of their future movements. They both agreed that it would be best to split up once again and for Ludlow to return to the passes in the Tsangpo valley and for Elliot to work the area around Nyoto Sama. They would reunite at Tsela Dzong in mid-October to formulate plans for their return to India.

Elliot's Journey

Elliot had had a splendidly profitable time on the passes on the main Himalayan range in the vicinity of Pe at the height of the flowering season and had returned with over 500 gatherings. From the Deyang La (14,176 feet), where he had spent a couple of weeks, he had made close on two hundred gatherings, including over forty rhododendrons, the most conspicuous plants on the pass, representative of over twenty species and varieties. With the exception of *Rhododendron chloranthum* (15193), another ally of *R. trichocladum*, none of them was new to the collections but many were interesting and important from a distributional point of view and in showing the variability, especially in flower colour, of well-known species. The flowers of *Rhododendron wardii*, for instance, varied from the most pale of yellows to a rich sulphur-yellow; sometimes they were quite free from any markings on the petals, sometimes the posterior one would carry small pinkish patches at the base and at other times a large purple blotch; usually the very young flowers, not fully opened, were of a clear yellow, but sometimes would be flushed in varying degrees with pink. *R. hirtipes* might be almost pure white, several shades of pink, or a mixture of white and pink, or even pale yellow; the posterior petal might be free of spots, or lightly or heavily speckled with purple and with or without a basal purple patch. The flowers of *R. cephalanthum* ranged from white through pale pink to dark pinkish-red, sometimes with a bright red tube. In *R. erythrocalyx* the red sepals were by no means a constant character for often they were a clear green, with no trace of any red pigmentation. With such variation in the field it is little wonder that there is so much among rhododendrons in cultivation.

Of primulas Elliot made few gatherings, less than twenty in fact,

but thirteen different taxa were represented, including the then little-known *Primula amabilis* (15137), the *violacea* form of *P. chionota* (15163) and the strange *P. advena* var. *concolor* (15180) which had been gathered on the 1938 expedition. For the rest, in Elliot's presses were meconopsis, orchids, fritillarias, cassiopes, irises, diapensias, diplarche, potentillas, pedicularis, and an interesting white-flowered greeny-eyed Kabischia saxifrage which Elliot had now found for the first time in se Tibet, *Saxifraga georgei* (15177). In fact Elliot's record from the Deyang La of this species which previously had been known only from Western China where George Forrest, after whom it was named, had found it in 1921, was not only the first record for Tibet, but the first record for the Himalayas. Since 1947 it hasn't been recorded from se Tibet but in 1949 Ludlow, Sherriff and Hicks were to record it on several occasions in Bhutan and it is now known to occur in Nepal. This wide distribution, from China to Nepal, is unique among the Kabschia saxifrages. And, unusual in a plant of so wide a distributional range, it exhibits but little variation and such as there is is due entirely to ecological conditions.

From the Deyang La Elliot had moved to the pilgrim pass of Budi Tsepo La where he had gathered, among much else, a new pedicularis with white and purple flowers which was later named *Pedicularis elliotii* (15242); *Primula genestieriana* (15240) which Ludlow and Sherriff had previously collected several times and which, when Kingdon Ward found it on the Doshong La, was named *P. doshongensis*; two forms of *Rhododendron doshongensis*, Ward's Doshong La discovery, one (15243) with the palest of pink flowers speckled with darker pink and the other (15247) with very dark pink heavily specked flowers; and a form of *Rhododendron cephalanthum* (15241) with sulphur-yellow trusses which had been discovered in Burma and named var. *nmaiense*.

Thence to the Doshong La (13,500 feet), Kingdon Ward's 'rhododendron fairyland', visited by Ludlow and Taylor in 1938; among the ten different rhododendrons Elliot gathered were the two forms of *Rhododendron cerasinum*, one with flowers of dark crimson (15279, 15280), the other with more or less white bells stained with cerise around the rim (15281) which Ward had named 'Coals of Fire' and 'Cherry Brandy' respectively; very much the same form of *R. parmulatum* (15291) which Ludlow and Elliot had found in Pemakochung but with the very pale yellow rose-tipped petals heavily speckled with purple; and a species new to the Ludlow and

92. *Saussurea laniceps*

Sherriff collections which Forrest had discovered on the border of Yunnan and SE Tibet in 1919, the creamy-white, purple-blotched *R. telopeum* (15306), close ally of the well-known *R. caloxanthum*, Farrer's discovery in NE Upper Burma in 1919 and 1920.

Shortly before meeting Ludlow at Shoga Dzong Elliot had spent ten days on the Nambu La (14,970 feet). As most of the rhododendrons had finished their flowering it was the herbaceous plants on which he concentrated, and among the primulas, the pedicularis, the asters, the meconopsis, the saxifrages, the fritillarias, the androsaces, the alliums, the cremanthodiums, the gentians, the swertias, the codonopsis, the saussureas, the ligularias and the other things, possibly the most interesting discovery was that of an aconite new to science which was named *Aconitum prominens* (15408) because the petals protruded prominently from out the deep blue very shallowly hooded sepal. This is the only occasion on which this very distinct plant has been found, its nearest of kin apparently being *A. polyanthum* which is seemingly restricted to the neighbourhood of Tongolo and Tatsienlu in Szechwan.

Ludlow around the Tsangpo valley

Having compared and discussed their collections it was now time for Ludlow and Elliot once again to separate, and on 25 July Ludlow and his party, with thirteen pack oxen, two ponies and four riding ponies, left Shoga and began a ten-day march down the great Gyamda Chu, now in full flood. It was a tiring and a tantalising journey because of countless transport changes due to the 'sadzi, sadzi' system of transport under which they were now operating and the constant fording of the swollen river. Taylor was much in Ludlow's mind for the extensive marshes of the Gyamda valley were rich in potamogetons and Ludlow gathered many on Taylor's behalf. He also gathered Taylor's barberry, *Berberis taylori* (14204) which was much in evidence on the hillsides and in the lanes bordering cultivation. The beautiful sky-blue *Onosma waddellii* (14189) had colonised the gravel and sand along the banks of the river whilst the same clear sky-blue of the open flowers of *Gentiana waltonii* (14195) coloured the nearby south-facing rocky hillslopes. On 31 July he reached Kyabden, scene of Taylor's strange illness of August 1938, and Dzeng on 4 August. Here, at the Gyamda-Tsangpo junction, the crops were ripe, cutting had commenced, the swallows were breeding in exactly the same rock face as in 1938 — and the Tsangpo was an amazing sight, a great muddy sea three or four miles wide. However, next day Ludlow reached the ferry, crossed on the tru, and reached the village of Tamnyen where he laid in supplies, gathered seeds of the new *Paeonia sterniana*

(14231) and left for the Deyang La on 6 August anxious to collect the autumn flora and such seeds as might be ripe.

And here, with the assistance of Ramzana and Tsongpen, during a break in the monsoon which gave them the most perfect of weather, he gathered copious material of exactly a hundred different taxa in exactly five days. Of course the great show of rhododendrons which had greeted Elliot at the end of May and in early June was long since past and the only rhododendron blooms now to be gathered were those of the pale lemon-yellow, rock-sprawling *Rhododendron trichocladum* (14244), of the dark damson *R. campylogynum* (14295) and of its cherry-pink form known as var. *cremastum* (14297) which Forrest had discovered on the Mekong-Salwin divide of NW Yunnan. Also long since past was the height of the primula season. Even so, seven species still were flowering; the creamy-yellow, orange-eyed *P. chionota* (14263 — as well as its violet-purple or pinkish-red form *violacea* (14262, 14264); those close allies, the rose-red *P. dryadifolia* (14271) and the pink *P. tsongpenii* (14270), the latter the new species found near Paka and on the Lo La in 1938; the golden-yellow candelabroid *P. morsheadiana* (14312) which Ward had discovered on the Doshong La in 1924; a fine vigorous form of the white-farinose, lilac-purple flowered *P. cawdoriana* (14318); *P. alpicola* (14319), its cream flowers sometimes flushed with pink; and the violet, copiously farinose *P. baileyana* (14320).

On the other hand it was fast approaching the peak of the flowering season of the gentians and with the sun to open their flowers they were a lovely sight on the grassy stony slopes and cliff faces. In all, Ludlow gathered eight different species, including the indigo-blue, white-tipped new species *Gentiana taylori* (14266) which had been discovered on the Kucha La during the 1938 expedition. Saxifrages, too, were much in evidence, splashing the rocks and screes with all shades of yellow; the one novelty among the collection of seven species was *Saxifraga thymoides* (14330). Other novelties included the deep violet *Aconitum tsariense* (14261) which Ludlow, with Sherriff, had discovered in Tsari in 1936 when it exhibited a degree of variation, sometimes carrying flowers of a rich blue-violet (2491), at other times white flowers stained with purple (2076); the yellowish-green *Pedicularis angustiloba* (14321) and a groundsel with yellow ray-florets and a dark brown-tipped green involucre, *Senecio atrofuscus* (14324).

Among the seeds Ludlow gathered on the Deyang La were those of *Omphalogramma minus* (14307) which for a time served to maintain this lovely purple-flowered plant in cultivation in gardens in Britain; those of *Primula dickieana* (13281) from which Mrs Renton

of Perth grew plants which were the first to flower in cultivation, in
1949; and those of *Lilium souliei* (13275). In a few weeks' time
living plants of the lily and of *Primula dickieana* would be sent to
Britain by air and would flower in several gardens in 1950.

Though novelties were few, the collections of Elliot and Ludlow
on the hitherto unexplored Deyang La had served to extend the
known distribution, within SE Tibet, of a great many species.

From the Deyang La Ludlow returned to Pe on 13 August,
whence he moved to the Doshong La two days later, camping not
on the site he and Taylor had occupied in 1938 but 1,000 feet lower,
on the spot Ward had pitched his tent in 1924, at the junction of
two valleys, one leading to the Doshong La, the other to the Pemo
La. Even before he began serious collecting he discovered a very
handsome aconite, with a broad short inflorescence of great purple
flowers, growing in the open glades of the conifer woodland. This
was the one and only time Ludlow found the plant in all his
explorations; it proved to be a species new to science which was
named *Aconitum elliotii* (14346).

On 16 August Ludlow's collectors were on the Doshong La 'but
did not do so well as I expected. They got no primulas, except
Primula valentiniana [14378]. Ramzana got a strange looking nomocharis [14363], greenish-white, but unfortunately could only find
one specimen. They also brought back a strange Cnicus (14359)
about 2 feet high, with one huge terminal flower . . .' The nomocharis was what is now known as *Lilium euxanthum* which Forrest
had discovered on the Mekong-Salwin divide in 1918 and again in
1921, whilst the so-called cnicus proved to be a cirsium and bears
the botanical name of *Cirsium eriophoroides* subsp. *bolocephalum*;
despite this cumbersome name this rather weird plant with its single
purple head, shaped like an old fashioned table-lamp complete with
globe, is a most attractive one which inhabited the snow-water-
irrigated screes and the grassy slopes. Even more weird than the
thistle was *Saussurea laniceps* which grew on boulder scree on the
summit of the pass, the entire plant, including the flowers, almost
completely enveloped in a dense tomentum of cotton wool (14426).

The following day 'I went up the pass myself — a climb of about
2,000 feet — and got one or two nice things including *Primula
falcifolia* [14390] and *P. morsheadiana* [14391]. *P. valentiniana*
and *P. dickieana* were very abundant. I got *Nomocharis souliei*
[*Lilium souliei* (14386)] and one or two nice saxifrages, also
Meconopsis simplicifolia but not a good colour [14383, the sky-
blue petals were marred by the purple-red stain at the base].
Of course one was assailed by the wind and the rain (some-

times snow and sleet) directly one reached the moorland above the tree level. But, on the whole, I suppose yesterday and today must be accounted good days for the Doshong La where it never seems to cease raining or snowing, day or night, year in and year out.' [37]

And on 18 August whilst the collectors were on the pass Ludlow 'stayed in camp and changed papers and separated dried specimens from those that are damp. It took me up to 2 p.m. to do this. We have four big presses full of flowers. The collectors brought back nothing of interest except a small white pedicularis with a purple spur [*Pedicularis bella* var. *holophylla* (14403)] from the cliffs on the crest of the main range where Tsongpen also found four specimens of *Primula* [*genestieriana* (14401)]. A gusty, windy, day with the usual drizzle. It is this perpetual drizzle, with gleams of diluted sunshine at intervals, which plants love and which accounts for their amazing variety and abundance on the main Himalayan range hereabouts.' [38]

Round about this time it wasn't only the rain which was never-ending. 'My menu seems to be curry and rice for all meals. Yesterday, Ramzana cooked [for me] enough curry and rice for 2 or 3 people. I had this curry last night for dinner, I had it for breakfast this morning, he hotted it up for lunch today and wanted to give it to me for dinner tonight. But I drew the line at this and asked for a roast cock (very old) as a change, intimating that the curry balance would be served up at breakfast tomorrow. But perhaps I malign curry and rice for I really don't know what I should do without it. If a sheep is lean and miserable, curry it; if a cock is old and stringy, which it nearly always is, curry it; if you have no meat but only vegetables, curry them. Curry is the solution to the traveller's needs in the Himalayas, and if you don't like curry then God help you.' [39]

From their camp on the misty Doshong La Ludlow and his collectors returned to the Tsangpo valley and on 28 August descended to Kyikar where the great river already has begun to bore its way through the Himalayas. From Kyikar there was a track moving south to the Nam La or 'Sky Pass' (17,140 feet) which is situated on a spur of Namcha Barwa and, like the Doshong La, leads to the 'Promised Land' of Pemako. Few travellers use the Nam La for it is a higher and more difficult pass than the Doshong La, but Ludlow was determined to go there and, having persuaded the villagers of Kyikar to carry the loads, the small expedition set out for the pass on 30 August. They climbed 800 feet or more to get away from the Tsangpo valley and then the track rose steadily through young larch and birch forest for another 1,000 feet until, rounding a spur, it entered picea forest. Still rising steadily, they eventually camped at 12,000 feet in a meadow of *Primula alpicola*. Not until the end of the following day did they reach the rhododendron moorland at the foot of the Nam La; they camped on the eastern shore of the Nam Tso where the rhododendron scrub was so thick that many plants had to be uprooted before a level enough surface for the camp could be made. From his tent door Ludlow could see the great peaks of Sengdam and Gyala Peri mirrored in the waters of the lake and from a nearby ridge Namcha Barwa and the Sanglung spur looked mightily impressive.

In South-Eastern Tibet, during the monsoon, photography of the great peaks is very much of a gamble for usually, day after day, the view may be obscured and only on the rare occasion may the veil of cloud and mist be lifted. Ludlow was lucky, for from the Nam La, for a brief space each morning, he was given a sight of the great guardians of the gorge.

And he was lucky with plants and made interesting observations on some of them. 'On the Nam La, near the lake, grew the trailing *Aconitum* [*stapfianum* (14470) with reddish-purple-hooded flowers, spreading luxuriantly over the dwarf rhododendrons]. The aconite is absent from the Doshong La only 5 miles away where another attractive aconite grows [*A. elliotii* (14346)] and the Doshong La aconite does not grow on the Nam La. All these Himalayan passes in SE Tibet seem to have their own particular flora. Another example is a huge yellow-flowered potentilla [*Potentilla arbuscula* (14466)] which abounds on this pass but is not to be found on the other passes further west.' [40]

Aconitum stapfianum may have been absent from the Doshong La but it had previously been recorded from the Kongbo district in 1924 by Kingdon Ward who found it on the Temo La. However these 1924 and 1947 records are the only ones from SE Tibet, the species having been described from material collected by Forrest on the Likiang range of NW Yunnan. Another very distinctive aconite with a branched inflorescence of greenish-white flowers faintly flushed with mauve (14487) which Ludlow now found on the Nam La had also been gathered in SE Tibet on one previous occasion, this time by Ludlow himself, with Sherriff, at Chayul Charme in 1936. It proved to be a new species, was named *Aconitum viridiflorum* and hasn't been found elsewhere.

Probably the Nam La plant which impressed Ludlow most was a dwarf delphinium no more than six inches high bearing large violet-purple flowers covered with golden hairs (14478). It was *Delphinium nepalense* and grew in gravel and scree at an altitude of 14,500 feet. Ludlow instantly saw that it was a perfect plant for the rock garden. Unfortunately seed was not available at the time, he was never able to return to the Nam La in late autumn to harvest seeds, and he saw it in no other locality.

From the Nam La he now descended to the big glaciers at the foot of Namcha Barwa above Tripé, where in 1938 Taylor had discovered the lovely *Primula aliciae* which Ludlow's collectors now failed to find, and spent the next month retracing his footsteps to Tse, re-ascending, for the seed harvest, the various passess he had worked on the downward journey. He reached Tse on 4 October and Elliot joined him there two days later, bringing with him 'the female takin skin and skull which the Trulung shikari had promised to get me, but though the skin was taken off well he had omitted to open up the hoofs and extract the cartilage from the ears and nose and to put on arsenical soap. In consequence decomposition had set in and the head was crawling with worms and stank to high heaven. Hair also came out in tufts from the head and reluctantly I had to decide to throw the skin away. A pity, because it would have completed the British Museum exhibit as we have a fine male and two calves.'[41]

Elliot in the upper Yigrong

Of course Elliot returned with more than a stinking takin skin. He had left Ludlow on 25 July and in the meantime, having travelled northwards over the Lochen La to the neighbourhood of Nyoto Sama in the Upper Yigrong, had gathered over 300 specimens, many of them fruiting ones, of such genera as cyananthus, codonopsis, campanula, adenophora, androsace, senecio, saussurea, cremanthodium (including the pretty pink *Cremanthodium palmatum* subsp. *rhodocephalum*), swertia, gentianella, gentian, salvia, dracocephalum, elsholtzia, pedicularis, pleurospermum, polygonum, clematis, cotoneaster, deutzia, potentilla, berberis, iris, allium, notholirion, fritillaria, and the 'Nyoto Sama' form of *Incarvillea mairei* (15614). Of the 'more aristocratic orders', as Ludlow called them, there were three lilies; *Lilium nanum* (15750), only one of many Ludlow and Sherriff gatherings of this lovely dwarf, which served to augment the stock of plants in cultivation, *L. wardei* (15806) and *L. tigrinum* (15707, 15736); there was *Fritillaria cirrhosa* (15609) the seeds of which served to reintroduce this desirable species into cultivation in Britain; there was the dainty *Streptopus simplex* (15818); there were five species of meconopsis; fifteen species of primula; a dozen species of rhododendron including *Rhododendron rufescens* (15669), the first time this pale pinkish-flowered antho-

pogon rhododendron, originally found in Szechwan, had been gathered in SE Tibet; eleven species of gentian and nine species of saxifrage including the deep yellow *Saxifraga implicans* (15570, 15582) which Ludlow and Sherriff had found near Sanga Chöling in 1936 (2061) and, with Taylor (6875), at Pangkar in 1938 and which was to be described as a new species in 1960, as was the bright yellow, maroon-spotted *Saxifraga tigrina*, based on a solitary collection (15805) which Elliot had made at Yumba. And there was another new aconite, a dark purple-flowered species, 2–3 feet tall, which Ward had first gathered in the Upper Yigrong valley in 1935, which Ludlow and Sherriff had also found in the hills north of Lhasa in 1942 (8770) and in 1944 (11085) and, with Taylor, on the Pasum La in 1938 (6883); Elliot recollected it on the Lochen La (15569) and in 1963 all these gatherings were described under the name of *Aconitum pseudosessiliflorum*.

Return to India

This was almost the end of the journey as far as collecting was concerned. The expedition left Tse on 12 October and returned to India by the circuitous and wearisome road via the Tsangpo valley which it had traversed the previous year. The shorter route, and the one Ludlow was anxious to take as it is incomparably richer botanically, via Tsari and Eastern Bhutan, was unexpectedly denied him at the last moment by the Political Officer in Lhasa. During the journey they daily augmented their collections by snatching such fruits and seeds, even plants, as appealed to them. For instance on 13 October they gathered an abundance of seeds of *Paeonia lutea* var. *ludlowii* (13313) which was already established in gardens in Britain from seeds of the 1936 and 1938 expeditions; on 16 October seeds of the orange-yellow, brown-spotted *Briggsia aurantiaca* (13320) and of the fine deep blue *Gentiana veitchiorum* (13321) both of which were successfully germinated in Britain; on 17 October seeds of *Cyananthus sherriffii* (13329) which served to augment the stock of plants in gardens grown from the 1936 and 1938 seeds, and plants of *Lilium wardii*; on 19 October fruiting samples of a local wheat as part of the collection of different wheats and barleys Ludlow was gathering for the British Museum, as well as two fruiting branches of two different species of clematis.

One lot of clematis fruits (13343) proved to be those of the fragrant nodding bell-shaped, yellowish-green *Clematis rehderiana* whilst the other (13342), those of a most desirable form of *C.*

93. *Cremanthodium palmatum*
94. *Streptopus simplex*

305

orientalis. In October 1950, Messrs W.E.Th.Ingwersen Ltd exhibited flowering specimens, grown from fruits of 13342, at the Royal Horticultural Society. The flowers, on peduncles four inches long, were bowl-shaped with four elliptic lemon-yellow sepals, almost as thick in texture as orange peel, with a central tuft of yellow stigmas surrounded by a mass of slaty-purple stamens. Many forms of *Clematis orientalis* had been in cultivation in the past but all had lacked the outstanding character of this particular form, the unusual thickness of the sepals. The plant was given the Award of Merit, since which time it has been known as 'Orange Peel'.

Though there were delays because of transport problems almost throughout the entire journey back to India the expedition succeeded in reaching Gyantse in time to hear Princess Elizabeth's wedding service on the BBC wireless programme on 20 November, and Kalimpong and the Sherriffs on 5 December. After an absence of fourteen months, Ludlow and his companions had returned with close on 4000 pressed plants in sixteen boxes as well as with four boxes of seeds and living plants. Subsequently the seeds were made up into 20,000 packets. 'This sounds a bountiful harvest and, as far

as herbarium specimens were concerned we had no cause to reproach ourselves . . . . But it would be wrong to give the impression that we collected seeds, or living material, of all the good garden plants we met with. This we most certainly did not do. A plant collector, I suppose, is never really satisfied with his harvest. Never does he completely garner all his spoils. The hazards and difficulties of seed collecting are many — the seed may be unripe and green, over-ripe and shed, eaten by grubs, grazed by cattle, destroyed by floods or buried 'neath the snow. But the real reason why we failed on this particular expedition to secure the seeds of so many desirable plants was due very largely to the twin obstacles of time and distance. Our passport gave us permission to spend but a year in Tibet, a woefully inadequate period considering the great distances we covered.' [42]

Though they didn't realise it at the time, this was to prove Ludlow and Sherriff's last Tibetan expedition.

Chapter Seven

Temperate and Alpine
BHUTAN

o

Ludlow and Sherriff had by this time officially left India, but before settling in Britain they determined to have one more expedition, 'our final fling' as Ludlow called it. In fact they planned two expeditions; Sherriff and his wife to spend a summer in the Mishmi hills of Assam whilst Ludlow would return to SE Tibet and especially to the Tsangpo gorge. However as both their plans for travel were refused, they turned once again to their good friends His Highness the Maharaja of Bhutan, and Tobgye, for permission to undertake further botanical exploration in Bhutan — and their wishes were granted. 'I, and Sherriff too, owe everything on this last trek of ours to Tobgye. Without his goodwill we hadn't the slightest hope of getting anywhere that was really interesting from a botanical point of view. Tobgye is one of God's good men. I have known him for two decades and I would trust him with my last rupee.'[1]

Even though permission for a summer's botanical exploration in Bhutan had been granted, Ludlow and Sherriff still decided to go their separate ways with the intention of working the whole of temperate and alpine Bhutan from west to east. Ludlow would concentrate his activities in Western Bhutan, Sherriff in Central Bhutan, whilst Betty Sherriff, with Dr J. H. Hicks who had joined the expedition as Medical Officer, would collect in Eastern Bhutan, They would all meet at Bumthang round about the middle of August to pay their respects to the Maharaja and to compare notes, before separating once again for the seed harvest.

Ludlow goes fishing (Map 19)

Thus it was that on 1 April Ludlow found himself once again at Sharithang — and a very different Sharithang from the encampment he had first seen during the latter part of May 1933. Then there was only one wooden building; now there were half a dozen, some for sleeping, some for cooking, some for servants' quarters, all very nice and clean and contained by a conifer fence. And unlike the time of his last visit the whole place was ablaze with the dark crimson flower-trusses of 3o-foot trees of *Rhododendron arboreum* (L s & H 16007)[2] and of *R. ramsdenianum* (16009) which abounded in the surrounding conifer forest and along the banks of the Sharithang Chu. Ludlow's companions were his three servants, Ramzana the faithful Kashmiri, the Lepcha collector Danon whom Ludlow hadn't employed since 1936, and Phak Tsering, a Sikkimese cook. All told there were thirteen pony, or mule, loads some of which would have to be left behind at Ha, or else sent on to Bumthang, for they couldn't go with him along the yak route he would have to follow after leaving Ha. Eastwards of Ha he intended to stay as high as possible under the eaves of the main Himalayan range. As no European had traversed the area the existing maps were quite useless. 'They are quite blank. I do not think that I have ever set out on a journey which is more blank on the map than this one. So if I do not succeed in finding many new plants I shall at least be able to insert a few diagrammatic rivers and ranges.'[3] To his immense satisfaction Ludlow was able to do both.

But first he had to cross the Kyü La and the Ha La, a mile separating the two, on the way to Damthang, and as he was in no hurry to reach his main collecting ground, he decided, for the next few weeks, to take a fishing holiday for he was anxious to test the success of Tobgye's endeavours to stock his waters with trout.

'There is a lake about 500 feet below the Kyü La which Tobgye has stocked with the trout sent from Kashmir in 1940. I did not test this lake but put up my rod and spent an hour on a lake about 1,000 feet below the Ha La on the east side. The surface of the lake was almost unruffled and I consider myself lucky to have got a couple of half pounders. I rose one or two good-sized fish and the little lake appears to be well-stocked. The fish, moreover, were in good condition. They are very dark in colour. In time Tobgye will have to net this lake and take out surplus fish otherwise they will degenerate with an increase in numbers.'[4]

Some four miles below Damthang (10,000 feet) he fished again. 'In the first pool I got into a two-pounder and all the way down to Ha Dzong (9,100 feet) I found fish numerous and in splendid condition. They are, I think, the most golden and handsome trout I have ever set eyes on and there is no doubt that the abundance of food accounts for this. The valley is full of springs which remain unfrozen in winter and must provide much larval life. In addition there are local fish which the trout finds an easy prey. I was broken three times by good fish I

Map 19. Ludlow in N.W. Bhutan

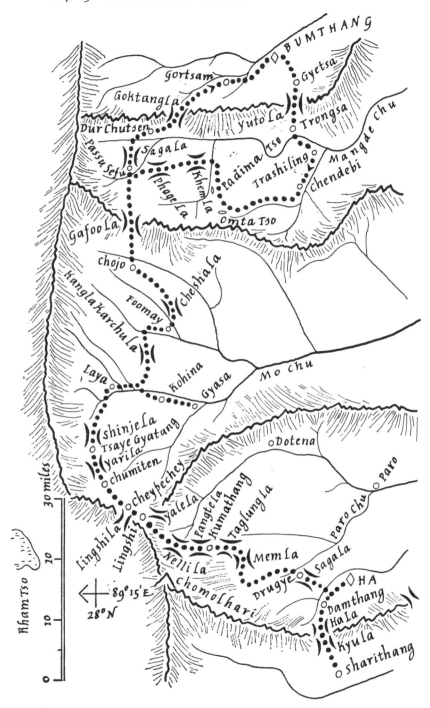

never saw. I kept 9 fish and put back 20, many of them $1-1\frac{1}{2}$ lbs. I got nothing bigger than two pounds but far bigger fish than these are in the river and report has it that there are fish of $8-9$ lbs. I can quite believe it for I saw several fish of $3-4$ lbs. The big fish are reported to be below the Dzong and trout are said to extend for at least 5 miles below Ha and probably a good deal further. I must investigate — I have plenty of time to do so. There is a very well constructed hatchery just above the Dzong in which there are 20 or so fish of $3-6$ lbs., and a number of smaller size. This hatchery was constructed under the supervision of Tendup [a member of the 1933 expedition] who was sent to Kashmir in the winter of 1939–40 and spent three months at the Harwan hatchery learning his job. He learnt it so well that he successfully transported 20,000 ova from Kashmir to Ha and it is entirely due to him that the Ha river has now, probably, the finest trout fishing anywhere in India. If Tobgye wished to commercialise his fishing (which he doesn't) I estimate that he could, by charging Rs 500/- a month and dividing his river into six beats, make Rs 18,000 a year out of licence fees. Many would pay Rs 500/- a month for such superb fishing. It would be cheap compared with comparable fishing anywhere [else] — say Scotland.'[5]

Ludlow spent most of the next four weeks fishing the Ha river between Ha and Chunzu Gompa, typing a paper on the birds of Kongbo and Pome, pruning Tobgye's fruit trees, and collecting plants. On one day, in the space of two hours, he caught eleven trout, on a Jock Scott, all in the junction pool of the Ha Chu and Lang Chu, at least six of them between 2 and $3\frac{1}{2}$ pounds. 'From this it is quite evident that the Ha Chu in its lower reaches is just stiff with trout.' Another morning he 'spent half an hour lying flat on a rock over a deep pool examining the fish in its depths. I saw four trout, one of which was well over 5 lbs. and the other three not far short of 4 lbs., cruising about. This river is a paradise for fishermen. I doubt if there is any river at home that can equal it in number and size of fish, and condition, per 100 yards.'

Not all the fish he caught were for himself and his companions for on 1 May 'Tobgye wanted four dozen trout sent in — two dozen to Gangtok and two dozen to Kalimpong. Owing to an increase in the volume of the Ha Chu it is becoming more and more difficult to catch trout with a casting net. The result of a whole day's catch with the net was six trout. So I went out in the morning to fish in earnest. I put on a new cast with a dropper as well as a tail fly and I donned my waders. Between 9 and 12 a.m. I had killed 17 trout and put back double this number. In the evening I fished for 1½ hours and caught or rather killed, another 10 fish. So in 4½ hours I killed 27 fish and must have returned 50. I got nothing big. I lost the only big trout — a fish of 4 lbs. — which I hooked. I was accompanied wherever I went by men with a bath-tub and bucket and the fish when caught were kept alive in the former till the evening when they were knocked on the head and packed in a box which a man took off posthaste to Gangtok. He travels all night and is due to reach Gangtok in 24 hrs!!'

As at Sharithang, so at Ha; since Ludlow's last visit in 1933 there had been many changes. There were now orchards of apples, pears, cherries and apricots; Rhode Island Reds strutted, and a pair of geese waddled, about the servants' quarters; there was more comfortable furnished accommodation for Togbye's guests as well as new quarters for the Political Officer's clerical establishment; there was a swimming pool some 50 yards by 25 yards, amply stocked with trout!; and there was an archery range for the locals to indulge in their favourite sport. The fruit trees had been brought from Kashmir and, not unexpectedly, many were suffering from the terrible Kashmir disease, St Joseph's Scale. Ludlow spent many hours pruning and spraying the orchards and lamenting the fact that the Bhutanese paid such scanty respect to their trees. Hillsides upon hillsides of trees were burnt to provide grass for the cattle and trees were deliberately ringed to kill them. Children, who usually looked after the cattle, also did a great deal of destruction by ringing the bark of *Pinus wallichiana*, generally in two places about half-way up the stem, to procure the bark's inner ring which was very sweet and which they ate. Naturally trees thus treated died and all round Ha the yellow dead and dying tops of *Pinus wallichiana* were a sad sight. On the other hand, on the right bank of the Ha Chu, where the trees had escaped the burning and the vandalism of the young, there were wonderful forests which caused Ludlow to contemplate

'the vast sums of money Bhutan could make if only she worked the forests nature has given her. Her population, instead of being poor and ill-clad, could have a far higher standard of living if only her rulers showed enterprise and judgment. There will come an awakening one day. The Bhutanese cannot be kept for ever in a state of ignorance and isolation. Bhutan could afford a population twenty times as big as the present one and with the Nepalese on her doorstep it is only a question of time before that vigorous and fertile race overrun the land.'[6]

Whilst Ludlow was thus occupied Danon and Ramzana were hunting for flowers. These, in the neighbourhood of the Ha Chu, were rather disappointing and the best they gathered was the pinkish-white *Rhododendron ciliatum* (16019), Hooker's discovery in the Sikkim Himalaya, a well-known species in cultivation and the parent of several clones which have received the First Class Certificate from the Royal Horticultural Society; a form of *R. cinnabarinum* (16027) with apricot-red to yellow bells which the local children ate with relish after first plucking away the stamens and ovary; a rosy-red shrubby form of *R. arboreum* (16026) with fawn indumentum on the lower surface of the leaves; apple-blossom-pink *R. virgatum* (16054); and finest of all, *R. griffithianum* (16068), a shrub no more than four feet tall whose large, loose, pink-suffused, white trusses were so fragrant that Ludlow at first mistook it for a member of the scented Maddenii series of rhododendrons. One of the most attractive plants, in fact it is one of the most attractive of all ericaceous shrubs, was *Vaccinium nummularia* (16056); growing in the wet ravines it was no more than two feet tall and was covered with red flower buds and cream, red-tipped, mature flowers. And certainly one of the most spectacular plants in this region was the pale pink-flowered, red flower-stalked *Bergenia ligulata* (16070) which at Chunzu Gompa literally plastered the rocks near the monastery.

Ramzana and Danon's most successful collecting was done south of Ha on the Tseli La (12,500 feet) where, in addition to several rhododendrons including the epiphytic, woolly-shooted, cream-flowered *R. pendulum* (16117) and a more pink than red form of *R. kendrickii* (16123), they found an abundance of the reddish-purple *Primula bracteosa* (16112), of the mauve-purple, yellow-eyed *P. tanneri* (16111, 16119) which had been among the first half dozen primulas Ludlow and Sherriff had collected in 1933, and of the rose *P. listeri* (16107), all of which had been gathered by Sherriff in Central Bhutan in 1937.

Ludlow in N.W.Bhutan

On 9 May Ludlow and his colleagues left Ha, crossed the Saga La (12,200 feet) to the north and descended to Drugye in the Paro valley where the Bhutanese did not seem so destructive of the native trees as in the Ha valley. Floristically the Saga La was disappointing and the only plant of any note was the lanky five-foot *Rhododendron lindleyi* (16184) which compensated for its ungainly habit by

95. The Ha valley
96. View from the Rip La

scenting the tsuga forest, low down on the Paro side of the pass, with its great cream, yellow-suffused trumpets. Although Hooker first collected this beautiful species in Sikkim it was actually described from a plant, raised by Mr Standish the Ascot nursery-man from seeds collected in Bhutan, possibly by Booth, between 1849–50. Many years later, in 1928 in fact, Kingdon Ward was to reintroduce the species into cultivation from the Delei valley in Assam.

At Drugye, where cuckoos called all day long and owls and nightjars all night long, and where black bear and barking deer were in the neighbourhood, the flora was again poor, three days of searching revealing little of interest except for a pale mauve-pink form of *Rhododendron virgatum* (16206) and the new *Paris marmorata* (16213) which Ludlow and Elliot had discovered in the Tsangpo gorge in April 1947, and the finding of which, at Drugye, now constituted a new record for Bhutan.

Ludlow was heading north for Lingshi Dzong, some forty miles away, and after three marches of roughly eight miles each which involved crossing the two passes, the Mem La (16,256 feet) and the Taglung La (14,470 feet), he halted at Kumatang (12,500 feet). Throughout these marches, and indeed throughout many future marches, he was constantly frustrated by lack of information about the route. 'It is quite impossible to obtain any information worth having about the route from the Bhutanese. They are either incapable of giving it, or they deliberately give false information, or not knowing anything about the route they concoct some story about it.' [7] He had never felt quite so 'in the blue' as on this particular trek. All he knew was that he was south of the main range, and not very far from it, and east of Chomolhari.

It was from near the snow on Chomolhari that in 1882, Dungboo, Sir George King's native collector, had gathered a primula clearly akin to *Primula sikkimensis*. In 1912 the same primula was found in the Chumbi valley and was named *Primula chumbiensis*. It is quite a handsome plant with reddish leaf-stalks and with up to half a dozen pale yellow flowers carried on a reddish scape 6–9 inches high. Until 1949 the species had been known only from SE Tibet. Ludlow now found it on the Mem La and by so doing recorded it for the first time in Bhutan. A week later Ramzana and Danon were to find more material of it near the Thimbu Chu above Naha; some plants (16327), growing in wet sand by the river, were extremely vigorous with fine large flowers, whilst others (16336), taken from dry ground, were much smaller and more like the Chumbi valley type.

Still another first record for Bhutan was the finding of the dwarf *Primula fasciculata* (16277) during the short halt at Kumatang. Kingdon Ward, on the Chungtien plateau of N W Yunnan, had been the first European to see its bright rose-red, yellow-eyed flowers, in 1913, after which others had recorded its presence in S W Szechwan, E Tibet and Kansu. No more than a couple of inches tall it is a denizen of boggy and marshy grassland and Ramzana found it on the Pangte La (15,357 feet), north of Kumatang, at an altitude of 13,000 feet. This was a new addition to the Ludlow and Sherriff collections and so were the specimens of the meconopsis Ludlow found, also on the Pangte La, and at the same elevation. 'I got *Meconopsis discigera* [16279] today, in bud . . . . It reminds me of *Mec. torquata* from Lhasa in one or two particulars. For example, the flower buds are closely adpressed together on the flower-stalk, then there are numerous persistent dead leaves of previous years attached to the base and it is obvious that it takes several (2 or 3 or 4) years to reach the flowering stage. In habitat it differs somewhat, being found along the banks of streams in the open, whereas *torquata* was always found on boulder scree, often overhung by boulders. Then again *torquata* occurs at 15,000 feet or more whereas *discigera* where I saw it was growing at not more than 13,000 feet at about the upper limit of the *Abies spectabilis* contour. Everything points to *Mec. discigera* being far more amenable in cultivation than *torquata*, alas, is.'[8] Fortunate is the monographer, in this case Sir George Taylor, who has his views on the relationship of a species in the genus independently confirmed by observations such as these. Ludlow never saw *M. discigera* east of the Paro watershed and, moreover, he never saw plants with fully opened flowers; those with buds about to burst being purplish-blue. In October he gathered seeds (17455) and young plants (17456) in the resting stage, and successfully reintroduced the species into cultivation where the first flowering specimens showed some variation in flower colour, some being of a good bright blue and others of an inferior slate-blue. However, unlike so many other meconopses, *M. discigera* has never made a very great impact on British gardeners. And unlike *Primula fasciculata* and *P. chumbiensis*, the finding of *M. discigera* did not constitute a new record for Bhutan for Cooper had recorded it there during his 1914–15 expedition.

From Kumatang they continued north through rolling alpine pastures and ascended the Pangte La, snow covered for the last 500 feet, finding tracks of the snow leopard and herds of burhel, some of them at least a hundred strong, and just below the summit

on the northern face the reddish-purple patches of another diminutive primula Ludlow hadn't seen in the wild before. This was *Primula waddellii* (16291), named after Major L. A. Waddell of the Indian Medical Service who travelled widely in the Himalaya and became an authority on matters Tibetan. He secured two scanty collections of his pretty little dwarf in SE Tibet in 1891. Further material was collected in Bhutan by Cooper, and in 1937 Sherriff had also taken beautiful specimens (3281) from the Black Mountain. Though never more than an inch tall, the reddish-purple, white-eyed flowers, with their deeply bifid petals, can be nearly an inch in diameter, and growing in masses, as it usually does, this primula can be a most attractive sight.

After the Pangte La there was still another pass to cross, the Nelli La (15,278 feet), before reaching Lingshi Dzong, and here, whilst stalking another herd of burhel, Danon found a fritillary new to him and to everyone else in the party. Inhabiting dry scree at 15,000 feet it was rather a strange looking affair with olive-brown leaves and with the same olive-brown pigment on the outer surface of the flower segments which were greenish on the inner face. It was a new species which later was named *Fritillaria bhutanica* (16296).

Making the easy descent of the Nelli La in a north-easterly direction Ludlow soon reached the dilapidated looking building on top of a bare hill about 500 feet above the headwaters of the Mo Chu, which was Lingshi Dzong. Straightway he saw an example of a phenomenon which had interested him greatly throughout his Himalayan travels and which now caused him to wonder — the evidence of a flood. 'The river coming down from the west which

drains the glacier snows from the main range has a scoured-out bed which has obviously been caused by a flood. On enquiry, I find that this flood occurred six years ago and was caused, like almost all Himalayan floods, by the bursting of a dam which had impounded a lake. In my travels in the Himalaya I have encountered many such cataclysms. There was the Kumdan ice dam which held up the Shyok river in 1929, the Hondar earth dam, also in the Shyok valley, which gave way in 1928. Then in Eastern Tibet there was the Yigrong flood of 50 or so years ago, and the Paka flood of about 18 years ago, both caused by the bursting of earthen dams which had given rise to lakes. And now here is a flood which also

97. *Lilium nanum* (above)
98. *Primula waddellii*

resulted from the bursting of a dam and the sudden liberation of the pent up waters behind it . . . . The dam was due to an ice fall from the main range damming up the outlet of one of the lakes at the head of the Lingshi stream. It swept away all the bridges on the Mo Chu . . . . All these floods scour out valleys at a rapid rate and as they are so frequent, geologically speaking, I am wondering if we appreciate correctly the magnitude of the effect of these floods in the past on the formation of valleys and gorges.' [9]

Though it was now approaching the end of May, Ludlow was too near the main range, and too high, and consequently in too cold an area, to find many of the plants for which he was hunting. Indeed it was so cold that in the night the milk froze into a lump of ice. Thus he sent Ramzana and Danon south, over the Yale La (15,180 feet) and into the Thimbu Chu valley, with instructions to descend as low as 11,000–10,000 feet and to collect everything that was worthwhile. In the meantime Ludlow stayed at the camp at Lingshi, was wakened every morning by Blood pheasants which saluted the break of day with their characteristic whistling screech, and thoroughly explored the area, mostly on the dzongpen's mule. Everything was very backward, even the birches and willows being still in bud. Even so, he made a few interesting finds. Growing at the foot of a great boulder on a glacier moraine, there was a solitary meconopsis with a single mauve flower. It was *Meconopsis primulina* (16310), originally described in 1896 from specimens gathered by native collectors in Bhutan and the Chumbi district of Tibet, and quite new to the Ludlow and Sherriff collections. During the next few weeks Ludlow found more of it and, with difficulty, collected its seeds in the autumn. Another meconopsis, *M. simplicifolia* (16313), a poorish form with reddish-purple flowers, was inhabiting the left lateral moraine of the Lingshi glacier, as were the dwarf violet-blue *Primula tenella* (16314) which Sherriff had collected at Tang Chu in 1937, the white, purple-flushed *Anemone rupicola* (16315) whose leaves were stained with reddish-purple on the lower surface, and, sprawling over the boulders, the large sulphur-yellow-flowered *Lonicera hispida* var. *setosa* (16312).

Ramzana and Danon descended the Thimbu Chu as far as Dotena, three long stages distant, and though they returned with a good collection of plants, 'did not do so well with primulas and rhododendrons — those two aristocratic orders — as I had expected.' Their collections which most impressed Ludlow were those of *Paraquilegia grandiflora* — 'quite the handsomest and largest flowered . . . . I have ever seen anywhere — as is its white form (16397).' Among the 'aristocrats' there was nothing new but there *were* trusses of *Rhododendron lanatum* (16324) yellow-flowered and with a russet indumentum on the lower leaf surface; large pinkish-red trusses of *R. hodgsonii* (16346) taken from 25-foot tall trees; rather poor trusses of *R. falconeri* (16371) from 30-foot trees — poor because, though they were of great size, they had been badly battered by the rain and hail and the colour wasn't so good, a pale cream with a tinge of pink; sprays of a form of *R. keysii* (16372) with tubular flowers yellowish at the apex and rich salmon-red towards the base, and of a splendid form of the fragrant white, pink-flushed *R. edge-worthii* (16378). Of primulas there was *Primula chumbiensis* (16327, 16336) which Ludlow had so recently found on the Mem La; the elegant *P. elongata* (16329) — a rich yellow form of it inhabiting the banks of the Thimbu Chu; and in the shade of the rhododendron forest, the magenta-red *P. geraniifolia* (16364) which Sherriff had gathered in 1937. For the rest, there was a specimen, in bud, of *Meconopsis discigera* (16399); a reddish-purple form of *M. simplici-folia* (16348); the reddish-purple *Cypripedium tibeticum* (16325); and, splashing the wet cliff faces with its lovely white flowers was a saxifrage which they were never to see again, which proved to be a new species, and which was later named *Saxifraga vacillans* (16352).

Ludlow and his colleagues left Lingshi Dzong on 2 June and travelled in a north-easterly direction into a valley where there were half a dozen houses and between twenty and thirty acres of cultivation before arriving at the small village of Cheypechey which lay at the mouth of a valley leading to the Lingshi La (17,200 feet) and which was two marches from the shores of the Rham Tso and the village of Rham (14,700 feet). Thence for some six or seven miles north-eastwards to Chumiten; and a further ten miles, almost due east and over the Yari La (14,950 feet), to Tsaye Gyatang; then the crossing of the Shinje La (16,070 feet) before turning in a south-easterly direction to Laya (12,150 feet), a scattered village of many houses on an open hillslope 500–750 feet above the upper Mo Chu, where Ludlow intended to halt for two or three days. From a plant point of view the marches from Lingshi were disappointing and quite obviously the area was too cold and dry for the type of plant for which Ludlow was searching. But if the plants were scarce the confounded ticks were in full force and were a constant plague. 'A blood thirsty country is Bhutan. Ticks, leeches, dim-dam flies, bugs, are all out for one's blood. And yet a day's march north as the crow flies these pests are unknown', wrote Ludlow from his camp a few miles from Laya.

The three days spent at Laya were hardly more productive of

plants than the previous ones though one or two rhododendrons were added to the collection including the small maroon-flowered *Rhododendron baileyi* (16442) which F. M. Bailey had discovered in 1913 on his journey to the upper reaches of the Tsangpo river; though he didn't collect specimens he did send seeds to the Royal Botanic Garden at Edinburgh whence they were distributed to other gardens and in this fashion introduced the species to cultivation in Britain.

Leaving Laya on 9 June Ludlow descended due south into the valley of the Mo Chu, through mixed forest of picea, tsuga and larch, to Kohina, a sort of half-way house between Laya and Gyasa Dzong at which latter place he halted for a couple of days. Between Laya and Gyasa the Mo Chu ran in a very steep and precipitous gorge and for most of the fourteen-mile march from Kohina to Gyasa Ludlow was some 3,000 feet above the river. And for most of the fourteen-mile march he was also in rhododendron country though most of the rhododendrons had already flowered; only *Rhododendron hodgsonii* (16494), two distinct forms of *R. cinnabarinum* — one with dull apricot flowers (16492), the other with red (16493 — var. *roylei*), and the pale sulphur-yellow *R. campylocarpum* (16495), all at 12,000 feet, were still blooming.

The Dzong was quite an impressive building standing beside a pretty pond with willows hanging over the waters. Unfortunately clouds of midges were also over the waters, and over everything else in fact, and were a source of great annoyance to the collectors who were forced to be always on the move. In these unpleasant conditions they added fifty different plant species to their collections during the two days spent at Gyasa. Pride of place was given to *Rhododendron rhabdotum* (16523) with its great cream, red-striped trumpets, and to the fragrant *R. maddenii* (16524) whose blush-pink flower-buds opened to pinkish-white; to *Dendrobenthamia capitata* (16513) for so long known as *Cornus capitata*, whose countless flower-bracts, plastering the many specimens which varied in height from 2–8 feet, coloured with creamy-yellow the open hillside where there was also an abundance of the white-flowered *Lyonia ovalifolia* (16526), a most handsome shrub of up to 10 feet; to a form of *Buddleia colvillei* with pink, white-throated, fleshy flowers (16514); to *Albizzia mollis* (16544), a tree up to 60 feet tall in the rain forest and a haze of yellowish-green, pink and white-stamened blossoms; and to *Cardiocrinum giganteum* (16545), the inside of the otherwise white trumpets suffused with pinkish-red towards the base.

For once fairly content with his haul of plants, but rather regretting, in spite of the midges, that he had to leave Gyasa so soon,

Ludlow returned to Kohina on 14 June and the following day continued to retrace his footsteps to a bridge over the Mo Chu. He then turned up the first big valley coming in on the left bank and ascended some 1,500 feet to Gyabna Thang, a charming encampment in a forest glade thick with *Primula sikkimensis*. Ludlow was tired and here would have rested for some days had he not been anxious to move eastwards over the Kangla Karchu La (16,630 feet) in the direction of Foomay. This journey of not more than twenty miles proved a difficult one for one of Ludlow's years and, with two halts, took four days. They left on 18 June on a track which almost immediately dropped several hundred feet, then rose steeply through the abies forest, and above the tree-line wound and twisted about in a most confusing fashion. 'We were in the clouds most of the time and the limited horizon added to our perplexity so that at times I didn't know in which direction we were moving. We skirted the shores of a number of small mountain tarns, and the path, always rough and often indiscernible, was continually ascending to a ridge and dipping down to a ravine. It is quite evident that this is going to be a high and difficult pass .... I felt the altitude more today than I have ever done before .... I had a very restless and sleepless night and again felt the altitude severely, I couldn't sleep because at frequent intervals I had to take long and deep inspirations. It rained most of the night and was miserably cold and damp. The poor coolies sleeping under rocks and cliffs must have had a foul time. I put up my spare tent for some of them.' [10]

'Every now and then, in the course of a journey, there comes a day when the march proves longer and more difficult than you anticipate and you reach your destination tired and weary and exhausted and wonder why on earth you leave the comforts of home and undergo such privations. Today's march fits into this category. I set out hopefully on Ramzana's and Danon's assurance that the ascent to the pass was a mere bagatelle and that the summit would be reached in [a couple of hours] of slow and easy climbing .... But the pass .... proved very elusive and we trudged for hour after hour until at last at 1.30 p.m. we reached the watershed .... We .... descended ever so gradually for three miles along a track as atrociously bad as the one we had ascended by. And then came the descent. Down we dived for a couple of hours until we reached the valley of one branch of the Po Chu having its origin in a series of glaciers on the main range .... But every painful experience generally has its complement of pleasure lurking somewhere and we must count ourselves fortunate in having had a fine — comparatively fine — day for the crossing of the Kangla Karchu La which is really a difficult pass and could prove formidable in bad weather . . . . I must confess that I was pretty "cooked" at the termination of today's march. At 64 it is a pretty severe test to climb 1,000 feet at any altitude above 15,000 feet and then to descend from 16,500 feet (approx.) along the roughest of tracts to 12,500 feet. However I managed it with a few groans and a few nips at the brandy flask. We got in at 4.30 p.m., having started at 5.30 a.m. The pleasure which pays for the pain of all this exertion, of course, is the discovery of something new to science, and the introduction of something new to the garden and, not least, the exploration of unknown country, or rather country unknown to Europeans, and blank on the map.' [11]

If during these last few difficult days he hadn't found any plants new to science he had found a few 'real aristocrats with the blue sap in their veins' — and none more aristocratic than some of the primulas, minute though some of them might have been; *Primula caveana* (16581), heavily dusted with farina, the pale yellow eye of the reddish-pink flowers surrounded by a ring of blue-violet; *P. soldanelloides* (16566) with the large solitary nodding glistening white bell; *P. sapphirina* (16576) no more than two inches high with 1 – 4 pendent lilac flowers; equally dwarf and with a beautiful pompon of hairs in the throat of the violet corollas, *P. pusilla* (16567) and *P. tenuiloba* (16579); a white form of the even dwarfer *P. concinna* (16593) whose normally pale pink flowers usually nestle among the tiny leaves; *P. glabra* (16577) and *P. kongboensis* (16588) with their small compact heads of pale mauve to reddish-purple flowers on a 1 – 4 inch scape — the latter a new record for Bhutan; and the rosy-red *P. waddellii* (16598) which Ludlow had recently found on the Pangte La. Apart from the primulas there was *Lilium nanum* (16574), the lilac-purple perianth segments streaked and spotted with a deeper colour; a few more plants of the blue-mauve *Meconopsis primulina* (16569); and 'a little gem of a saxifrage with a huge red-brown calyx and yellow corolla and yet a midget of 4 or 5 inches high' — *Saxifraga lychnitis* (16599).

Having crossed the Kangla Karchu La, Ludlow reached Foomay where he halted for the best part of a week, resting a little, drying his collections and watching the birds, whilst Danon and Ramzana searched the glacial reaches of the Pho Chu for plants, returning with very little that was of great interest. Much their best find was a dwarf potentilla no more than six inches high, with a one-inch yellow flower, with a thick sericeous indumentum covering the plant and with a dense mat of foliage concealing the stem. This splendid little plant proved to be a new species which Ludlow himself later was pleased to describe, and name, as *Potentilla bhutanica* (16623).

They had to wait until they reached Chojo Dzong on 29 June, after crossing the Chesha La (15,040 feet), before becoming really enthusiastic about the plants. Then 'both Danon and Ramzana returned with a meconopsis (16704) which I think must be new. It is a plant with a very large rose-pink flower borne singly on a scape which varies in length from 18—30 inches. Half way up the stem is a whorl of 3 or 4 cauline leaves more or less sessile. The basal leaves are lanceolate and petiolate. The stem and leaves are covered with long soft fawn-coloured hairs and each leaf at its apex has a little tuft of these hairs. The most important fact connected with this plant, apart from its loveliness, is that it is perennial and gives off lateral basal shoots. I think this meconopsis will prove amenable in cultivation. It occurs abundantly, locally, on cliffs and in boulder scree. In addition to this new? meconopsis D. & R. brought back *Meconopsis bella* (16703) and a white variety of *Primula caveana* (16695). I got a pretty salvia and also a primula which I think must be *Primula bellidifolia* (16685) growing underneath a rock exposed to the sun but sheltered from the wind and rain (unless the wind and rain come blowing in violently from the NE). It was a drizzly day but nevertheless it is a red-letter day if the pink meconopsis proves to be new.' [12]

Later Ludlow took several photographs of the lovely poppy. 'It grows chiefly on boulder scree, but also on cliffs. It is a most attractive plant and, if new, will create quite a sensation in horticultural circles. Danon and Ramzana went to an adjoining valley and found the pink meconopsis in even greater numbers than in the original valley I went up today. They say the valley for great distances was just pink with the blooms. Danon reported that in an area of two feet in diameter he counted twenty-six flowers, and it must be remembered that each plant bears only a single scape and that each scape bears only a single terminal flower.' [13] Danon and Ramzana were as excited by the pink meconopsis as was Ludlow. 'They are indefatigable in scouring the hills, in all weathers, for plants. They take pride in their work and the discovery of a primula, meconopsis or some other aristocratic plant new to the collection gives them the greatest pleasure. I do not wish for a better pair of collectors than these two.' [14]

Leaving Chojo Dzong on 9 July they met with the poppy once again during the crossing of the Gafoo La (16,900 feet). 'On a boulder strewn moraine the new pink poppy was growing to perfection. Some of the blooms were enormous. The flowers look you straight in the face and do not hold their heads down like *Mec. simplicifolia* and many other meconopses. I noticed that this poppy always seems to choose a position facing north . . . . This poppy inhabited this boulder strewn moraine for about half a mile and then vanished. We saw none higher up though the ground seemed suitable. Under the boulders where the meconopsis was growing, and sometimes mixed up with it, was a Rotundifolia primula which is . . . . *Primula caveana* (16791).'

They were soon to learn that boulder-scree and boulder-strewn moraine was not the only habitat of their exciting poppy-wort, for, moving still further eastwards and in crossing the Saga La on 14 July, 'we suddenly came upon the new pink meconopsis in great abundance, and here it grew not at the base of boulders, or on cliffs, but just mixed up with *Meconopsis paniculata* in the grassy stony shrubby valley-bed. There is no reason as far as I can see to prevent this poppy growing well at home — at least it shouldn't prove any more difficult to raise than *paniculata*. And what a lovely sight it is in full bloom. I saw one clump, about 2 feet in diameter, with 30 flowering scapes. *Mec. horridula* was growing with these other two poppies.'

Ludlow's 'new pink meconopsis' was not new. It was *Meconopsis sherriffii* which Sherriff had discovered on the Drichung La in SE Tibet in 1936 and which Ludlow knew well, having described it as the gem of the entire 1936 collections. It is remarkable that apparently he didn't once think of Sherriff's pink poppy when he was collecting, photographing and writing about his one, which was, of course, a new Bhutan record for *Meconopsis sherriffii*. Though Ludlow was later to harvest pounds of seeds, the plant was not to prove as amenable in cultivation as he had supposed it would. However it was entirely appropriate that those who were to solve most successfully the problems of its cultivation in Britain were to be George and Betty Sherriff in their garden at Ascreavie in Scotland.

By 16 July Ludlow had reached Dur Chutsen (10,800 feet), south-east of the Saga La, and whilst he was halted there a letter from Sherriff arrived with the news that he was at nearby Namdating and suggesting that they meet to discuss their progress. Ludlow

Meconopsis simplicifolia

proposed that they rendezvous at the hot springs equidistant to both of them and thence wind their way to Trongsa. Sherriff duly arrived at the hot springs on the 18th — to be followed a couple of hours later by a messenger from Hicks bringing the sad news that Betty Sherriff had fallen from her pony, had broken the left humerus near to its articulation with the shoulder, had had slight concussion, and that it was necessary for her to be X-rayed in Calcutta. Thus Sherriff and Ludlow now decided to make their way to Bumthang where they hoped to meet Mrs Sherriff and Hicks in some ten days time.

During their journey to Bumthang Ludlow and Sherriff were fortunate to find a lily (19490) at Gortsam, some ten miles from Bumthang, which certainly pleased Sherriff as much as the pink meconopsis. Ludlow simply recorded the fact of the finding; 'It occurred in one place at about 12,000 feet where the hillsides were rocky and grassy. We got only one flower — a blood red corolla about 2″ long. We saw hundreds of plants, but they were all in fruit. We dug up a number of bulbs and Sherriff's collectors will come back in the autumn and get more.'[15] Sherriff's reaction was a more excited one. 'The pick of the bunch came . . . . with the finding of a new lily. How lucky this was can be guessed when one thinks that only one flower was seen and that that happened to be almost exactly where I stopped to put flowers in the press. We would almost certainly all have passed that flower and the many plants with no flowers, had it not been for these coincidences. It is a new lily; it must be, isolated like this in Central Bhutan. It is not a startling flower, rather tubular, 2½″ long, and brownish-red, one-, or in one case only, two-flowered. A new lily is what we have wanted all these years and now we have it . . . . Altogether a very fine day to remember.'[16]

The lily was indeed new and was appropriately named *Lilium sherriffiae* after Betty Sherriff who had forestalled her husband and his friend, by two months, by finding one bulb and one flower (20658) near Lao, one march from Shingbe and two marches from the Me La, on 23 May. And she described the flower more accurately than did her colleagues, 'maroon with inside chequered with gold', and this is the only lily which is known to have a chequered perianth comparable to the patterning in various species of colchicum and of fritillaria. In October a good supply of bulbs and seeds was procured, under the number 19490, from which this unusual plant was introduced into cultivation and figured in the Botanical Magazine (New Series tab. 141).

With the Sherriffs and Hicks (Map 20)

It is necessary, now, to return to the beginning of April. The Sherriffs and Hicks had arrived in Bhutan on 26 March and, riding three of the mules of the Maharaja of Bhutan up the valley of the Mangde Chu, had reached Kinga Rapden (5,800 feet) on 2 April, there to be straightway involved in a week of entertainments with His Highness. 'When about three miles short of Kinga Rapden we were met by two of H. H's. men and given tea and saffron rice. H. H. also sent us three very fine ponies, richly caparisoned with gold brocade saddle cloths, to ride in on. The next excitement was meeting the two dancers with their two-man orchestra of sarnais. This was about two miles from camp, but they played and danced the whole way in. How they manage that I can't think — uphill, downhill, over muddy patches, small streams, and so on, the music and dancing continued. They are dressed in very bright clothes, the dancers in white skirts and many coloured top half clothes, with a little brightly coloured crown adorned with a gold ornament. The sarnai players are in red flannel. This all caused the greatest excitement locally and our procession was joined by everyone, men, women and children, and also by all the coolies carrying our loads. We reached camp after a pause to get scarves ready in case the Maharaja and two Maharanis were there to meet us. But they were not, so we had a rest here till H. H. came at 3.0 p.m. The camp is a magnificent place with three large tents for us, an EP tent for reception and another huge tent for our kit, besides a kitchen with a good chula made. Everything is laid on and all our wants catered for. At 3.0 p.m. H. H. arrived. He rode up and got off his pony 30 yards from the door where we all went to meet him . . . . He has aged greatly. But he is still the pleasant thoughtful man he always was to meet. He remembers all the old Political Officers and asked after all by name. Bailey particularly. He seems to have been H. H's. particular friend.' [17]

'We had lunch with H. H., on 3rd, in great ceremony with the dancers to meet us and all the accompanying tamasha. But meals with H. H. are always private. No servants are allowed in the room except to serve and then only for the minimum time . . . . On 4th H. H. proposed a picnic but it rained and misted and we put it off. An army of servants had already set out with food etc., but they were recalled and the unwanted

99. Dancers at Kinga Rapden

food sent to us in enormous bowls.' [18]

'H. H. sent to say he would come up after our lunch at about 12.30 p.m. and would like to have a friendly talk and stay to dinner. His usual hour for dinner is 9.0 p.m. but we could not manage that and compromised at 7.0 p.m. He arrived a little before 12.0 and left at 8.0 p.m. so we had a pretty hard afternoon, but still a pleasant and I hope a profitable one for all concerned. He led me away after lunch to have a talk and we talked for fully two hours or more . . . I told H. H. about the awful present state of the pine forests between Biyita Sam and here. The forests are only of old fully or more than half grown trees. Young trees barely are to be seen. All are burnt when the grass is burnt annually. I suggested either shutting off whole areas, never to be burnt, or shutting areas for 5 or more years at a time. The grass is only burnt to get fresh grass for cattle grazing. When I asked H. H. how many cattle there were in the 25 miles of forest we passed through, he said "at most 300". I hope something may be done . . . . H. H. was undoubtedly interested.' [19]

'H. H. has entertained us fully and we have entertained him too. We were down on 6th to watch Bhutanese dancers, male and female, and next day we had H. H. up for a good long session, 1.0 p.m. till 5.30 p.m. He was then in great form reminiscing mostly about his Calcutta visit and various doings of Political Officers Gould, Williamson, Bailey and Bell. He would very much like to see Bailey again.' [20]

'We had a long and pleasant session with H. H. on 8th. He came up again at 3.0 p.m. and we had dinner at 7.0 p.m. and he left at 8.0 p.m. . . . . We had games with the two Maharanis and H. H., and the greatest success was animal grab which so amused H. H. that we had to stop for fear he would do himself some harm through overlaughing. Today H. H. came up again to say goodbye at 9.30 a.m. and we were off at 10.0 a.m. The road was pretty steep, straight up behind Kinga Rapden, to the east. There is now a good deal of cultivation for about 2,000 feet, then the path goes along the edge of the forest and so to a large area of grass. Round about here there were a number of rhododendrons. *Rhododendron arboreum* [18620 — bright red form] is common, as is *R. grande* [18660 — a tree up to 40 feet with great trusses of pale yellow, purple nectar-pouched, bells] and some epiphytic — like *R. edgeworthii* and *rhabdotum* which were not in flower. *Magnolia campbellii* [18648 and up to 80 feet high] was a very fine sight indeed, in

Map 20. The Sherriffs and Hicks in Central Bhutan

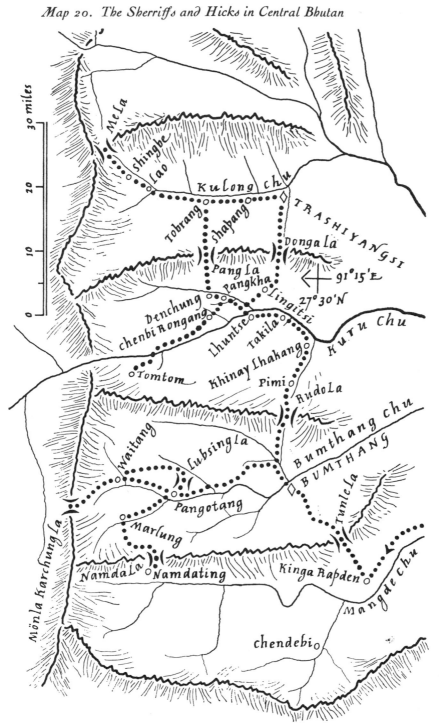

30 miles

20

10

0

Me La

Shingbe La

Kulong chu

Tobrang

Shabang

Dongala

TRASHIYANGSI

Pang La
Pangkha

91° 15' E

Denchung
chenbi Rongang

Lingitsi

27° 30' N

Lhuntse

Takila

Kuru Chu

Tomtom

Khinay Lhakang

Pimi

Rudola

Waitang

Lubsingla

Bumthang chu

BUMTHANG

Mönla Karchung La

Pangotang

Tunle La

Marlung

Namda La

Namdating

Kinga Rapden

Mangdechu

chendebi

full bloom [cream] and very common. Of primulas we found *Primula bracteosa* (18644) almost over and another I don't think I have seen before [18669 — the rich rose or deep pink *P. listeri*, of which Sherriff had made the first record for Bhutan in 1937]. It is really lovely to get away on the job again.' [21]

The Sherriffs and Hicks were now heading north-east for Bumthang whence they would strike eastwards over the Rudo La (12,600 feet), scene of the finding of *Primula dickieana* var. *aureostellata* in 1933, to Pimi, thence to Khinay Lhakang (8,000 feet). Shortly after leaving Pimi there 'was certainly the finest thing we have yet seen, a rhododendron (18720) which I cannot make into anything in the rhododendron book. This is a tree of 20–30 feet with a nice, clean, smooth, thin, flaking bark and lax trusses of 3 or 4 enormous flowers. When they are in bud they are a good pink and even when first full out they are pale pink but go white later. The curious thing is the absence of glands on any part but the ovary. I hope that this may be a new one. It is a real beauty, should be hardy, and we are practically certain to get ample seed. Next came several trees of *Prunus cerasoides* var. *rubea*, magnificent trees up to 60 feet high. This we have now found in Pome (Gompo Ne), E Bhutan (Gamri Chu), and here.' [22] It is rather strange that Sherriff didn't recognise his rhododendron as the pure white one he had collected at Chendebi during his trip to Dungshingang in 1937 — *Rhododendron griffithianum*; possibly, like so many others who have become interested in rhododendron taxonomy, he was misled by the glandulosity, or otherwise, of the ovary.

From Khinay Lhakang they marched up the Kuru Chu to Takila (6,000 feet) which Ludlow and Sherriff had visited in 1933 when it was called Tangmachu and where they had found a fine form of *Lilium nepalense*; thence to Lhuntse Dzong (4,520 feet), just beyond the junction of the Kuru Chu with the Khoma Chu. This was the point at which the party had planned to split; Sherriff would move almost due north up the Khoma Chu, whilst Mrs Sherriff and Hicks would travel in a south-easterly direction to Trashiyangsi Dzong before marching north to north-east, up the Kulong Chu, to Shingbe and the Me La, the route Ludlow and Sherriff had taken in 1933. They would all endeavour to reunite at Bumthang round about the middle of August.

With Sherriff in central Bhutan

On leaving his companions Sherriff moved on to Pangkha and there halted for three days for it was still too early for flowers and time had to be put off. A good deal of the time was spent in the region of the Rip La (c. 9,600 feet) and a rather disappointing time it was even though the finding of the rich mauve-pink usually fimbriate-petalled *Primula normaniana* (18740) which Sherriff had found on two or three occasions in SE Tibet did now constitute a new record for Bhutan. The only other primula of interest — Sherriff never ranked it highly from an ornamental point of view — was the mauve *P. listeri* (18749) which plastered the wet moss on the rocks beside a stream in the only bit of rain forest there was, for most of this country seemed to be bared for grazing.

On 26 April Sherriff crossed the Rip La and spent five days, again not very profitable ones, at Denchung (7,300 feet) in the Khoma Chu basin. *Rhododendron lindleyi*, here almost entirely a ground-growing plant and not epiphytic, was very common, and so were other rhododendrons, in a practically inaccessible spot on a high cliff face across the river. 'This place is a 45° rock slope with

> no soil on it in places whilst in others some trees and shrubs have managed to find an anchorage. Obviously every now and then anything on the slope just shoots off to the bottom. It is quite impossible to climb the smooth slippery rock but we went up the between bits with the odd tree and bamboo and I found it very hard work indeed. There were mases of *Rhododendron edgeworthii* (18777) and a little higher up of *R. griffithianum* [which Sherriff now immediately recognised] . . . We had to go on up about 2,000 feet and then down another way. I hope I have, at any rate, some good photos of these two species.'[23]

He had indeed taken excellent photographs and that evening wrote equally excellent field notes for his two gatherings.

> '18776. *R. griffithianum*. Tree 20–25 feet with grey bark which peels off in thin flaky reddish pieces. Corolla white, filaments white, anthers pale brown. Style cream, with pale green or pink glands. Ovary green. Calyx pale pinkish-green. About the commonest rhododendron in this valley, between 6–8,000 feet, but always almost impossible to get, as it grows chiefly on cliffs in places similar to *R. edgeworthii* (18777). Unopened buds are red.

> '18777. *R. edgeworthii*. Shrub 3–7 feet. Corolla always slightly tinged pink. Buds rose, opened flowers gradually becoming white. Filaments white, anthers dark brown. Style pale pink. Calyx pink to green-pink. Bud scales rich rose. Very fragrant indeed. A beautiful sight, always on cliffs or over-hanging river.'

How fortunate is the rhododendron taxonomist to be given field information such as this.

The Maharaja had loaned Sherriff a very good man, Ngudup Namgyel, with full authority to do anything Sherriff asked of him. Ngudup was anxious to learn all he could from Sherriff, not only about plants, but about other things as well. 'We discussed Delhi today and got a bit mixed up as Ngudup was under the impression that Delhi was in England. I have told him I will teach him Hindu when we reach Nashima. He also asked if, when I went to London, I had to pay my respects to the King, or, if when I was in Calcutta I had to obey the King's orders. In Bhutan the Maharaja is everything. One can see this by the way it is impossible to stop the making of camps. H.H. ordered it; it must be done; it does not matter whether I want it or not.'[24]

On 2 May Sherriff recrossed the Rip La heading west, and then north, for Chenbi Rongang (6,500 feet) and the finding of a primula shortly after leaving Denchung prompted another splendidly detailed field note.

'18806. [Primula]. 7,000 feet. 2.5.1949. Not yet in full flower. Only one plant found with a few flowers open, and this shows two whorls, and a third starting. Corolla deep purplish-red with a yellow eye. Tube greenish-red. Calyx dark crimson. Pedicels green; bracts green. Leaves darker above than below. Petioles very pale green. Main veins in leaves very prominent below and very pale green. Hairs everywhere white. Root-stock when cut is the colour of beetroot. Growing in dense rain forest, on steep slopes.' And the following day Sherriff added 'Seems much more common at Chenbi Rongang, Kuru Chu, Kurted, but still not in full flower. Habitat same but altitude from about 6,500 feet to 7,500 feet.'

This was the first time either Sherriff, or any of his colleagues, had seen this very distinct species which therefore was new to the Ludlow and Sherriff collections; it was *Primula mollis* which had been discovered originally about 1852–53 by T.J. Booth who gathered ripe seeds from which the plant was introduced into cultivation and figured in the Botanical Magazine in 1854. In Bhutan it was again found by J.C. White in 1905, and by R.E. Cooper in 1915, but is also present in Assam, in Upper Burma where it appears to be more common than anywhere else, as well as in NW Yunnan.

This was really the only plant to interest Sherriff for the next two weeks during which time he marched north-west, almost as far as Tomtom, before retracing his steps to Pimi which 'in my memory

will always be a foul place. I won't see it again and I don't want to.'
He could never forget the foul swampy camp and the leeches of
1933.

But Pimi had at least one redeeming feature. It was within easy
reach of the Rudo La (12,600 feet) and approaching the pass there
was *Rhododendron rhabdotum* (18877) in abundance in the rain forest;
it was usually an ungainly straggly shrub and sometimes epiphytic,
but always bore heads of 2–5 magnificent fragrant flowers. And on
the east side of the pass, much wetter than the west side, there were
many other rhododendrons in plenty. There was the pale creamy-
yellow, dark-spotted *R. triflorum* (18881) at 9,500 feet. There was
the rose form of *R. arboreum* (18882) up to 30 feet tall, at 10,500
feet. And in the next 2,000 feet there was *R. hodgsonii* (18884),
30-foot specimens with large compact rose-pink to pale pink
trusses; *R. glaucophyllum* (18887) no more than two feet tall,
covering the rock faces with rose-pink flower-buds and very pale
pink, darkly spotted mature flowers; the white, reddish-brown-
spotted *R. pendulum* (18888) trailing its woolly young shoots over
the cliff faces; *R. cinnabarinum* (18889) varying greatly in colour,
sometimes a good yellow, sometimes salmon, sometimes red-salmon;
two quite distinct colour forms of *R. lanatum*, one (18890) pure
yellow-cream with a varying degree of red spotting, the other
(18890A) distinctly pink; a purple-mauve to blue-mauve form of *R.
wallichii* (18898); and the creamy-yellow, red-speckled *R. wightii*
(18899). 'How pleasant to see alpines again after all the low muck
I've seen for two weeks or more. *Primula calderiana* (18895), *P.
elongata* (18896) and *P. smithiana* (18900) were taken. *P.
sikkimensis* only in bud. We also got *Diapensia* [*himalaica*]
(18891), but saw no *Bryocarpum*. There was no time to go off
the path though .... This was a refreshingly good day for
flowers again.' [25]

Not until Sherriff reached Pangotang (12,100 feet) on 24 May,
five marches in a north-westerly direction from the Rudo La, did he
have a similar 'refreshingly good day' with his flowers. 'We got
16, including the yellow primula taken on the Rudo La
[*Primula elongata* (18955)] and also what may be *P. barnardoana* [18946], *P. calderiana* [18947] purple, *P. calderiana*
[18954] white, *P. atrodentata* [18957]. Funnily enough. *P.
sikkimensis* doesn't even show leaf, yet on the Rudo La the
flowers were just opening on the 18th. *Rhododendron setosum*
[18956] and *R. anthopogon* [18949] are just in flower while *R.
campanulatum* [18945 — a white, red-spotted form] is really
beautiful in full flower. *R. cinnabarinum* (yellow) also very

prolific and *R. campylocarpum* common . . . . Obviously we are nearer the real thing now . . . .'[26] The specimens gathered under 18946 more than ever convinced Sherriff that *Primula barnardoana* (placed by the taxonomists in the section of the genus known as Rotundifolia) and *P. elongata* (in the section Nivales) were merely forms of one another and that these two sections could well be amalgamated; certainly he believed that no sharp line of demarcation could be drawn between them. Later Sherriff found more of the white form of *P. calderiana* sometimes growing quite separate from, and sometimes mixed with, the usual deep purple form; and sometimes he found colonies with all shades between white and purple.

Sherriff's luck at Pangotang continued. 'I went to the nearest cliffs with Ngudup. There we found a small daphne with reddish flowers [*Daphne retusa*, 18973], an androsace newly out and very rich rose [*Androsace strigillosa*, 18974], a fine cushion saxifrage (18972) [a new species with rich yellow flowers later named *Saxifraga sherriffii*], a small berberis [*Berberis parisepala*, 18975] and one or two other little things. But what pleased me most was that I saw a lot of another primula, completely dried out and not started in any way to grow. It was on cliff ledges, or under jutting out rocks. There is a great deal of it and it is certainly a Soldanelloideae section primula. I guess it to be *Primula eburnea*. It will not be in flower for two months I'm sure.

'Today I took some of the dormant Soldanelloid primula off the cliff opposite the camp and packed them in an Oxo tin, sealed the tin and packed it off to go to Taylor by air mail. I have asked him not to plant all but to examine some to see what really happens when one of these primulas is really dormant. Nothing may come of this, but it will be interesting if they get through, still dormant. They were packed dry, on a dry day, or at a dry time anyway, and *may*, by being sealed, remain dormant.'[27]

The sealed tin reached Taylor, in London, almost a month later. From the small centre-bud of each plant radiated long brittle, wiry, reddish-purple roots and every plant looked completely dried and quite lifeless. When Taylor crushed a bud between finger and thumb most of it crumbled to dust except for a minute hard green centre; and when such a bud, with its attendant roots, was placed overnight in water in a shallow dish by next morning the roots had

100. Camp at Denchung

become quite soft and pliable and the bud had expanded sufficiently to distend the dead scaly leaves and to disclose the green germ within. The stimulus of water and room temperature had restored an apparently dead plant to one of great vigour. The remaining plants were sent to various expert cultivators in Britain and, though some complained that it was too much to be asked to bring the dead to life, they all had their plants in full flower in the space of a month. In the meantime Sherriff had been far out in his estimate as to when the plants on his cliff would be in flower. Not two months later, but only sixteen days later, the cliff was covered not, as he had suspected, with the cream flowers of *Primula eburnea*, but with the blue of *P. umbratilis* (19128) — but not nearly such big and strong plants as he had collected on previous occasions in Central Bhutan.

Sherriff decided to halt at Pangotang, 'as we have such a hell of a place built for us [under the orders of His Highness].' He himself was enclosed in a fir zareba which was useful for there was frequently a strongish wind, whilst the servants had a palatial wooden hut, big enough for dozens of them. The camp was at the junction of two rivers, that from Waitang and the north, and another from the west. One day he made an exploratory visit to Waitang (13,500 feet) and found the same phenomenon that he had witnessed in the Rinchen Chu in July 1937, the suspected hybridisation of *Primula calderiana* (19000) with *P. strumosa* (19001) with the resultant offspring in all variations of colour from white to powder-blue to violet to yellow.

On the whole, Pangotang, from a flower point of view, rather disappointed Sherriff and Kantanang (12,900 feet) to the south-east and over the Lubsing La (c. 14,500 feet), whence he moved on 2 June, was even more disappointing. 'This place seems completely dead. I don't know where the flowers are. We've gone high and we've gone low, and we can't find anything. I went to Gormotangka [a few miles north-east] today and looked at some most promising cliffs. But there was nothing to be seen, and not even signs of anything. I wish I had never come up this way at all, but had stuck somewhere [else], perhaps the Donga La ridge running up between the Me La and Singhi Dzong. With difficulty now, we are getting 2, 3, 4 or 6 specimens in a day and none of these is interesting. *Meconopsis sinuata* [19052] is common here and represents the beauty and interest of the flora pretty well. It is a miserable little thing which rarely has more than one flower open at a time — and the colour is purple-blue at that.' [28]

336

Concluding that he had been a little too early, at Pangotang, for those plants in which he was particularly interested, and in any case being anxious to return to the cliffs whereon he had found the desiccated Soldanelloid primula, Sherriff returned there on 13 June —and with much better results. The dried up cliff primula, *Primula umbratilis* (19128) was now in full flower and so were several other species. 'Got *P. capitata* [19163] (one only), *P. tenuiloba* [19167] and *P. walshii* [19172] (very few) and saw the finest lot of *P. caveana* I have ever seen. It is as common as dirt here. Lots of *P. bellidifolia* [19181] beside the *caveana* in places .... Obviously we are into things now — all of a sudden really.'[29]

Thus optimistically Sherriff returned north again to Waitang on 17 June. He had a miserable pony which shied at everything and in some way managed to loosen the girth to the extent that the saddle slipped and Sherriff was thrown. Luckily he wasn't in a dangerous spot above the river and only fell amongst smooth boulders; the main damage done was the smashing of his reading glasses but fortunately he carried a spare pair. In less than a month, exactly the same accident was to befall Betty Sherriff, but with much more unfortunate results.

At Waitang, where Sherriff halted for some ten days, *Primula umbratilis* (19183) and *P. caveana* (19177) were two of the commonest plants, and on several occasions, growing close by the normal pale purple or lilac form of the latter, Sherriff found a few plants with pure white flowers (19175) some of which he marked for collecting later in the season. True *P. calderiana* appeared to be absent and many of the presumed hybrids between it and *P. strumosa* had by this time cast their flowers. However, one day he marched in a north-westerly direction to the Monla Karchung La (17,442 feet) on the main Himalayan range and was quite certain that he was about to meet *P. calderiana* once again for he could detect its unpleasant fishy smell up wind for at least 100 yards distant. Greatly to his surprise he was greeted not with masses of purple *P. calderiana* but with masses of the rich yellow *P. strumosa* (19204) which, for the first time, he realised had exactly the same odour as *P. calderiana*. He was now firmly convinced that these two species were very closely allied, much more closely than the monographers had supposed, and that they hybridised freely. At Pangotang, *P. calderiana* and its albino form was the dominant plant, and at the Monla Karchung La, *P. strumosa*; half way between the two stations, at Waitang, the coloured hybrids, white, powder-blue, violet, cream, yellow and the rest, had taken possession.

Thus the days passed; back once again to Pangotang and then

north-west to Marlung for an exploration of the passes on the
Tibdey La range; Sherriff always in search of his beloved primulas
and of Ludlow's 'other aristocrats with the blue sap in their veins',
always hoping to find new species, always making shrewd observations. He crossed the Namda La, the divide between the Bumthang
Chu and the Mangde Chu, to Namdating, and 'one thing was
immediately noticeable, *Meconopsis horridula* [19436] this side
[west] of the pass, is twice the size of those on the other side, and
the colour is a fine blue, instead of the miserable dirty colour on the
other side. It is curious and most marked.' [30] And at Namdating the
cliffs were covered with a mixture of *Primula umbratilis* and a fine
blue form of *Meconopsis bella* (19437). He had searched for this
meconopsis in the extensive area in the Bumthang Chu—Trongsha
Chu divide without finding a single plant. 'It would be interesting
to know why *Mec. bella* is so happy here and completely absent only
a few miles away over the divide.' [31] It was while Sherriff was in
this area that he heard of Ludlow's near presence at Dur Chutsen
whither he hastened, to learn of his wife's unfortunate accident. He
and Ludlow made their way to Bumthang where they were joined
by Mrs Sherriff and Hicks on 26 July.

With Mrs Sherriff and Hicks to the Me La

With Tsongpen as their chief collector, Mrs Sherriff and Hicks had
parted from Sherriff at Lhuntse Dzong on 22 April and in somewhat
leisurely fashion had travelled first south to Lingitsi (6,500 feet)
and then east over the Donga La (12,500 feet) to Trashiyangsi
Dzong (5,800 feet) and Chorten Korra where they had halted for a
few days. Whenever they pitched camp Hicks' medical services
were in demand. Some patients, such as several blind, he couldn't
help; others he could, with M & B and other drugs. And on at least
one occasion his skill as a surgeon was put to the test in ususual
surgical conditions, as Betty Sherriff recounted on 28 April. 'More

> patients appeared after lunch including a small boy with a cyst
> the size of an egg dangling from an ear. His father was
> pathetically anxious for him to have attention and was very
> pleased when John Hicks said he would tackle it. By this time
> the wind was blowing hard and the tent flapping and billowing
> about—not what one would consider very suitable conditions
> for an operation. However the instruments were boiled up in a
> dekchi and the table set in readiness and the small boy installed
> in a camp chair with the father in attendance. J.H. gave a
> local anaesthetic and very deftly and well cut away the cyst

101. The Marlung Chu

. . . . The father was thrilled when the unsightly growth bounced on to the cotton wool he was holding! He is to bring the lad up to Lao in ten days' time to have the stitches removed.'

After so valiant a performance on Hicks' part under such difficult conditions it was entirely appropriate that a 6 feet tall mahonia which he now found should prove to be a new species and should be named *Mahonia hicksii* (20200).

From Chorten Korra it was their intention to journey due north to Lao, only three marches distant but they would halt on the way, thence north-east to Shingbe and the Me La. This was a route well known to Ludlow and Sherriff who had covered the ground on two occasions, and at different times of the year. In 1933 they had made the journey at the end of July and early in August, whilst in 1934 they had marched from Shingbe south to Chorten Korra at the end of August and during the first week in September. Whilst both journeys had been immensely profitable from a flower point of view Ludlow and Sherriff were only too well aware of the fact that they must have missed many of the earlier flowering species. Betty Sherriff, Hicks and Tsongpen would now try to fill the gaps in the collections for they would have six weeks — part of May, the whole of June, and the beginning of July — in which to explore the region of the Me La, the aptly named 'Pass of the Flowers'.

They left Chorten Korra on 30 April, marched easily through a good deal of cultivation, mostly rice-fields which the villagers were ploughing, to Shapang (6,500 feet), thence to Tobrang (7,500 feet) and to a delightful camp where they halted for the best part of a week. They were surrounded by rhododendron forest, but forest with so dense an undergrowth of bamboo as to be almost impenetrable except after much cutting and then much crawling on hands and knees. *Rhododendron griffithianum* (20220) deep pink in the bud, both the pink (20583) and the crimson (20586) forms of *R. arboreum*, the crimson *R. neriiflorum* subspecies *phaedropum* (20582) which Sherriff had recorded for the first time in Bhutan in 1937 at Chendebi, and the orange and yellow form of *R. keysii* (20581) — all were dominant and gave much colour to the landscape. Tsongpen was quite an expert at hunting for, and finding, primulas and now he gathered the geranium-like-leaved *Primula normaniana* (20589) although only a few of its pinky-mauve, frilled, flowers were fully open. On the Pang La (14,000 feet), directly west of Tobrang, a few days later, it was again found in considerable quantities although still it was only just beginning to flower. On the other hand, Sherriff had collected it in full flower, a little to the south-west, two weeks previously — the first record for Bhutan. It rather seemed that Mrs Sherriff and Hicks would be a little early for good primula collecting on the Me La.

And indeed at Lao (9,200 feet), the next halt on 10 May, though *Primula bracteosa* was now in immature fruit, *P. whitei* (20619) was still in flower and growing in a remarkable fashion on the moss covered trunks of trees and over mossy rocks. Here again, the rhododendron and bamboo forest, especially in the region of the great waterfall behind their camp, was only penetrable when Tsongpen had cleared a path with his kukri. Tsongpen's efforts were rewarded by his finding, at 10,000 feet, 20-foot trees of the smoky-blue or purple-mauve, round-trussed *Rhododendron niveum* (20620), one of Hooker's Sikkim Himalayan discoveries. The specimens which Tsongpen gathered were quite new to the Ludlow and Sherriff collections, and the species were never seen by them again.

The cliffs in the region of the Trashiyangsi Chu were the favoured habitat of many more rhododendrons; the tubular-flowered form of *Rhododendron glaucophyllum* known as *tubiforme* (20613, 20623), the flowers shading from deep rose-pink almost to white, and sometimes crimson-spotted; the orange-belled form of *R. cinnabarinum* (20622); one form of *R. lanatum* (20628) the yellow flowers vermilion-speckled and the lower side of the leaves with a palish fawn indumentum, and another (20648) with orange bands running along the yellow petals and with an orange-brown felt below the leaves; a form of *R. pendulum* (20627) the red-speckled white flowers shaded with pink; *R. wightii* (20641, 20642) varying from pale to a deeper yellow, always red-spotted and sometimes crimson-blotched as well. The cliffs and their rhododendrons, as well as the waterfalls losing themselves in spray so long and sheer was the drop, were a splendid spectacle, as was, at one point where the valley suddenly broadened and the river ran smoothly, a large colony of a magnificent form of the purple, golden-eyed, very farinose *Primula calderiana* (20625). And as a sure indication that they were still too early for primulas, a little north of Lao, in the direction of Shingbe where there was a considerable amount of snow, the white *P. vernicosa* (20637) and *P. dickieana* in its reddish-purple (20268) and pale yellow (20330) forms, still carried their flowers.

Had Ludlow been with the party without a doubt he would have called the weather execrable, and would also have had a name for the leeches which were present in their millions. Because Ludlow had not once found a leech on any of the hundreds of birds he had shot in leech infested jungles it had occurred to him that this might

be connected with the oil on the feathers. Thus, now, the experi-
mentally-minded Hicks removed the oil-gland from a Blood
pheasant, squashed it, and was mortified to see that the leeches
moved over the oil and were in no way contained. Only saline kept
them at a comfortable distance. But in spite of the weather, in spite
of the leeches, Mrs Sherriff, Hicks and Tsongpen persevered with
their plant hunting in the region of Lao and the Trashiyangsi Chu
and on the last day at the Lao camp, 23 May, were rewarded with
the finding of the new *Lilium sherriffiae* (20658) which Ludlow and
Sherriff were also to find, near Bumthang, some two months later.

They all left Lao on 24 May for Shingbe. For most of the way
they were never far from the river, marching mostly through
rhododendron, abies and juniper forest, often through rocky defiles,
and then through broad wet marsh sheeted with the magnificent
Primula calderiana. Three or four miles below Shingbe they met with
the fast melting snows and the river raced down between great
banks of yellow and pink rhododendrons; it was a glorious sight
with, in the background, dark conifers silhoutted against the gleaming snows. After some difficulty they found a comfortable camping
site at Shingbe for their five tents round which the coolies dug deep
drains. The coolies also buckled to and built a fine shelter of abies
planks and branches which would be much used for the next six
weeks for drying flowers and pressing-paper, as well as clothing.

Their collecting at Shingbe began auspiciously — and in somewhat surprising circumstances. During the night of 25 May, Betty
Sherriff dreamt that her husband walked into her tent, stood
beside her campbed, and gave her instructions for collecting on the
following day. She was to seek out below the camp a small track
leading to the Me La; to follow the track for about three miles until
it bifurcated; to take the right hand fork and walk some 300 yards
to a large rock mass. On the far side of the rock she would see a
poppy she hadn't found before. As Sherriff left the tent he turned,
shook a finger at her, and said 'Be sure you go'. The next morning,
at their usual 5.00 a.m. breakfast, when she told Hicks and Tsongpen the substance of her dream, they were both very sceptical and
urged her to keep to their original plan of collecting in a particular
valley they called the Glacier Valley. But the dream, and especially
the shaking finger, had been so vivid and the instructions so clear
and positive, that she determined to leave the rest of the party to
seek her dream poppy. She had no problems; she found the track

102. *Codonopsis convolvulacea*
103. *Aconitum fletcherianum*

easily; she found the mass of rock easily; and behind it she found a glorious blue meconopsis which she hadn't collected before, a form of *Meconopsis grandis* (20671). Hicks, at first unbelieving, returned to the spot the next day and took several photographs. More important, in the autumn he succeeded in collecting seeds of what is certainly one of the supreme gems of the Ludlow and Sherriff collections; when plants from these seeds flowered in some Scottish gardens they became known as 'Betty's Dream Poppy'. Mrs Sherriff wrote to her husband telling him of her experiences; the letter had to be carried first for seven days to His Highness the Maharaja and then for a further eight days to Sherriff. She asked him if, on that particular night of 25 May, he had been thinking of *Meconopsis grandis*, or of his wife. After several weeks, she received Sherriff's reply: 'Neither!'

There was still a good deal of snow on the north face of the valley leading to the Me La at the end of May, the south face being almost clear, and it was in the region of the melting snows that Betty Sherriff found her first primulas. Sherriff had so impressed on his wife the desirability, nay the necessity, of searching for, and gathering, as many species of primula as possible that she was now almost as interested in these plants as was he, and reacted with equal delight whenever she found one, as on 28 May two days after finding her 'Dream Poppy'. 'A good day for flowers as we got two primulas! *Primula macrophylla* [20679] hiding under rhododendron bushes just where the snow had melted, and the most adorable little primula, growing on a soaking wet rock, which looks as if it must be *P. jigmediana* [20681]. Such a dainty little flower, lilac-blue with a maroon eye, only about $1\frac{1}{4}''$ high. It was a great thrill finding it.' [32] She had indeed found *P. jigmediana*, last gathered on the Me La, and for the first time, by Ludlow and Sherriff on 5 August 1933, the same day on which they had also collected *P. macrophylla*. Rather remarkably these primulas on the Me La thus had a flowering season of at least ten weeks.

Sherriff had already made one effort to introduce *P. jigmediana* into cultivation during his first expedition to the Me La in 1933, and although a few plants had flowered in 1936 they had quickly been lost. Now he would make a second attempt. Seeds and roots were gathered by Hicks in September (21187) and flown to Britain. Although Mrs J. B. Renton of Perth succeeded in flowering one or two plants the following year they were soon lost and thus, sadly, this second effort of Sherriff's to introduce this minute gem to gardens in Britain was no more successful than his first.

Unfortunately the Me La primulas weren't very easily found for there was far too much snow which began about a mile and a half below the pass. And although it was rather dramatic plodding along through the snow with dazzling snow peaks on either side and with a radiant blue sky, it was also very bad for the plimsoll-shod feet and Betty Sherriff's chilblains became very troublesome. However they all persevered and gradually, as the days passed, they rediscovered Ludlow and Sherriff's findings of 1933. Hicks more than once had explored the valley directly behind the camp, the valley he called 'My valley'. Here the snow was no obstacle; it was a plod first through abies forest, then through thick rhododendron scrub, until the valley opened up and offered grassy slopes, rhododendron and juniper scrub, screes and great cliffs, for exploration.

It was in this valley, on the cliff faces, that Tsongpen found the glistening white-petalled, black or dark purple-sepalled *Primula soldanelloides* (20698). 'It's always a thrill when a new primula [new to the collection in this case] is found and worth all the slogging around.'[33] And no doubt an especial thrill when they discovered a primula not seen by Ludlow and Sherriff on either of their visits, as happened on four occasions the most notable of which was on 17 June. 'A big day in our collecting as between us we got three new primulas — at least new to our collections. Hicks went up "His" valley and climbed high up over a grassy path and came back with what we think are *P. cawdoriana* (20369) .... and *P. tenuiloba* (20370). The former is a very pretty primula — I hope it is *P. cawdoriana* as that will mean a new record of it in Bhutan. Tsongpen and I went up to the Me La .... and crossed over the river [which] was well over our knees. We then climbed up a snow filled nullah straight up to some high and promising looking cliffs. Tsongpen had to cut steps in the snow as it was very steep and slippery. When we got above the snow we found .... lots of *P. jonardunii* [*dryadifolia*] and we also saw *P. soldanelloides* glistening white and remote on a precipitous rock face. We collected a wee pale blue-violet primula (only a few specimens seen in flower and not many plants) cushion-like in habit, which we think may be *P. muscoides* (20743), a small blue lily which was flowering practically as the snow melted around it [*Ypsilandra yunnanensis* (20741)], a charming dwarf cassiope [*Cassiope selaginoides* (20739)] .... We have decided that we must open our second and remaining bottle of sherry tonight to celebrate three primulas.'[34]

Primula 20743 was indeed *Primula muscoides*, a white form of which Sherriff had recorded for the first time in Bhutan in 1937; 20743, growing at the foot of a dripping wet cliff, at 14,000 feet, had pale blue-violet, white-eyed flowers. On the other hand, 20369 was not, as they all thought and hoped that it would be, *P. cawdoriana*, which apparently has never been found in Bhutan. But it was a species which they now recorded for the first time in Bhutan and which was quite new to the Ludlow and Sherriff collections — *P. wattii*. Like so many of its relatives among the Soldanelloid primulas it is a most beautiful species, usually no more than six inches high, with a compact head of 5–10 widely cup-shaped blue-mauve flowers the petals of which are irregularly toothed, sometimes almost fringed, and the eye of which is conspicuously white-farinose. Discovered in 1877 on the Chola range in Sikkim, it is known to occur sporadically in Eastern Sikkim and Ward had also recorded it in the Assam Himalaya and at Seingku Wang on the Burma-Tibet frontier. Periodically it has been introduced into cultivation — it flowered for the first time in Britain in 1912 at the Royal Botanic Gardens at Kew and Edinburgh — but has always been short-lived for it rarely ripens seeds in cultural conditions. On this expedition seeds and roots were gathered (21178) in September in an endeavour to reintroduce the species into cultivation and although plants persisted for some years, gaining the Preliminary Commendation from the Royal Horticultural Society when exhibited by Mrs Knox Finlay in 1957, they were lost soon afterwards.

Return to Bumthang

The time was fast approaching when they must leave Shingbe and the Me La and begin their return journey to Bumthang. Quite clearly June had been too early to study the flora of the 'Pass of the Flowers' and their haul of Ludlow's 'aristocrats' had not been great. They had obeyed Sherriff's instructions regarding the search for his favourite primulas and though they had gathered all the species Ludlow and Sherriff had found in 1933 and 1934, they had found only four which had been missed on the two previous journeys; *Primula wattii*, *P. muscoides*, *P. glabra* (20355) and *P. macrophylla* var. *macrocarpa* (20736).

They left Shingbe on 5 July for Tobrang on a track which for the most part was a tearing river. It was a long trek of over ten hours during which they gathered sixty specimens including the epiphytic *Rhododendron rhabdotum* (20489) and the white, pink-flushed *R. camelliiflorum* (20488), as well as the yellow, fleshy-flowered *R. micromeres* (20825), the yellow *Meconopsis villosa* (20817) and the creamy-yellow form of *Lilium nepalense* known as *concolor* (20843).

At Tobrang they halted for two days, chiefly to give aid to an ailing Tibetan who, with his wife and child, the latter also very ill, as well as with a donkey and a puppy, had been lying in a rough shelter, thick with flies, for the last three weeks, too weak to move. He and the child were suffering from acute dysentery and worms. Hicks treated his patients with worm medicine and with M & B, before he and Betty Sherriff marched westwards in the direction of the Pang La (14,000 feet) which Ludlow and Sherriff had crossed in the middle of August in 1933. Mrs Sherriff firmly believed that no one had used the pass during the last sixteen years, so difficult was the journey, first through shoulder-high bracken which had to be cut and then through thick rhododendron and fir forest, constantly hopping from boulder to boulder along a muddy and slippery track, and ascending and descending by old notched trees which were used as ladders. Not until 10 July were they safely over the pass and had reached Denchung.

Here they made plans to journey north up the Khoma Chu in the direction of Narimthang where, in August 1933, Ludlow and Sherriff had gathered *Primula eburnea, P. dryadifolia* and many other lovely plants. On asking the village headman to provide the necessary transport they were told that the Khoma Chu valley was full of evil spirits and that it would be safe to attempt a northwards journey only after animal sacrifices had been made. This they refused to do and set out northwards after a day's rest at Denchung. Betty Sherriff was riding a small pony which one of the villagers had found for her, a welcome change after so many weeks of walking with bad chilblains. A small boy, carrying a large parcel of plimsolls which had been sent by His Highness, was leading her pony along an extremely narrow and rough path and was clearly in difficulties. To assist him, Betty Sherriff relieved him of his parcel. Not more than a mile from Denchung the saddle, owing to a loose girth, swung round under her pony and she, holding the plimsoll parcel, couldn't fling her arms round the pony's neck as she automatically would have done. Instead she was shot from her pony and down the hillside, hitting a rock and breaking her arm in two places. Naturally the expedition to Narimthang had to be abandoned and all returned to the Denchung camp where the headman said 'I told you so'. Hicks of course treated the invalid with the resources at his disposal but the next two weeks were very painful for Mrs Sherriff. However, she, and the rest, safely reached Bumthang on 26 July and the entire expedition was then united.

And once again the expedition received the hospitality of His Highness the Maharaja; and when not being entertained, and

347

entertaining, attended to their plant specimens and developed photographs. And Ludlow pondered on one of the most interesting ornithological problems of Bhutan, the presence in the Bumthang area of the Tibetan, or Black-rumped, magpie (*Pica pica bottanensis*) which both he and Sherriff had seen on several occasions in Tibet. It was not only present around Bumthang but occurred at Gyetsa to the west and at Bumdang Thang to the east; Sherriff had seen it on the Rip La north-west of Lhuntse in the Khoma Chu valley and thought that quite probably it could also be found in the Kuru Chu valley. There was no evidence of its occurrence elsewhere in Bhutan. How had this colony of Tibetan magpies in the neighbourhood of Bumthang originated ? How had it survived in a region of wet forests and heavy rainfall, so different from its much drier Tibetan habitat, for over a hundred years ? Had it migrated over one of the Himalayan passes, or had travellers brought it to Bumthang ? It was quite impossible to say. The only point on which Ludlow was pretty certain was that as the magpie had been described in 1840 from a specimen taken from an unrecorded locality in Bhutan, Bumthang probably must represent the type locality.

One other matter was causing Ludlow a good deal of thought; at the age of 64, how on earth was he to surmount the numerous passes on his return journey to India ? His problem was solved for him by His Highness who generously gave him the use of his own riding yak, to take him as far as Lingshi, if necessary.

The autumn flowers and the seed harvest

On 1 August the Sherriffs left Bumthang, Mrs Sherriff to return to India for an X-ray of her arm, Sherriff to accompany her as far as Ha whence he would return to his collecting grounds for the seed harvest. Three days later Ludlow and Hicks also left in the their quest for seeds, Hicks for the Me La, and Ludlow first for Chendebi. Before leaving they were received by His Highness on the archery ground and given the usual tea. The Maharaja told Ludlow that he would never forget Ludlow's kindness to his son in Kalimpong, and thanked him, not for the first time, with touching and simple sincerity. Ludlow, much moved, replied that this was probably the last time he would meet His Highness; that he had spent half of the last twenty-five years of his life either in Tibet or Bhutan; that these had been the happiest years of his life; that the recollection of them would be the solace of his old age. With that, he rode west to Gyetsa with the main intention of moving north to Passu Sefu and to seeds of *Meconopsis sherriffii*.

From Gyetsa (9,525 feet) he crossed the Yuto La (11,200 feet)

348

the rolling pine-clad slopes on the east side of the pass giving way to steep hillsides clothed with oaks and deciduous forest on the west. Then the descent to Trongsa Dzong (7,125 feet) and its dense semi-tropical jungle with leeches abounding, followed by the march south-west to Trashiling and the collecting of the new species *Luculia grandifolia* (17040) with its tassels of large pure white flowers, which Sherriff had first found on the Chungkar cliffs in 1934. Thence to Chendebi (8,400 feet) and to the same camping ground he and Sherriff had used in 1933, there to halt for a day to attend to the seventy or more different gatherings of plants he and his collectors had taken in the last three days.

On 8 August he began his journey northwards up the Rinchen Chu collecting seeds, and later on, bulbs, of *Lilium nepalense* (17081) which form has proved to be more hardy in cultivation, in some gardens, than other forms, and reaching the Omta Tso (14,200 feet) two days later. At his camp, some 400 feet above the lake, he intended to halt for the greater part of the next two weeks. There was little point in reaching Passu Sefu before the beginning of September at the earliest so Ludlow planned to move slowly, making halts on the intervening passes.

The area was a very wet one. 'The rain is terribly persistent. It just pours, and pours, and pours, relentlessly. It is very depressing. But the plants love it and were it not for the relentless rain there would not be the lovely plants.' [35] And lovely plants there certainly were, some of them now in fruit. In the alpine grassy pastures, between 14,000 and 16,000 feet, there were large quantities of *Primula uniflora* (17108) which Sherriff had recorded for the first time in Bhutan in 1937 and which Ludlow hadn't seen in the wild before; legions of *P. strumosa* (17105) still in magnificent yellow flower, and of *P. elongata* (17119) now in fruit; great cushions of lilac *P. muscoides* (17127); sheets of purple *P. calderiana* (17131) intermixed with *P. sikkimensis* on the shores of the lake; both the purple and white forms of *P. tsariensis* (17140, 17146); countless spikes of bright pink *Polygonum vacciniifolium* (17136) plastering the rocks and boulders; an abundance of *Codonopsis nervosa* (17125) amidst dwarf fruiting rhododendrons, the great cup-shaped flowers bluish-white without, veined with reddish-purple within; the delightful large-flowered, finely cut-foliaged *Delphinium muscosum* (17102); the pale lemon-yellow *Aconitum orochryseum* (17122) which A. F. R. Wollaston had first found on Mount Everest in 1921 and which now was a new record for Bhutan; and at least three saxifrages new to science, the port-wine-red *Saxifraga rubriflora* (17099), the yellow *S. lepida* (17095) which has since been found by Stainton, Sykes and Williams in Nepal, and the pale lemon-yellow form of *S. thiantha* known as var. *citrina* (17104). Unfortunately Ludlow's collecting staff became seriously depleted through an accident to one of Danon's feet which became so swollen and painful that for a time he couldn't put it on the ground. He had to return, with another collector, Pompoli, to Bumthang for treatment, so that Ludlow was left only with Ramzana to continue his journey, first eastwards by the Padima Tso and then north up the Mangde Chu to Passu Sefu, on 22 August.

They had to move very leisurely over the two passes between the Padima Tso and Passu Sefu, the Khem La (14,800 feet) and the Phage La (16,280 feet), because their yaks, carrying the baggage, were very slow and made heavy weather of the ascents. Thus there was ample time to search for plants — and they were handsomely rewarded. At the Padima Tso they gathered a beautiful cremanthodium (17177) with olive-green involucral bracts and pale yellow ray florets; at 15,000 feet it was growing amidst the dwarf rhododendron scrub and proved to be a new species which Ludlow described, and to which he gave the name *Cremanthodium bhutanensis*.

And on the two passes they were greatly attracted by several aconites, all of great merit, growing in rock crevices amongst boulder scree, sometimes on steep alpine pastures, between 13,500 and 15,000 feet. On the Khem La there was the incomparably beautiful *Aconitum fletcherianum* (17198), a few inches high with very large bright violet flowers, the two lateral segments edged with white. He and Sherriff had first discovered it — and only one specimen of it — in 1933 but had refound it on several later occasions. 'Friar Tuck' was the name Ludlow gave to it. And 'Little Tich' was the name he gave to the midget, violet-purple-flowered *A. pulchellum* (17204), widespread in Yunnan, Szechwan, SE Tibet, Upper Burma and Sikkim, which he found on the Phage La, not far removed from another dwarf aconite (17203) which he and Sherriff had first seen on the Cho La in SE Tibet in 1934 and which proved to be a new species with flowers varying in colour from deep blue-violet to pale purple, sometimes tinged with red, and which was named *A. parabrachypodum*. Ludlow was keenly aware of the desirability of introducing these dwarf gems into cultivation in Britain. But how ? Flowering so late in the season, during August and September, it is unlikely that their seeds ripen before they become shrouded in snow in October. Ludlow's only chance was to take away living plants with him in the hope that they might set seeds on the journey, and in any case survive to be flown home from Calcutta. Thus he dug up the fibrous roots of

Aconitum fletcherianum and the tubers of the other two species. These were finally flown to Britain where *A. fletcherianum* and *A. pulchellum* certainly flowered for a year or two, before dying without setting seeds.

In camp at Passu Sefu on 1 September, Ludlow's first task was to secure seeds of the 'plant of incomparable merit', *Meconopsis sherriffii*, which he did that same day. And on the 7th he gathered an even greater amount under the number 17231. Plants were growing at 13,800 feet, along with *Meconopsis paniculata*, *M. simplicifolia* and *M. horridula*, in a flat valley bed where lonicera, potentilla, salix and dwarf rhododendrons flourished amongst rocks, boulders and grasses. Once again Ludlow could see no reason why the pink poppy should not prove as amenable in cultivation as its companions. Certainly the seeds of 17231 germinated well when sown early in 1950 and magnificent flowering specimens, grown by Mrs Knox Finlay, were one of the highlights of the Chelsea Flower Show of 1951 where they gained the Award of Merit. Unfortunately, although in nature, and certainly in one garden, Sherriff's poppy is a true perennial — and some of the plants Sherriff flew home had the remains of flower stems and had produced offsets after flowering — in cultivation it has mostly tended to be monocarpic and has died after flowering. Only Mrs Sherriff has cultivated it successfully over the years and still maintains a good stock in her garden at Ascreavie in Scotland where it is truly perennial.

For some years both Ludlow and Sherriff had been worried about the political situation in Bhutan and in SE Tibet, and the announcement on 3 September, which he heard on his small wireless, that the Chinese Communist Army had reached Sining and had announced that in due course it would proceed to 'liberate' Tibet which was an integral part of the Chinese Empire, prompted an unusually long entry in Ludlow's diary that evening. 'Poor old Tibet. I wonder if

America, Britain, or India, will take up the cudgels on her behalf, or just stand still and watch her gobbled up. Tibet can do nothing by herself of course. With Communist China at her back door India will find herself in a very unenviable position. In fact there will now be a North East Frontier as well as a North West Frontier. It seems to me that India would be well advised to help Tibet repel the Communist armies. She could do this with a powerful air-force based on Lhasa, and a scorched-earth policy along the route of the invader. Troops and transport columns would be very vulnerable from the air on the bare Tibetan plateau. And the Indian air-force has experience of landing and taking off at high altitudes, e.g. in

Ladakh. There is an excellent natural aerodrome near Sera on the Lhasa plain. India would have to make the road over the Nathu La motorable. This could be done without much effort as the alignment is already laid down. And Pangda Tsang has recently improved the road from the Nathu La to Phari out of all knowledge. Jeeps could traverse it already and very little extra work would be needed to improve it so as to take lorries. From Phari onwards, the only major obstacles are the Kamba La and the Tsangpo crossing. Of course all this political trouble knocks on the head any chance I had of going to the Tsangpo gorge [next year].'

On 14 September when Ludlow and Ramzana left Passu Sefu and travelled west to the Gafoo La (16,900 feet) neither Danon nor Pompoli had returned from Bumthang. Danon, especially, was a great loss as he was a fine collector and knew his plants even better than did Ramzana. Not until 26 September, whilst Ludlow was halted at Foomay, six marches to the south-west of the Gafoo La, was Danon fit enough to rejoin the expedition. In the meantime Ludlow and Ramzana crossed the Gafoo La and marched on to Tranzo and Chozo Dzong, where early in July Ludlow had found a fine incarvillea (16722) the large gloxinia-like flowers of which were pinkish-red, white-streaked in the throat and creamy-white in the lower half of the tube. It was the Chinese *Incarvillea mairei* which he had collected in s E Tibet at the beginning of June in 1947. He now found the plant, its leaves having turned bright red in the autumn, in fruit, and collected seeds under the number 17250. As with the Tibetan plant there was some variation in the resultant seedlings and an almost stemless form with vivid carmine-pink flowers appropriately has been named 'Frank Ludlow'. Happily this fine cultivar is well established in cultivation — in some gardens and in at least one nursery, at any rate.

For the next month Ludlow and his companions retraced their footsteps to Ha, over the ground they had covered during May and June, in their quest for seeds, bulbs, tubers and plants. Frequently the conditions were anything but ideal for their task, as was the case on 29 September during the march from Foomay to Gyabna Thang, the half way stage between Foomay and Laya. 'It snowed last night and when we awoke in the morning there were 3–4 inches of snow on the ground — sufficient to hide all the nice little primulas such as *glabra, pusilla, soldanelloides, waddellii, tenuiloba* of which I hoped to collect seeds. One of the most important plants of which I wanted seeds (and fruiting specimens [for the herbarium]) was the small *Meconopsis primulina*. When we arrived at the place where it grew not only was the ground covered with a carpet of snow but the snow which had ceased to fall in the early morning began to fall again. We knew roughly the area in which the plants were for we had marked the spot with stones in June last. I suppose there must have been 100–150 plants in flower in an area 60 yards square. But when it came to finding these plants (only about 6″–8″ in height when in flower) in the dried up seedling stage under a carpet of snow we found it a very difficult task indeed. There were four of us, Ramzana, Danon, Pompoli and myself, and after 1½ hours' groping under the snow, with a blizzard raging the while, we managed to resurrect 15 fruiting specimens only two-thirds of which possessed seeds [17338]. Such are the difficulties, or one of them, with which the seed hunter is beset. And the horticulturists at home, when they receive a mere pinch of seed, will probably remark on the niggardliness of the donor, little realising with what toil and labour and discomfort that little packet was obtained.

'The rest of the day was one of hard foot slogging through snow and sleet most of the time along the most execrable of tracts — a track so ill-defined that we often lost ourselves in the mist and cloud and had to circle about for lengthy periods to find the footprints of our transport which was ahead. It was one of the weariest marches I have ever accomplished, and I reached camp footsore and wet and weary half an hour before darkness fell. Danon, who has been marching without a halt since he left Bumthang on the 16th, was very exhausted and his foot still gives him much pain. It snowed, sleeted and rained the whole live-long day — a day in which we had hoped to reap a rich harvest of seeds.' [36]

Ludlow was able to increase his pinch of seeds of *Meconopsis primulina*; he collected a tablespoonful (17426) in the Yare La, and at Lingshi Dzong four fat packets of it (17446); 'I am very pleased with the result'. And no doubt desperately disappointed with the behaviour of his precious seeds in cultivation. Though some germinated there is no record of plants having become established anywhere. The seeds he gathered of *Meconopsis discigera* (17455) fared better and this species is still in cultivation; although monocarpic it usually sets seeds before dying and thus can be renewed.

Thus the weeks passed, Ludlow reaching Ha on 21 October with the largest collection of dried plants, over 1500 numbers, and of seeds, over 250 collections, he personally had ever gathered. He was quite certain that both Sherriff and Hicks, whose arrival he now awaited, would have been equally successful.

Sherriff had left Bumthang on 1 August and had accompanied his wife as far as Ha. There, Mrs Sherriff had left for India and Sherriff had returned to Bumthang to make a last journey north, as far as Waitang, for the autumn collecting. The return journey to Ha occupied the whole of August and he enriched his collections by some hundred and thirty gatherings. He was especially pleased to gather once again, at Ha, the very lovely *Codonopsis convolvulacea* (19620) which he had first collected in 1937, under 3568, near Paro, and which he had missed there, on his present trip. Since 1937 the recollection of its bright vinca-blue flat flowers with their wine-red ring and large rusty-brown globular stigma had been a very vivid one; he regarded it as a finer plant than *C. vinciflora* and one which would be a great attraction in gardens at home. At Ha he marked several plants for seed collecting in October when he would also gather tubers which, 2 – 4 inches under the ground, were like potatoes and of much the same size, sometimes over two inches in length. Naturally on the Ha La he visited the site of *Meconopsis superba* which he had first gathered in 1933. He found most beautiful plants of it growing amongst potentilla, dwarf rhododendrons and bright pink sheets of *Polygonum vacciniifolium* (19616) which was sprawling over the rocks above the limit of the fir forest. The meconopsis (19617) was not yet with ripe seeds but the splendid plants were over six feet in height and the thick hard stems had carried over thirty flowers. Sherriff was equally impressed by the very large root which, about six inches below the ground surface, was as much as 15 inches in circumference before it divided into smaller roots all of which exuded a thick yellow juice when the surface was cut.

Sherriff considered himself lucky to reach Bumthang again alive, for on 29 August, riding his pony on the march from Chendebi to Trashiling, he reached the scene of an accident on a bridge over the river. Five minutes earlier a man with three ponies had crossed the bridge whose sides had been badly scoured. Two of the ponies had made the crossing but the third had broken one of the long planks of the bridge and in some remarkable fashion had fallen through the gap, the breadth of one plank, into the chasm below. Sherriff would have crossed this same bridge, riding his pony, and believed that the plank would certainly have given way under him.

On 5 September Sherriff began his month's trip to the north of Bumthang, to the ground he had covered during the last part of May, the whole of June and the first half of July; before leaving he gave Danon, Ludlow's collector, penicillin injections and a course of M & B for his injured foot. There were still plenty of flowers to collect including, much to Sherriff's delight, more splendid material of *Codonopsis convolvulacea* (19674). There were saxifrages. There were gentians, including a new species with silvery-edged leaves closely imbricated in four rows and with remarkable flowers partly of a wine-purple and partly of a very pale blue, which was later named *Gentiana emodii* (19721); Sherriff sent to Britain, by air, living plants which, unfortunately, did not survive the journey. There was *Delphinium muscosum* (19725) which he had first found in 1937 and which both he and Ludlow had refound several times during this present expedition. There were several aconites; the new blue-violet species from the 1934 expedition, *Aconitum para-brachypodum* (19708); *A. orochryseum* (19719) which Ludlow had recorded for the first time in Bhutan only the previous month, this time with almost white flowers instead of the usual yellow; and his great favourite, the delightful *A. fletcherianum* (19734).

And there was one of the most beautiful plants Sherriff had ever seen, Kingdon Ward's *Wardaster lanuginosus* (19764). Ward had discovered this interesting link between the floras of the Eastern Himalaya and Western China in 1921. He had found it near Muli in sw Szechwan, had christened it his 'flannel-leaved aster', and had described it as 'a particularly striking species with chubby heads of violet flowers cuddling down amongst the foliage'.[37] In 1936 Ludlow and Sherriff had recorded it (2001) for the first time in se Tibet at 16,500 feet on the Traken La and thus more than five hundred miles due west of Ward's original locality. And in Central Bhutan Sherriff had extended its range another hundred miles still further westwards by recording it first at Marlung in July (19415) and now, in September, at Waitang (19764) where he found it both in fruit and in fine flower, perfuming the hillside very strongly and very pleasantly with verbena. Living plants of 19764 were sent to Britain where they flowered at the Royal Horticultural Society's Gardens at Wisley, and at Mr R. B. Cooke's garden at Corbridge, Northumberland. Unfortunately, as with so many of these densely felted high alpines, wardaster proved to be very intractable in cultivation, and although plants did set seeds in Mr Cooke's garden this most desirable alpine plant was quickly lost to cultivation.

Although Sherriff succeeded in harvesting a fine lot of viable seeds of the wardaster he wasn't always so lucky with those of other plants. By the end of the first week in September most of the seeds he was anxious to collect were barely ripe and many of those that were ripe were infested 'with caterpillars or little bugs of some kind'. There was a fine salvia of which he wanted seeds; he found plenty of unripe ones but 'as soon as they [the fruits] showed any

colour at all a bug got into them'. The seeds of *Primula soldanelloides* had all been eaten by these 'bugs'. From among the capsules of
P. umbratilis he collected nearly seventy caterpillars and every single
capsule of a beautiful large swertia was inhabited by three or four
red worms. He was anxious to gather seeds of the rich deep
Polygonum griffithii (19715), and to this end examined several
thousands of old flowers, and harvested only fifteen seeds. But 'bugs'
were not the only problem. Almost as great a menace were the yaks
which had demolished an aster which at Pangotang at the end of
June had covered a great area. Again in the Pangotang area he had
seen hundreds of *Notholirion bulbuliferum* (19292) which, at the
beginning of July, was only in flower-bud. Now, in September, when
he returned to the spot to collect bulbs, and perhaps seeds, for some
time he could find no signs whatsoever of a single plant. However
by searching closely in the low-growing potentilla which grew
everywhere he discovered a few leaves of the notholirion — the yaks
having eaten everything else. However he succeeded in digging up a
hundred good-sized bulbs, some fifty smaller ones, and in finding at
least 1500 bulbils and by so doing reintroduced into Britain this very
beautiful plant which, as grown in the garden at Logan in Wigtown-
shire, Scotland, received the Award of Merit from the Royal
Horticultural Society in 1953.

To add to Sherriff's problems towards the end of September
coveys of Snow partridge (*Lerwa lerwa*) became very vociferous
and some of his men maintained that this was a bad sign and that it
heralded much snow and rain. 'They were quite right. It has barely
stopped raining for over three days and yesterday and today
have been as bad as we've had all year. Last night it sleeted,
and this morning snowed hard, even down here [Marlung] at
13,500 feet. And it has not let up raining hard and snowing all
day long. I had not expected this I must confess. We had two
high altitude primulas to get today. One, I gave up altogether
— *Primula concinna*, white. The other, *P. soldanelloides*, we had
marked with a small cairn of stones . . . . I tried to find it but
rain had removed all the stones and snow hid the primula. Up
there, there were 3"–4" of snow. We eventually found a little
of what I think is *P. soldanelloides* in a cliff face a little distance away.' [38] It was the same awful weather which had hit
Ludlow on his quest for seeds of *Meconopsis primulina* and which had
prompted the same kind of depressing entry in his diary on the same
day of 29 September.

104. *Meconopsis bella* with *Primula umbratilis*

Under all these circumstances Sherriff, anxious to gather a record harvest of seeds, was rather disheartened, until a letter from Ludlow, written on 20 September, gave the news that his friend had collected five pounds of seeds of *Meconopsis sherriffii* ! 'That in itself is a pretty good collection to take back from this trip because it will almost certainly grow well from this low altitude seed. On the Drichung La it was found at from 15,500–17,000 feet as far as I remember, whereas this is mixed up with *Meconopsis paniculata, simplicifolia* and *horridula* all of which are in cultivation at home.' [39]

Sherriff need not have been dissatisfied or depressed with his results for, when he reached Bumthang on 7 October, he had close on 180 gatherings of seeds, some 25 rooted plants, a goodly number of bulbs and tubers, as well as close on 1500 dried plants for the herbarium.

Hicks, who reached Bumthang on the same day, had also had a successful trip. He arrived with nearly 150 gatherings of seeds. Among them were some thirty primulas, including *Primula dickieana* var. *aureostellata* (21266), *mollis* (21239), *xanthopa* (21230), *soldanelloides* (21197), *jigmediana* (20681), *dryadifolia* (21185), *wattii* (21178), *macrophylla* var. *macrocarpa* (21172) and *normaniana* (21038); there were about twenty rhododendrons, including *Rhododendron rhabdotum* (21257); there were six meconopsis — *Meconopsis grandis* (21069), *bella* (21124), *sinuata* (21131), *simplicifolia* (21162), *paniculata* (21168) and *horridula* (21198); and several gentians including the new species *Gentiana taylori* (21246) first found in 1938 in SE Tibet, and now, gathered on the Me La, constituting a new record for Bhutan.

Sherriff was particularly pleased with the refinding of *Primula xanthopa* for he had been anxious for Hicks to collect further material of it from the Donga La, its only known locality where he and Ludlow had gathered it during their first joint expedition in 1933. Hicks was successful in obtaining flowering specimens (21010) as well as seeds and living plants (21230). The living plants flourished abundantly in several gardens in Britain for a year or two, and some gained a Preliminary Commendation when exhibited by Mrs Knox Finlay at the Royal Horticultural Society in 1950; the impression, in horticultural circles, was that this species would prove to be very amenable in gardens. Unfortunately, with one exception, all plants died without setting viable seeds. The exception was in the garden of Mr and Mrs John Renton of Branklyn, Perth, Scotland where, for at least twelve years, the charming species was maintained in cultivation.

In addition to his seeds and living plants, Hicks, with the assistance of Mrs Sherriff and Tsongpen of course, had prepared nearly 1500 herbarium specimens which included several new species, one of them, a crimson-flowered lousewort ultimately being named *Pedicularis hicksii* (21099).

During his trip Hicks had tended the sick on many occasions and had performed several operations of a minor nature. But at Bumthang he, with Sherriff, was the witness of an operation he would not have ventured to have done. He described it all on 12 October.

'I have just watched an operation for the cure of cataract performed by a Tibetan woman. Most of the doctoring here consists of such procedures as placing red hot irons on the patient's head but in this case the operation was quite a scientific one though entirely different from ours. She makes an incision or rather a mere puncture a sixteenth of an inch long through the sclera, inserts a flat gold probe across and behind the iris and then scoops the cataract away in one dexterous movement. She does not remove it — she says this cannot be done — but leaves it somewhere out of the way in the medial part of the eye. It is all done in five minutes and really nicely if I am a judge of surgery. There is no anaesthetic but the old priest upon whom she was operating never flinched. Neither is there any antisepsis but this is compensated for by speed, simplicity and a "no-touch" technique. It was all carried out in the weird surroundings of a Buddist temple with the pictures of various gods leering down upon us from the walls. There was no "jardu" or magic ceremony about it except perhaps the "medicine" with which the eye was drenched and this was made from saffron, musk and bear's liver. And there was an assistant who tapped two stones together by the patient's ear but this was really to attract his attention and cause him to deviate his eye. I was intrigued at the discovery that the West had not the monopoly of inventing surgical operations.

'The woman's story is that she is doing this as a sort of penance. When she was young she was travelling with her father's caravan — for he was a trader — when it was attacked by robbers and she was knocked down. Her father's servants got the better of their assailants and caught one and put out his eyes. This act the father, in time, came to regret and when he asked a Lama how he could obtain forgiveness he was told that he must give his daughter to a life of curing blindness. So the girl was taken to Lhasa and there the chief physician taught her the secret which was known to only four people. When she had practised scores of times on animals' eyes she was at last declared competent and allowed to embark upon her life of healing.

'Now where fiction ends and fact begins in this story I cannot tell, but here was the woman, looking well-dressed and prosperous, here was her word that she had done over eighty cases and here was the undeniable fact that she had very neatly just done her eighty-seventh. What is more, she had had only eight failures — so she said, and who in England is a good enough liar or a good enough surgeon to make such claim for his workmanship.'

Sherriff was equally impressed. 'I would not have believed it had I not actually seen it done.'

Once again, during their five days at Bumthang, Sherriff and his friend Hicks were given the generous hospitality of His Highness the Maharaja. During the course of his expeditions Sherriff had now stayed with His Highness for almost three weeks 'and he has made me and my friends his guests all the time we have been in Bhutan. He has fed us all, and our servants; he has provided saddles for us, sent our mail each way and generally looked after us all. Could hospitality ever be greater ? In return Betty and I now have to take Tasho [his son] home and we will have a job looking after him, no doubt. He will be with us for about five months.'[40]

Sherriff and Hicks left Bumthang on 13 October and reached Ha on the 24th, there to find Ludlow awaiting them. Between the three of them they had amassed beautiful material of five thousand gatherings for the herbarium, as well as some six hundred collections of seeds, bulbs and tubers, the largest collection ever made from Bhutan during the course of seven months. In addition there were ninety-three lots of living plants which Sherriff proposed to fly to London and which were now packed in three loosely woven bamboo hampers, three tiers of plants in each hamper, the individual plant clumps wrapped in moss so that once in Britain they could be forwarded to the gardens specified by the collectors with the minimum of disturbance. Sherriff would make a special journey to Calcutta to ensure that the plants were properly handled and placed on the plane, whilst Dr Taylor, who would arrange for import permits and satisfy the customs and disease control authorities,

105. Village near Lhumbe Dzong
106. View from the Lubsing La

would also meet the freighter at London Airport, divide the collection and despatch each share to the recipients Ludlow and Sherriff had nominated.

The plants left Calcutta by BOAC on 11 November and Taylor was at London Airport when the plane touched down on the afternoon of the following day. Within an hour the three hampers were on their way to the Natural History Museum where they remained over the weekend, and on Monday 14 November were unpacked at Chelsea Physic Garden. On the whole the plants' had travelled quite excellently though some were rather dry and a few, such as *Gentiana emodii* (19721), *Diapensia himalaica* (19146) and *Diplarche multiflora* (21152) were quite dead. The losses weren't really surprising considering the length of the journey, and the rigours of the journey, from their cool wet alpine haunts of Bhutan, through the high temperatures in the steamy lower parts of the country to Kalimpong and then between Kalimpong and Calcutta when the temperature no doubt would be in the region of 100°F.

In the consignment there were over thirty primulas including such beautiful rarities as *Primula xanthopa* (21230), *P. jigmediana* (21187), *P. uniflora* (19836), *P. soldanelloides* (19420, 21197); two aconites, *Aconitum fletcherianum* (19771, 17198) and *A. pulchellum* var. *pulchellum* (17204); *Meconopsis sherriffii* (16704); *Wardaster lanuginosa* (19764); three orchids, *Cypripedium tibeticum* (19123), *Spathoglottis ixioides* (16920), as well as a yellow-flowered species which was never identified; *Delphinium muscosum* (19375); *Paraquilegia grandiflora* (16356); *Streptopus simplex* (21044); *Androsace globifera* (19309) and *A. haemispherica* (19404) — every single plant one of Ludlow's 'aristocrats with the blue sap in their veins'.

With the living plants was a quantity of lily bulbs, including those of the new species *Lilium sherriffiae*, of notholirion bulbs, and of tubers of one of the most successful introductions, the twining *Codonopsis convolvulacea* (19620, 19674).

At that time this effort of Ludlow and Sherriff to introduce, by air, living plants to cultivation in Britain probably was the greatest one of its kind and was a fitting climax to their years of magnificent collecting. However, time has shown that their endeavours were not as successful as they had hoped. Many of the plants made but a fleeting appearance in British gardens, flowered and then were gone. And over twenty years later, most of those which have survived have maintained but a tenuous hold on cultivation.

Thus ended the sixth joint expedition of these great collectors. They returned to Britain in 1950, Ludlow to spend the remaining years of his life working on the Ludlow and Sherriff collections in the British Museum, Sherriff to farm in Angus in Scotland where, at his home at Ascreavie, near Kirriemuir, he and his wife transformed a wilderness overrun by rabbits into a Himalayan garden of surpassing beauty where grew, and where still grow, primulas, meconopsis, gentians, lilies, rhododendrons as in their native haunts. When he had spoken to His Highness the Maharaja of Bhutan for the last time, Ludlow had said that the recollection of his happy days in Bhutan and Tibet would be the solace of his old age. And indeed they were, for he lived, working among his plants, to the ripe old age of 86 years, dying on 27 March 1972. Sherriff was not so fortunate; he died at Ascreavie on 19 September 1967, in his seventieth year.

'Knowledgeable and resolute plant collectors whose rewarding journeys in the Eastern Himalaya and in South Eastern Tibet have added greatly to our knowledge of the flora and natural history of these regions and whose strenuous efforts on hazardous expeditions have enriched our gardens by the introduction of plants of outstanding merit.' Thus was volume 174 of the *Botanical Magazine* dedicated to those two friends and outstanding characters who, strangely, were known to each other, very simply, as Ludlow and Sherriff.

NOTES

Chapter One
1. Ludlow, 26.IV–5.V.33. All such notes refer to the dates of entry in the collectors' diaries.
2. All collectors' numbers, unless otherwise stated, refer to those of Ludlow and Sherriff (L & S) but to avoid repetition the initials henceforth are omitted.
3. Ludlow, personal communication.
4. Ludlow, personal communication.
5. Ludlow, personal communication.
6. Ludlow, 24.VI.33.
7. Sherriff, 11.VII.33.
8. Ludlow, 16.VIII.33.
9. Sherriff, 17.VIII.33.
10. Sherriff, 18.VIII.33.
11. Cooper in *Notes Roy. Bot. Gard. Edin.* XVIII, 81(1933) — a judgment expressed before he became familiar with Ludlow and Sherriff's numerous primula gatherings of later years.
12. Sherriff, 21.VIII.33.
13. Sherriff, 23.VIII.33.
14. Sherriff, 1.IX.33.
15. Sherriff, 3.IX.33.

Chapter Two
1. Ludlow, 22.VI.34.
2. Kingdon Ward in *J. R. H. S.* 74, 334 (1949).
3. Sherriff, 1.VII.34.
4. Sherriff, 10.VII.34.
5. Sherriff, 14.VII.34.
6. Ludlow, 23.VII.34.
7. Sherriff, 23.VII.34.
8. Sherriff, 6.VIII.34.
9. Sherriff, 8.VIII.34.
10. Ludlow, 16.VIII.34.

11. Sherriff, 22.VIII.34.
12. Sherriff, 24.VIII.34.
13. Ludlow, 30.VIII.34.
14. Sherriff, 24.IX.34.
15. Ludlow, 4.X.34.
16. Sherriff, 2.X.34.
17. Sherriff, 3.X.34.
18. Sherriff, 5.X.34.
19. Ludlow, 5.X.34.
20. Ludlow, 6.X.34.
21. Sherriff, 6.X.34.
22. Sherriff, 7.X.34.
23. Ludlow, 8.X.34.
24. Sherriff, 20.X.34.
25. Sherriff, 7.XI.34.

Chapter Three
1. Sherriff, 9.III.36.
2. Sherriff, 9.III.36.
3. Ludlow, 12.III.36.
4. Ludlow, *Ibis*, 45 (1944).
5. Sherriff, 27.IV.36.
6. Sherriff, 29.IV.36.
7. Sherriff, 8.V.36.
8. Sherriff, 18.V.36.
9. Sherriff, 7.VI.36.
10. Sherriff, 3.VI.36.
11. Sherriff, 11.VI.36.
12. Ludlow, 15.VI.36.
13. Ludlow, *Himalayan Journal* X, 11 (1938).
14. Ludlow, 29.VI.36.
15. Ludlow, 26.VII.36.
16. Ludlow, *Himalayan Journal* X, 14 (1938).
17. Sherriff, 13.VI.36.
18. Sherriff, 18.VI.36.
19. Sherriff, 22.VI.36.
20. Sherriff, 24.VI.36.
21. Sherriff, 7.VII.36.
22. Sherriff, 14.VII.36.
23. Sherriff, 3.VIII.36.
24. Sherriff, 10.VIII.36.
25. Ludlow, 26.VIII.36.
26. Sherriff, 16.IX.36.
27. Ludlow, *Himalayan Journal* X, 15 (1938).
28. Ludlow, 6.X.36.

Chapter Four
1. Sherriff, 3.V.37.

2. Sherriff, 14.v.37.
3. Sherriff, 11.v.37.
4. Sherriff, 14.v.37.
5. Sherriff, 15.v.37.
6. Sherriff, 21.v.37.
7. Sherriff, 22.v.37.
8. Sherriff, 22.v.37.
9. Sherriff, 22.v.37.
10. Sherriff, 15.vi.37.
11. Sherriff, 21.vi.37.
12. Sherriff, 22.vi.37.
13. Sherriff, 23.vi.37.
14. Sherriff, 24.vi.37.
15. Sherriff, 3.vii.37.
16. Sherriff, 4.vii.37.
17. Sherriff, 5.vii.37.
18. Sherriff, 10.vii.37.
19. Sherriff, 12.vii.37.
20. Sherriff, 14.vii.37.
21. Sherriff, 23.vii.37.
22. Sherriff, 24–31.vii.37.

Chapter Five
1. Ludlow, *Himalayan Journal*, XII, 1 (1940).
2. All collectors' numbers on this expedition are those of Ludlow, Sherriff and Taylor (LS & T).
3. Sherriff, 23.iv.38.
4. Sherriff, 24.iv.38.
5. Ludlow, *Himalayan Journal*, XII, 3 (1940).
6. Sherriff, 25.iv.38.
7. Sherriff, 25.iv.38.
8. Sherriff, 25.iv.38, written on 27th.
9. Ludlow, 27.iv.38.
10. Sherriff, 2.v.38.
11. Sherriff, 3.v.38.
12. Sherriff, 30.v.38.
13. Live plants of both these primulas were flown to Britain in 1938 and, although *P. calderiana* still retains a hold in cultivation, it is doubtful if *P. chamaethauma* now survives.
14. Sherriff, 7.vi.38.
15. Forms of *P. ioessa* were first named varieties of *P. sikkimensis*
16. Sherriff, 14.vi.38.
17. Sherriff, 15.vi.38.

18. Sherriff, 22.vi.38.
19. Sherriff, 4.vii.38.
20. Sherriff, 8.vii.38.
21. Sherriff, 18.vii.38.
22. Sherriff, 18.vii.38.
23. Sherriff, 18. vii.38.
24. Sherriff, 21.vii.38.
25. Ludlow, 31.vii.38.
26. Ludlow, 2.vi.38.
27. Ludlow, 15.vi.38.
28. Taylor, *J. R. H. S.* 72, 139 (1947).
29. Described by Taylor and first recorded by R. L. Harley in 1926; Harley regarded it as a yellow-flowered variety of *M. simplicifolia* although suggesting that it might possibly by a hybrid between that species and *M. integrifolia* as both were growing in close proximity in his garden. It was raised from seeds saved from Bailey's form of *M. simplicifolia* and flowered for the first time in 1925.
30. Taylor, *J. R. H. S.*, 72, 139–40 (1947).
31. Taylor, *J. R. H. S.*, 72, 141 (1947).
32. Taylor, *J. R. H. S.*, 72, 142 (1947).
33. Ludlow, 13.vii.38.
34. Ward, *Riddle of the Tsangpo Gorges*, 106 (1926).
35. Ludlow, *Himalayan Journal*, XII, 10 (1940).
36. Ludlow, *Himalayan Journal*, XII, 10 (1940).
37. Taylor, *J. R. H. S.* 72, 143 (1947).
38. Ludlow, 11.viii.38.
39. Ludlow, *Himalayan Journal*, XII, 12 (1940).
40. Sherriff, 13.viii.38.
41. Sherriff, 22.viii.38.
42. Taylor, *J. R. H. S.* 72, 168–9 (1947).
43. Ludlow, *Himalayan Journal*, XII, 15 (1940).
44. Ludlow, *Himalayan Journal*, XII, 15–16, (1940).
45. Ludlow, *Himalayan Journal*, XII, 16 (1940).

46. Sherriff, 16.x.38.
47. Sherriff, 5.x.38.
48. Sherriff, 18.x.38.

Chapter Six
1. 'Although transport and supplies in Bhutan and Tibet were cheap, expeditions on the scale we organised were not run without incurring considerable expense. Occasionally we received grants from funds at the disposal of the British Museum (Natural History) and members of the expedition at times contributed according to their means, but it was Sherriff who defrayed the greater part of the costs. Without his financial help our efforts would have been far more restricted and our collections much more modest.' (Ludlow, *J.R.H.S.*, 93, 17 (1968).)
2. Ludlow, *Gard. Chron.*, 141, 670 (1957).
3. Ludlow, *Gard. Chron.*, 141, 671 (1957).
4. All collectors' numbers on this expedition are those of Ludlow, Sherriff & Elliot (L S & E).
5. Ludlow, *Gard. Chron.*, 142, 70 (1957).
6. Ludlow, *Ibis.*, 93, 550–551 (1951).
7. Ludlow, *Gard. Chron.*, 142, 309 (1957).
8. Ludlow, 15.I.47.
9. Sherriff, 16.I.47.
10. Ludlow, *Gard. Chron.*, 143, 14 (1958).
11. Ludlow, 31.I.47.
12. Ludlow, *Ibis*, 93, 575 (1951).
13. Betty Sherriff, 20.II.47.
14. Kingdon Ward, *The Riddle of the Tsangpo Gorges*, 233–234 (1926).
15. Sherriff, 23.II.47.
16. Ludlow, *Ibis*, 93, 561 (1951).
17. Ludlow, *Ibis*, 93, 550 (1951).
18. Ludlow, 20.III.47.
19. Betty Sherriff, 31.III.47.
20. Ludlow, 27.IV.47.

21. Ludlow, *Gard. Chron.*, 143, 102–103 (1958).
22. Ludlow, *Gard. Chron.*, 143, 103 (1958).
23. Ludlow, 2.V.47.
24. Ludlow, 4.V.47.
25. Ludlow, *Gard. Chron.*, 143, 103 (1958).
26. Ludlow, 11.V.47.
27. Ludlow, 25.V.47.
28. Ludlow, *Gard. Chron.*, 143, 196 (1958).
29. Ludlow, *Gard. Chron.*, 143, 196 (1958).
30. Ludlow, *Gard. Chron.*, 143, 283 (1958).
31. Ludlow, 24.VI.47.
32. Ludlow, 4.VII.47.
33. Ludlow, 5.VII.47.
34. Ludlow, 7.VII.47.
35. Ludlow, 7.VII.47.
36. Ludlow, 11.VII.47.
37. Ludlow, 17.VIII.47.
38. Ludlow, 18.VIII.47.
39. Ludlow, 23.VIII.47.
40. Ludlow, 31.VIII.47.
41. Ludlow, 7.X.47.
42. Ludlow, in *Gard. Chron.*, 143, 425, (1958).

Chapter Seven
1. Ludlow, 27.III.49.
2. All collectors' numbers on this expedition are those of Ludlow, Sherriff and Hicks (L S & H).
3. Ludlow, 27.III.49.
4. Ludlow, 2.IV.49.
5. Ludlow, 3.IV.49.
6. Ludlow, 15.IV.49.
7. Ludlow, 15.V.49.
8. Ludlow, 18.V.49.
9. Ludlow, 22 & 23.V.49.
10. Ludlow, 18 & 19.VI.49.
11. Ludlow, 20.VI.49.
12. Ludlow, 30.VI.49.
13. Ludlow, 2.VII.49.
14. Ludlow, 8.VII.49.
15. Ludlow, 22.VII.49.
16. Sherriff, 22.VII.49.
17. Sherriff, 2.IV.49.
18. Sherriff, 3.IV.49.

19. Sherriff, 7.IV.49.
20. Sherriff, 9.IV.49.
21. Sherriff, 9.V.49.
22. Sherriff, 16.IV.49.
23. Sherriff, 27.IV.49.
24. Sherriff, 29.IV.49.
25. Sherriff, 18.V.49.
26. Sherriff, 24.V.49.
27. Sherriff, 26 & 28.V.49.
28. Sherriff, 5.VI.49.
29. Sherriff, 16.VI.49.
30. Sherriff, 16.VII.49.

31. Sherriff, 17.VII.49.
32. Betty Sherriff, 28.V.49.
33. Betty Sherriff, 8.VI.49.
34. Betty Sherriff, 17.VI.49.
35. Ludlow, 12.VIII.49.
36. Ludlow, 29.IX.49.
37. Kingdon Ward, *Romance of Plant Hunting*, 198 (1924).
38. Sherriff, 29.IX.49.
39. Sherriff, 3.X.49.
40. Sherriff, 12.X.49.

Indexes

Italic folios refer to plates